U0293174

OFFICE BUILDING NOW Ⅱ

现代办公建筑 Ⅱ

邱凌云 周冬梅 黎健 译

高迪国际出版有限公司 编
大连理工大学出版社

图书在版编目 (CIP) 数据

现代办公建筑 . 2 : 汉英对照 / 高迪国际出版有限公司编 ; 邱凌云 , 周冬梅 , 黎健译. —大连 : 大连理工大学出版社 , 2014.1
ISBN 978-7-5611-8214-7

Ⅰ . ①现… Ⅱ . ①高… ②邱… ③周… ④黎… Ⅲ . ①办公建筑 - 建筑设计 - 作品集 - 中国 - 现代 Ⅳ . ① TU243

中国版本图书馆 CIP 数据核字 (2013) 第 216005 号

出版发行：大连理工大学出版社
（地址：大连市软件园路 80 号 邮编：116023）
印　　刷：上海锦良印刷厂
幅面尺寸：240 × 320mm
印　　张：22
插　　页：4
出版时间：2014 年 1 月第 1 版
印刷时间：2014 年 1 月第 1 次印刷
策划编辑：袁　斌 刘　蓉
责任编辑：刘　蓉
责任校对：王丹丹
封面设计：邵嘉盈

ISBN 978-7-5611-8214-7
定价：348.00 元

电话：0411-84708842
传真：0411-84701466
邮购：0411-84703636
E-mail:designbookdutp@gmail.com
URL:http://www.dutp.cn

如有质量问题请联系出版中心：（0411）84709246 84709043

PREFACE Ⅰ

序言一

Baik Eui-hyun
Design Partner
Junglim Architecture Co.,Ltd
白义铉
双林建筑事务所设计合伙人

Office Architecture as a Residing Space

In office design, architects spend most of their time creating spaces with a high building efficiency ratio, mass, shape, skin, and spacious lobby. In particular, they spend most of their time in morphological attempts, such as tilting, twisting, and sharpening, to stand out from the crowd.

Recently, there has been a rising interest in eco-friendly architecture without the distinction of public and private areas and attempts at diverse architecture are being made. The "glass curtain wall" office building, which is a symbol of modernization, is being called "Hippo eating energy" and is being singled out as an energy waster. Various regulations for eco-friendly architecture are still focusing on energy consumption and architects are focusing on this concept as well. With experience in eco-friendly design processes, which focus on facilities, equipment, and materials, I asked myself a most basic question, "What are the important planning factors in office design?"

Natural lighting and natural air ventilation, excluding direct sunlight on computer-based office environments, are not only the most fundamental and important design factors but also the ones that determine energy performance and residence quality in office architecture. Particularly, these factors act as a planning unit for the entire building in large-sized offices.

Planning for a lobby, elevator hall, rest area, and conference space that function as a communication hub between internal and external spaces is an important factor in making an office special. A common use space equipped with high-tech facilities represents an enterprise and creates user confidence.

A comfortable toilet and convenient pantry are factors that impress and satisfy users.

There are many attempts to create office architecture as a residing space through the introduction of locker rooms and shower rooms for bicycle commuters, lactation rooms for pregnant women, privacy-considered lounges for female employees, sports facilities, and food & beverage facilities. Some offices offer hotel-style services. These offices are often called "Premium Offices" and receive higher rental fees. Officetel (office + hotel), a residing and office space, has long been vitalized as office architecture in Korea.

As a result, I believe that an attempt at investment to increase residential performance can enhance property value and act as a benefit in renting. To Korean people, who are famous for working many hours, offices are a residing space where they spend more time than at home, particularly in the architecture design area. It seems to me that architects should understand offices as a residing space considering users' comfortability and take greater concerns and efforts.

作为居住空间的办公建筑

对于办公室设计，建筑师把大多数时间花在空间设计上，力图创建一个既具有高层建筑效率比，又具有一定的体块、形状、皮肤和宽敞的大厅的独特空间。特别是，他们大部分时间都在尝试建筑物形态的变化，如倾斜、扭曲、锐化等，并通过这些特殊的造型使其设计的建筑物能从其他建筑群中脱颖而出。

最近，我对环保建筑兴趣渐浓，无论是公共建筑还是私有住宅，并尝试研究不同风格的在建建筑。"玻璃幕墙"办公大楼，这种现代化建筑的象征，却被称为"河马式的大胃王"，并因此被选为能源浪费建筑。环保建筑的各种规定一直关注的是能源消耗，建筑师也在关注这个概念。以我在环保设计流程中积累的经验来看，环保重点关注的是设施、设备和材料。于是，我问自己一个最基本的问题："什么是办公室设计最重要的规划因素？"

自然采光和自然通风以及避免阳光直射在基于计算机的办公环境中，不仅是最基本和最重要的设计因素，同时也是决定办公建筑能源效益和住宅质量的重要因素。特别是，这些因素是作为大型办公楼整座大楼的一个规划单元来进行设计的。

大堂、电梯厅、休息区和会议区，作为内部空间和外部空间的联系枢纽，起着至关重要的作用，规划好这些区域就会使你的办公空间别具一格。一个配备高科技设施的公用空间不仅代表着一个企业的形象，同时也让用户充满信心。

舒适的卫生间和便利的食品室可以给用户留下深刻的印象，满足用户的需求。

作为居住空间的办公建筑在设计上有许多尝试，有为自行车通勤者修建的更衣室和淋浴房，为产妇设置的哺乳房，出于隐私考虑而为女雇员设计的休息室，还有配备的体育设施和餐饮设施，不一而足。一些办公楼还提供酒店式服务。这些办公楼通常被称为"高端办公楼"，会收取更高的租赁费。Officetel（办公+酒店），一个居住兼办公的空间，作为韩国的办公建筑，长期以来一直方兴未艾。

因此，我相信，尝试在提高住宅性能方面进行投资既可以提高房地产价值，也能在出租中获利。对于以长时间工作著称的韩国人来说，办公室是一个居住空间，他们在此花费的时间远远多于呆在家里的时间，特别是对那些在建筑设计领域工作的人来说尤其如此。在我看来，建筑师应该明白办公室也是一个居住空间，要更多地考虑到用户的舒适性，并在设计中对此给予更多的关注和努力。

序言二

PREFACE Ⅱ

Jiří Hejda
Partner
DaM Architectural Office

吉里・海伊达
DaM 建筑事务所合伙人

In our opinion, office buildings changed a lot in the past years (or decades). A great deal of these changes may be attributed to the recent technological development – office work completely changed over the past 20–30 years; and buildings seemed responding to these changes rather slowly at the beginning...

Together with the great development of IT and increasing mobility and flexibility of office workers increases also the demand on a flexible workspace and different functions blend more and more. The same as "home office" has been becoming more popular, that is working from home in close contact with the firm via IT networks we feel that the opposite development has been taking place – parts of homes have been becoming part of offices – firms develop their work environment with greater respect to their employees and their comfort. And so workplaces, meeting rooms and relaxation areas mingle transforming to a compound. It was quite natural that this development also affected architecture of buildings. Buildings are designed as a harmonic complex not subject to only a straightforward economical calculus. Through a character and design of workplaces firms have more often been expressing their corporate individuality.

Another strong aspect is the increasing pressure put on following the environmental friendliness of buildings in the past ten years. To our satisfaction it has become also a commercial article and so clients approach us requesting this type of buildings; it is a pure joy to design a building as an intelligent complex with the perspective of long decades of a lifespan and not only with respect to an instant profit. An increasing popularity of environmental certificates such as LEED clearly proves it.

Yet one basic rule applies here the same as to any other type of a building – the elementary and defining factor for the quality of the final build is the client and his wishes and requirements. Where a good brief and the desire to push the final quality at least a short step ahead are missing even the best architect can do nothing about it. Yet in the opposite situation when an architect meets his client mentally a symbiotic team is established where parties do not fight each other but on the contrary, they multiply the quality of their work by mutual stimuli and this way unique pieces of architecture may be built.

在我们看来，办公建筑在过去的几年（或者几十年）改变了很多。这些众多的变化可能是最近的科技发展引起的——办公室工作在过去的 20 至 30 年间完全改变了，而建筑似乎才开始慢慢地应对这些变化……

随着 IT 技术的快速发展以及办公室员工流动性和灵活性的增大，有必要增加一个灵活的工作空间，让不同功能越来越多地融合在一起。于是，“家庭办公室”已经变得越来越流行，员工们在家办公，并通过 IT 网络与公司保持密切的联系。我们也觉察到事情已经朝反向发展了——家的一部分已经逐渐变成办公室的一部分——公司改善其工作环境均以更加尊重员工的意愿和注重他们工作环境的舒适性为基础。因此，工作场所、会议室和休闲区进行混合，转变为一个复合区域。很自然地，这种发展也影响着建筑物的风格。各式建筑在设计时不应只考虑其成本核算，而更多的是要考虑到将它与周围的建筑融为一体。当然，公司也可以通过对工作场所独具匠心的设计来表达企业的个性，展示企业的魅力。

另一个显著的特点是，在过去的十年中伴随着开发环境友好型建筑后不断增加的压力。让我们满意的是，环境友好型建筑已然变成了一种畅销品，越来越多的客户需要这种类型的建筑；设计这样的建筑是一种纯粹的欢乐，因为它是一种长达几十年寿命的智能型建筑，设计时更多考虑的是环保节能，而不会只考虑即时利润。越来越普及的环保认证如 LEED 认证显然证明了这一点。

一项基本规则适用于此，也同样适用于其他任何类型的建筑——对建筑质量的评估标准，最基本的同时也起决定性作用的因素，是客户以及他的愿望和要求。如果连一份良好的职责和推动成品质量向前一小步的愿望都缺失了的话，再好的建筑师也无济于事。相反地，当建筑师以满足客户的心理需求为宗旨，建立一支共生团队，团队成员不是互相攻击而是互相鼓励，他们的工作质量会成倍提高；唯有如此，独特的建筑作品才可以创作出来。

PREFACE Ⅲ

序言三

Michal Leszczynski
Partner
Grupa 5 Architects

米哈尔・雷克钦斯基
Grupa 5 建筑事务所合伙人

In the past years we have seen how the comfort of the employee has a direct impact on the results of his work. Appropriate workplace area, some form of working privacy, common space, all these factors have an impact on productivity of an employee, a workgroup, and consequently performance of the entire company.

Similarly, the level of the technology and installation provided in the building, adequate ventilation systems, providing good lighting, not to mention the necessary equipment at telecommunication systems, make the employee can better focus on work, increase his productivity.

For this purpose the office building features for the possibility of a recovery at work such as inner garden, reading room, dining room or cafeteria. Buildings are set in a neighborhood of the green areas, parks. Work becomes a human-friendly environment. New office buildings allow you to come to work by bicycle, park it and take a shower.

Another important element is the ability to identify with your company, your workplace. The building's architecture is of significant importance, the shape and form of the building, the materials from which it is made, entrance hall and its interior make the employee will be identified with their work or not.

More and more office buildings are designed as buildings for rent. Addressed to a typical tenant they are fitted in a similar standard. Buildings have an optimal technology. In a world of widely available standard challenge for the architect is to propose interesting and original design solutions. The architect is to convince client to the proposed solutions and at the end to ensure that the construction phase has been properly executed.

Efficient and effective design generates savings in management costs. Initially, it requires incurring increased costs of implementation. But at operational stage you get measurable savings. Sustainability and environment certification become another standard. Lower water use, improvement of the ventilation and heating systems, rainfall collected in a storage tank for rainwater intended for re-use, all these solutions improve the quality and comply tenants' requirements.

What is the future development of office buildings? I think the evolution of the architecture of the office building will focus on the individual human being, and a friendly working environment.

在过去的几年中，我们已经看到员工工作环境的舒适度高低直接影响着其工作成果。适宜的工作区域、某种形式的私人工作空间、公用空间，这些因素无不影响着一名员工或一个工作组的生产效率，并因此影响整个公司的业绩。

同样，技术水平是否先进，配套设施是否精良，也直接影响到工作效益。完善的通风系统，良好的照明设备，特别是必要的通信系统设备，可以使员工更好地专注于工作，提高工作效率。

因此，办公楼在设计时必须涵盖诸如内部庭院、阅览室、餐厅或食堂等功能区域。办公楼的选址最好毗邻绿地和公园。这样，员工可以在一种人性化的环境中工作。新办公大楼允许你骑自行车上班，停好自行车后还可以冲个澡，以最佳的状态投入工作。

另一个重要的元素是大楼的设计必须与你的公司及你的工作场所相匹配。大楼的建筑风格具有非常重要的意义，因为无论是外部造型还是内部结构，以及材料的选择、门厅的设计、内部的装修等，都会直接决定员工们能否融入工作环境。

越来越多的办公楼设计为出租之用。一个典型的承租商也会提出如上述的类似标准。因此，大楼的设计应该有一个最优技术方案。世界上建筑标准的使用非常广泛，对建筑师的挑战是他必须提出有趣而原创的设计解决方案。建筑师必须说服客户接受其提出的解决方案，以确保施工正常进行。

高效而实用的设计能够节省管理成本。最初，它会导致实施成本的增加。但是在往后的具体运营阶段，可节省不少开支。可持续性和环境认证成为建筑的另一个评价标准。减少用水量，改善通风和供暖系统，收集雨水并重新利用，所有这些解决方案不仅提高了办公建筑的性价比，也顺应了承租人的需求。

办公建筑的未来发展趋势是什么？我认为，办公建筑的演化将更注重个体的人以及友好的工作环境。

CONTENTS

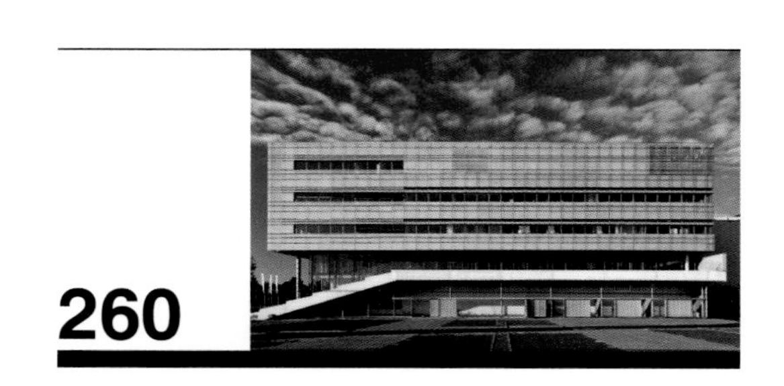

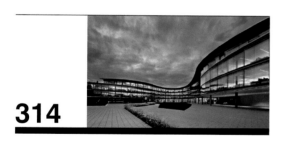

Filadelfie 办公楼

FILADELFIE OFFICE BUILDING

ARCHITECT
DaM spol.s r.o., Jan Holna, Petr Šedivý, Richard Doležal, Petr Malinský, Lenka Kadrmasová

ASSOCIATE ARCHITECT
Michaela Čechová, Jana Hanzalová, Monika Pejsarová, Gabriela Šatrová, Jitka Šindelářová, Mária Urbanová, Roland Vančó, Kateřina Víděnová, Jindřich Ševčík

FIRM
DaM

LOCATION
Praha, Czech Republic

AREA
35,000 m²

PHOTOGRAPHER
Filip Šlapal

The new office tower Filadelfie became in 2010 a part of the large newly urbanized site in Prague 4 – Michle. At the same time, a revitalization of a disused space between streets (Želetavská and Baarova) took place turning the site into a park.

Although the Filadelfie office building is located in a relatively urbanized environment, the designers designed the shape to maximize the detachment from surrounding buildings. The shape of a cross diagonally oriented following the adjacent communications allowed to keep enough of free space and to approach the design as a solitaire building. The building is placed without any base onto a grassed plane, which surrounds Filadelfie while transforming to the mentioned park.

The basic concept of a typical office floor is a modified letter H. The main communication spine is located in the central part of the disposition, from which the four administrative wings are stretching. The opening of the cross shape is providing both premium levels of interior lighting and panoramic views to any place along the facade.

The variability of the office floor disposition allows dividing the floor into four independent units, or utilization of the whole floor for single tenant. Staircases in the two larger wings enable inner vertical connection for a tenant using several adjacent storeys.

Distinctively shaped volume of seventeen office floors is changing in the parterre into six sub-terrain storeys filling the whole available site area. Thanks to the terrain morphology the first basement is partially exposed towards street Želetavská, thus creating a portal for building main entrances. Retail areas are situated on this floor. The other five basements are accommodating parking and building technology facilities.

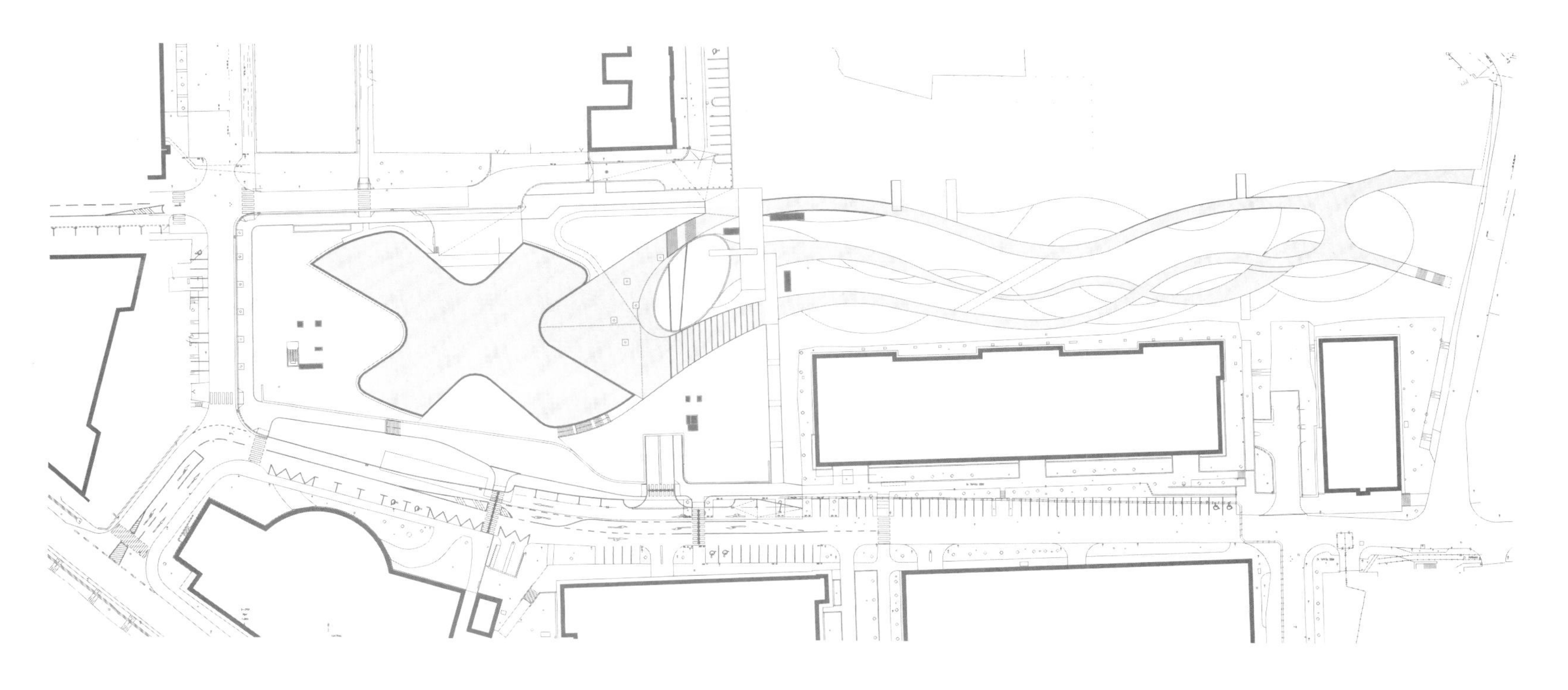

Situation

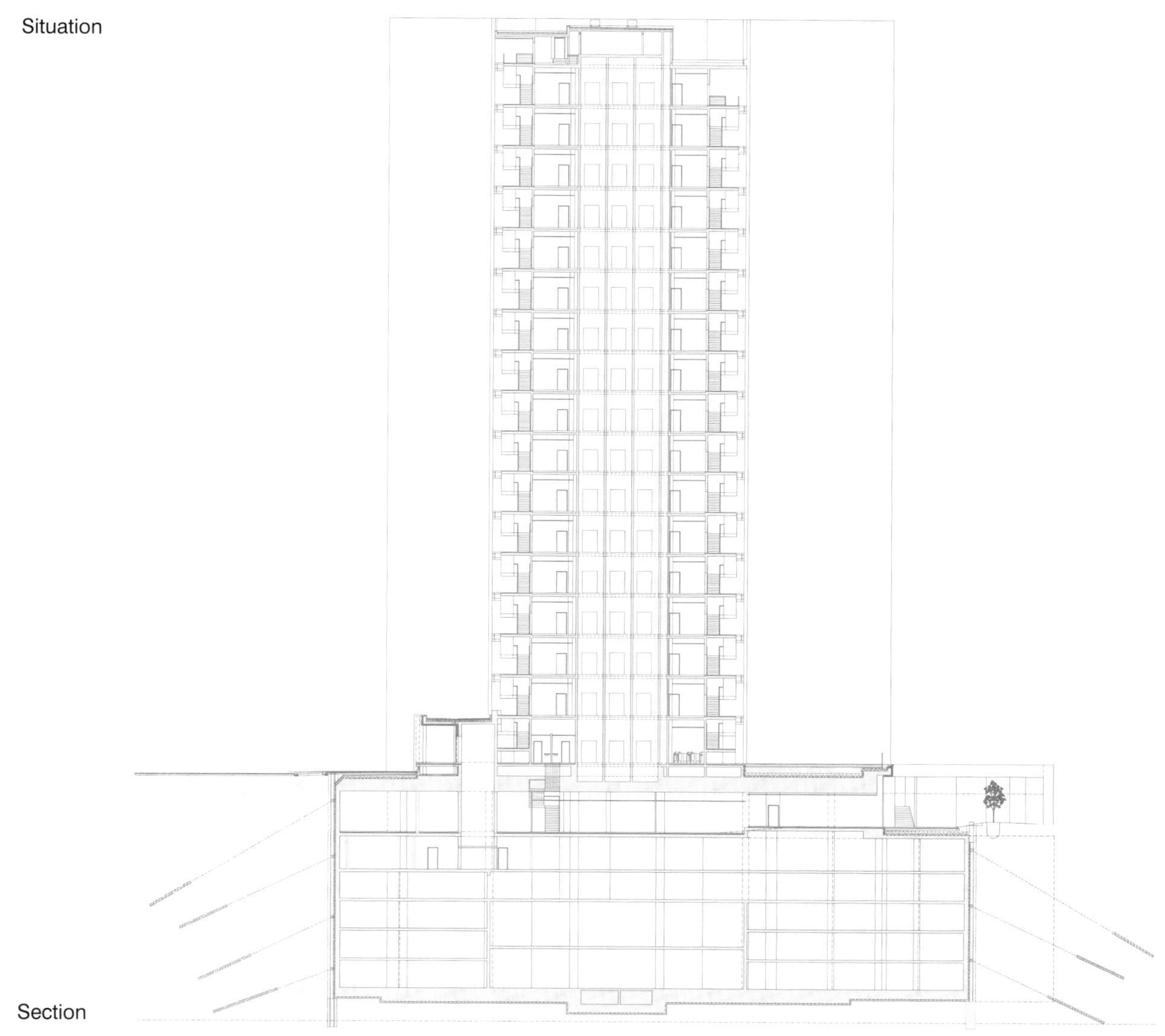

Section

新建的 Filadelfie 办公楼在 2010 年成为布拉格市第 4 辖区的 Michle 区大型新兴城市网的一部分。在修建大楼的同时，还将位于 Želetavská 和 Baarova 两街之间的废弃空间改造成了一座公园。

虽然 Filadelfie 办公楼处于高楼林立的城市化环境中，但其独特的外部造型让它从周围众多的建筑物中脱颖而出。外形呈四叶螺旋桨状，内部通信设施互相连接，这样的设计使建筑物有足够的自由空间来进行灵活布局，也诠释了它作为一座纸牌状建筑应有的风格与特点。这座大楼没有任何底座，直接建于草坪之上，草坪环绕着 Filadelfie 大楼，形成上文提到的公园。

典型办公室楼层的基本设计理念基于字母 H 的变形。主要的通信枢纽位于大楼的中部，四个行政区像翅膀一样从中间向四个方向伸展。四叶螺旋桨外形设计不仅确保了室内光线充足，还可以保证沿着各个立面观赏到市区的全貌。

每一层楼的办公室都由灵活的模板分隔而成，这些模板可以按照需要把一层楼分成四个独立的办公单元，也可以将隔板拆除形成一个大的单元供单户租用。两个较大翅膀区各设了楼梯通道，可以将租户租用的相邻层的内部垂直空间连接起来。

除了十七层楼高的形状独特的大楼外，在花坛的位置还设计有六层地下空间，占据了全部的规划用地。由于地形独特，地下一层的上半部分露出了地面，对着 Želetavská 街，形成了大楼的一个主要入口。零售店坐落在这一层。其他五层用于停车和修建机械设备用房。

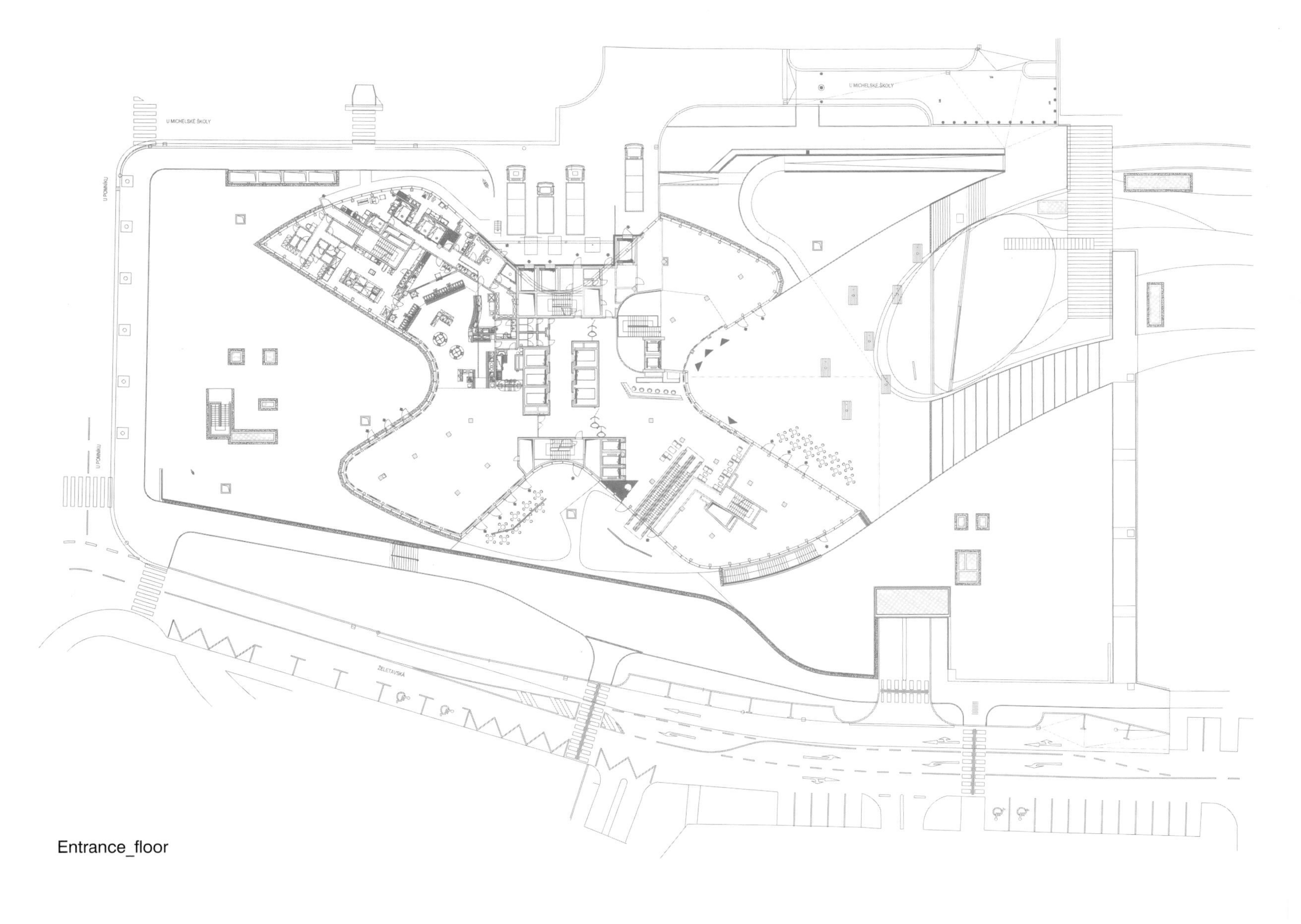

Entrance_floor

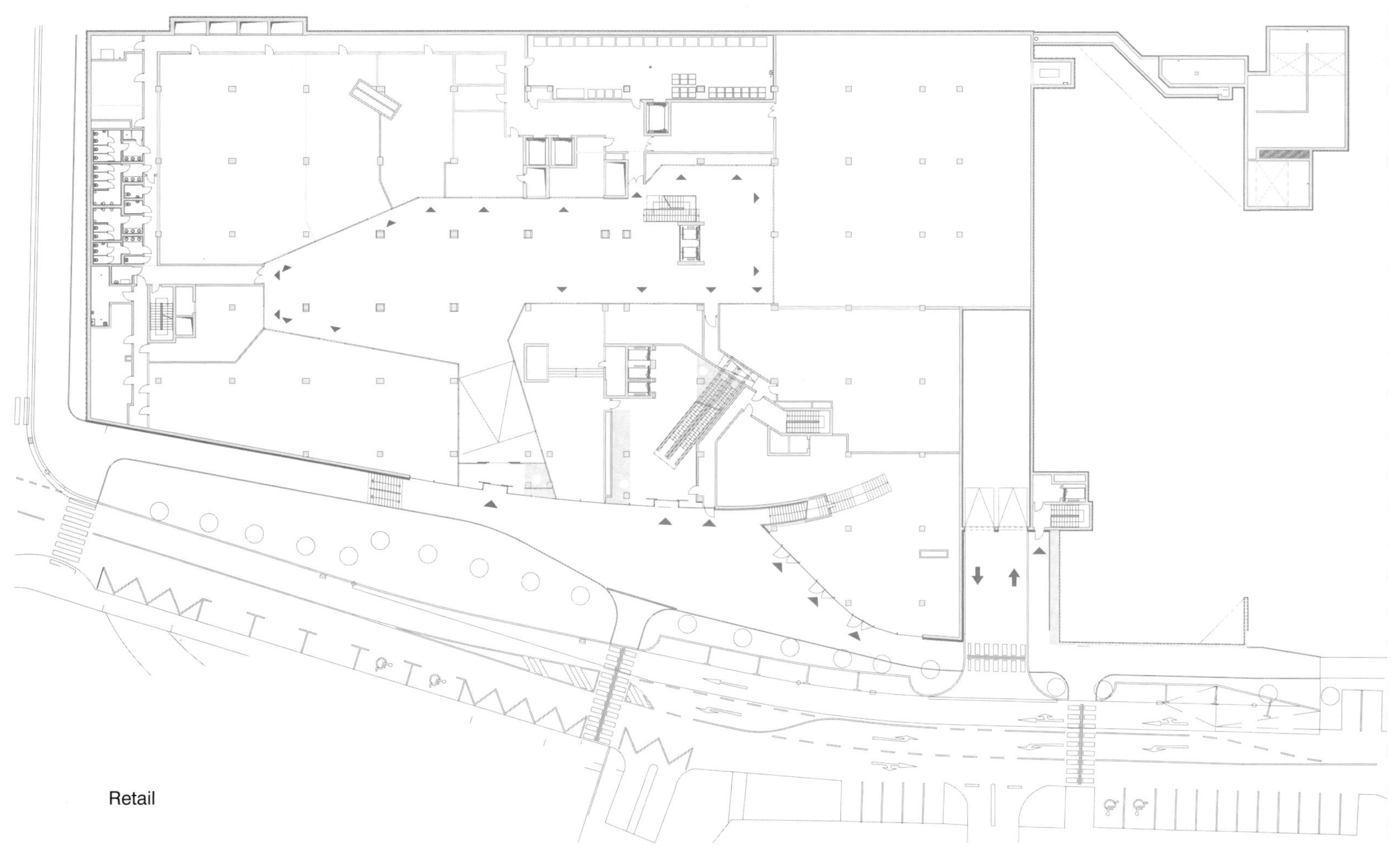

Retail

RETAIL
FILADELFIE

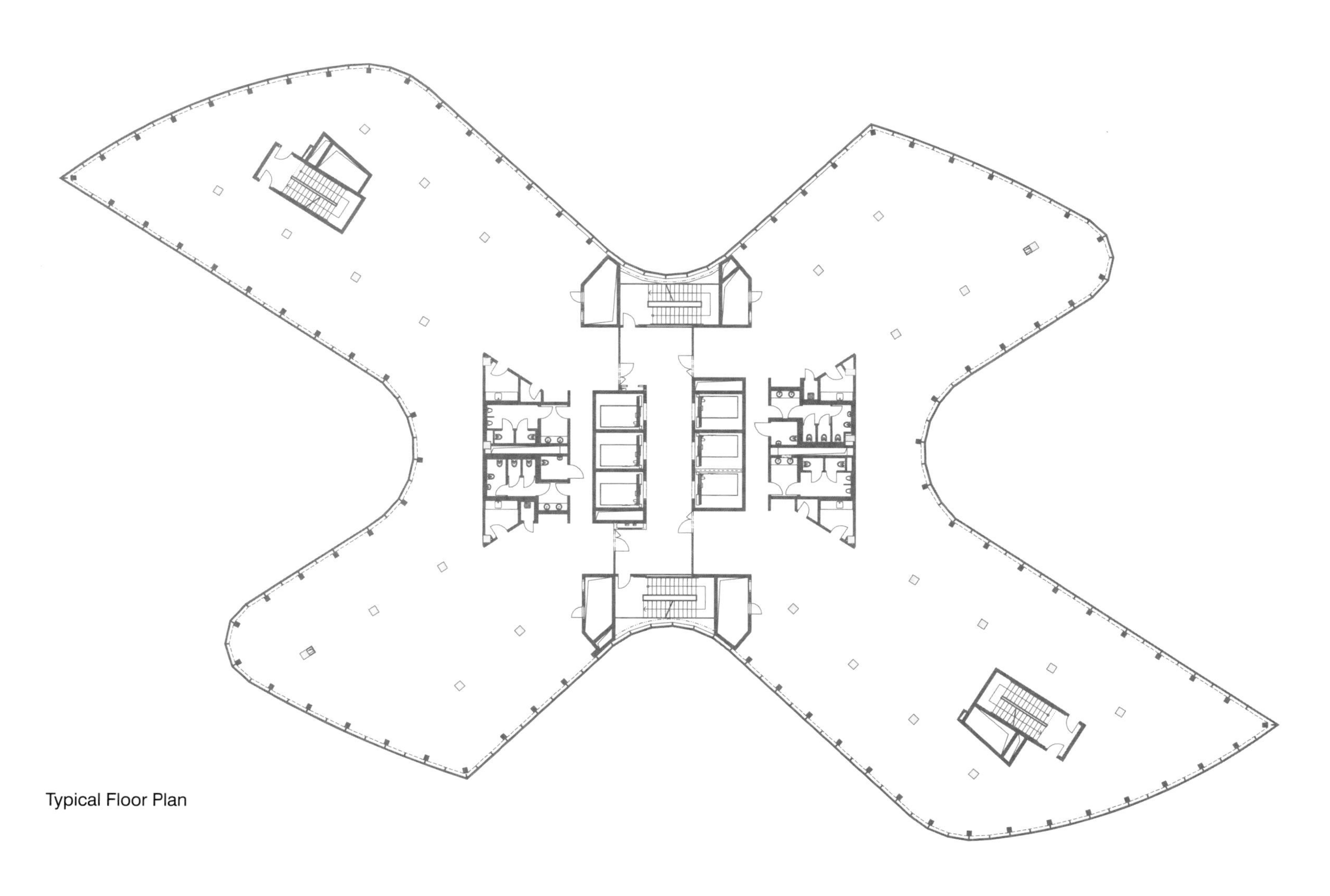

Typical Floor Plan

REZIDENČNÍ PARK BAAROVA
BAARŮV PARK
WWW.BAAROVA.CZ
BYDLENÍ

FILADELFIE

FILADELFIE
BBCENTRUM FILADELFIE
TERASA

巴拿马银行大厦

BANK OF PANAMÁ TOWER

ARCHITECT
Herreros Arquitectos, Mallol y Mallol

PROJECT DIRECTOR
Jens Richter

PROJECT TEAM
Gonzalo Rivas Zinno, Carmen Antón, Joanna Socha (HA), Amilton Jaramillo (M&M), Ruben Taboada (M&M)

CLIENT
Banco de Panamá

LOCATION
Panamá

AREA
35,395 m²

PHOTOGRAPHER
Fernando Alda

BancoPanamá
ANCON

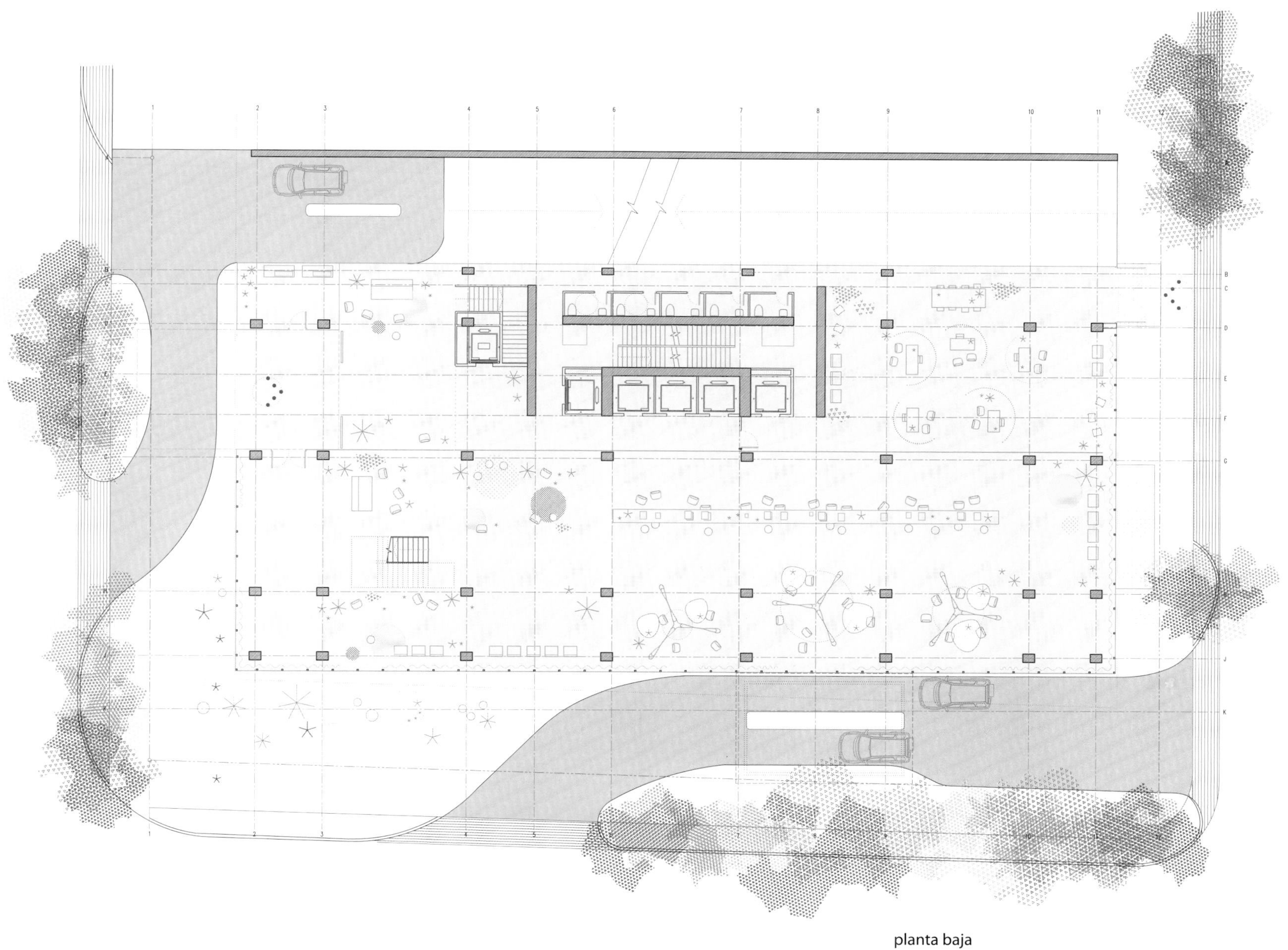

planta baja

Programmatically, the tower section is organized as a piling up of four separate buildings. The lowest first part relates to the city with the bank offices and its counter hall on the ground and mezzanine floor above which a six-storey parking is located. The remaining three building parts are accentuated as prisms of different sizes adapting to different positions, offsetting each with the one below and hereby creating a series of terraces, with different orientations and views over the sea and the historical city center.

Each prism can be read as a small canonical office building with its sky lobby with shared uses by all tenants: a restaurant, a gym and a club for meeting and relaxing. On its up most part, a representative hall provides stunning views all over the city. The building envelope is resolved with a glass curtain wall supplemented of different color tones and transparencies that enhances a vibrant effect of randomness in the facade, caused not solely by apparent effects of opacity and reflection but also by the scalar ambiguity of the standardized elements that never coincide with any obvious horizontal line of the building. This prevents "reading" the program distribution from the exterior while introducing a diversity of elevations of the horizontal facade elements as seen from the inside. The proposed system incorporates naturally all particular solutions such as the railing of the terraces, air intakes of the technical rooms or the integrated ventilation openings of the parking levels.

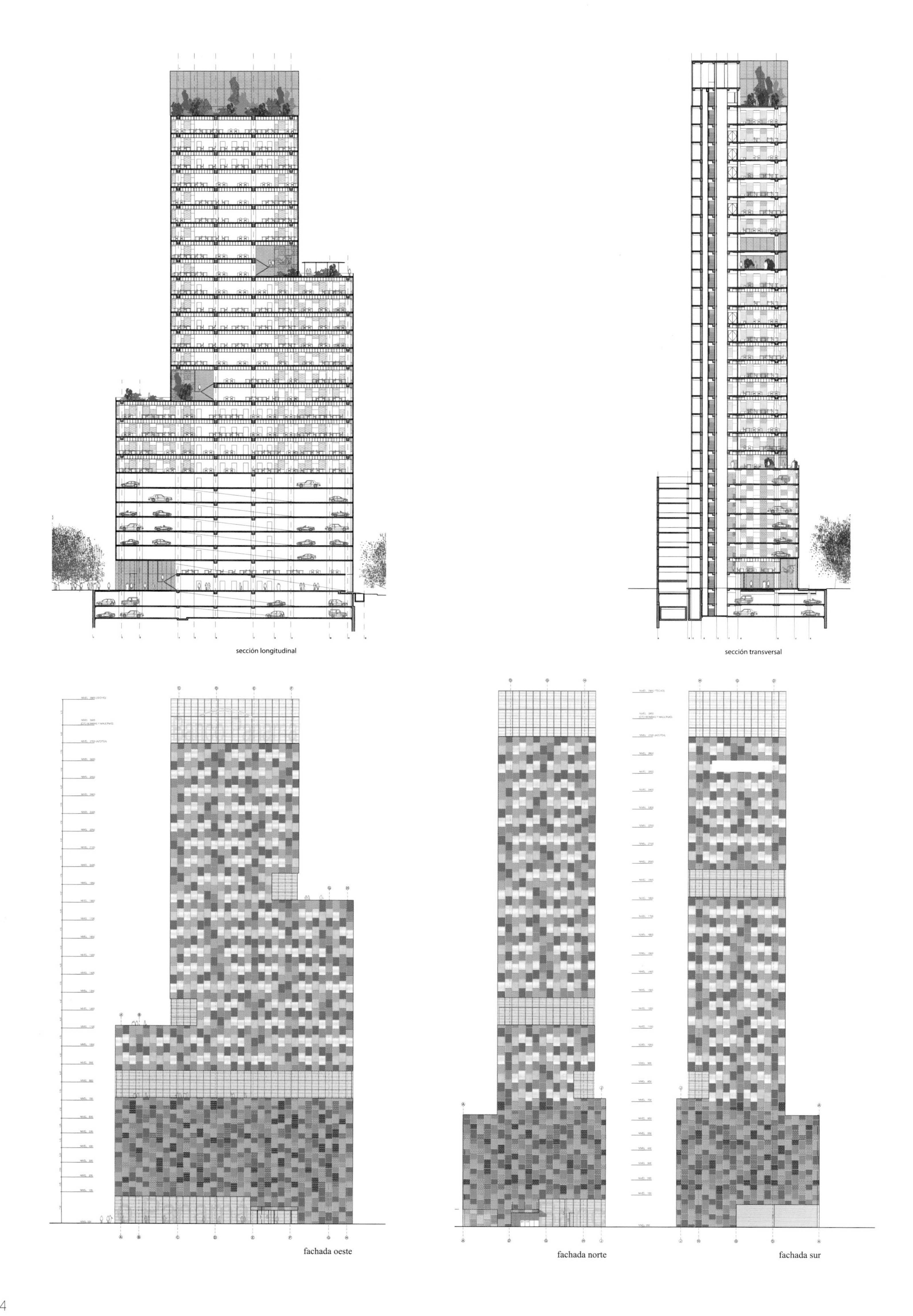
sección longitudinal
sección transversal
fachada oeste
fachada norte
fachada sur

BancoPanamá

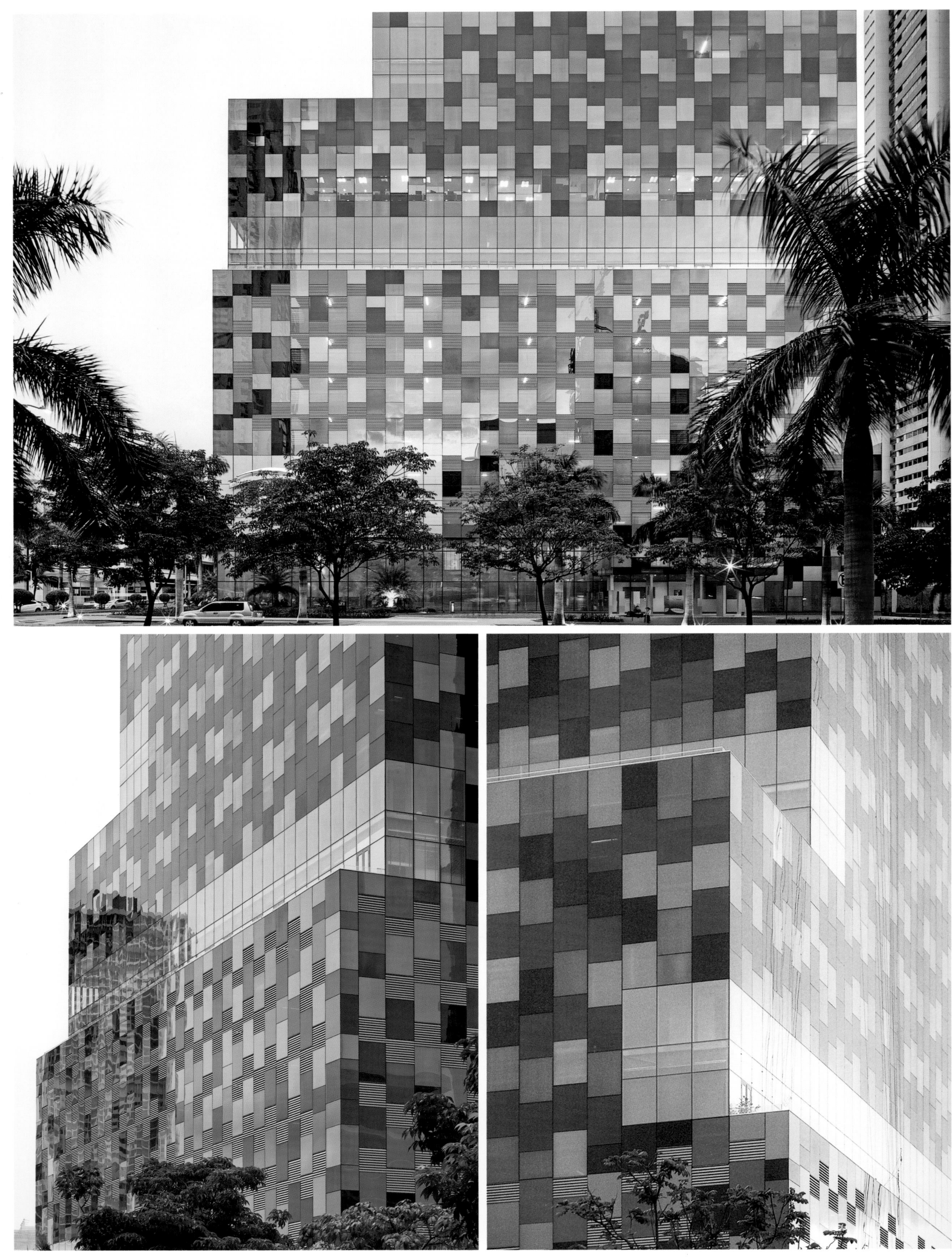

按照设计规划，塔楼部分由四个独立的建筑叠加而成。最下面的建筑包括银行办公室、地面营业厅和一个夹层，夹层上面是一个六层的停车场。其他三座建筑像三个大小不一的棱柱体，处于大楼不同的位置，由于上面的建筑相比底下的都要向内缩进而形成一个个露台，有些朝向大海，有些面对历史名城中心，形成非常美妙的景观。

每个棱柱体可以看做是一座标准的办公大楼，各有各的大堂，连接着所有租户共享的区域：餐厅、健身房以及休闲和会议会所。在大楼最顶端有一个别具一格的大厅，在此可以饱览全城美景。大楼的外围结构是玻璃幕墙，呈现出不同的色调和透明度，增强了立面的随机性与活力感；这种随机性不仅是因为不透明性与透明玻璃上的映像之间的鲜明转换，也因为标准构件的安放基本都很随意，并未与建筑物的任何一条明显的水平线条相吻合。这样的设计可以避免从外部“解读”大楼的功能布局，而只能从大楼内部观察到室内极大的空间灵活性。该规划还包含了所有特定的解决方案，小到露台的栏杆设计，大到技术房的进风口或停车层的综合通风口的设计等等。

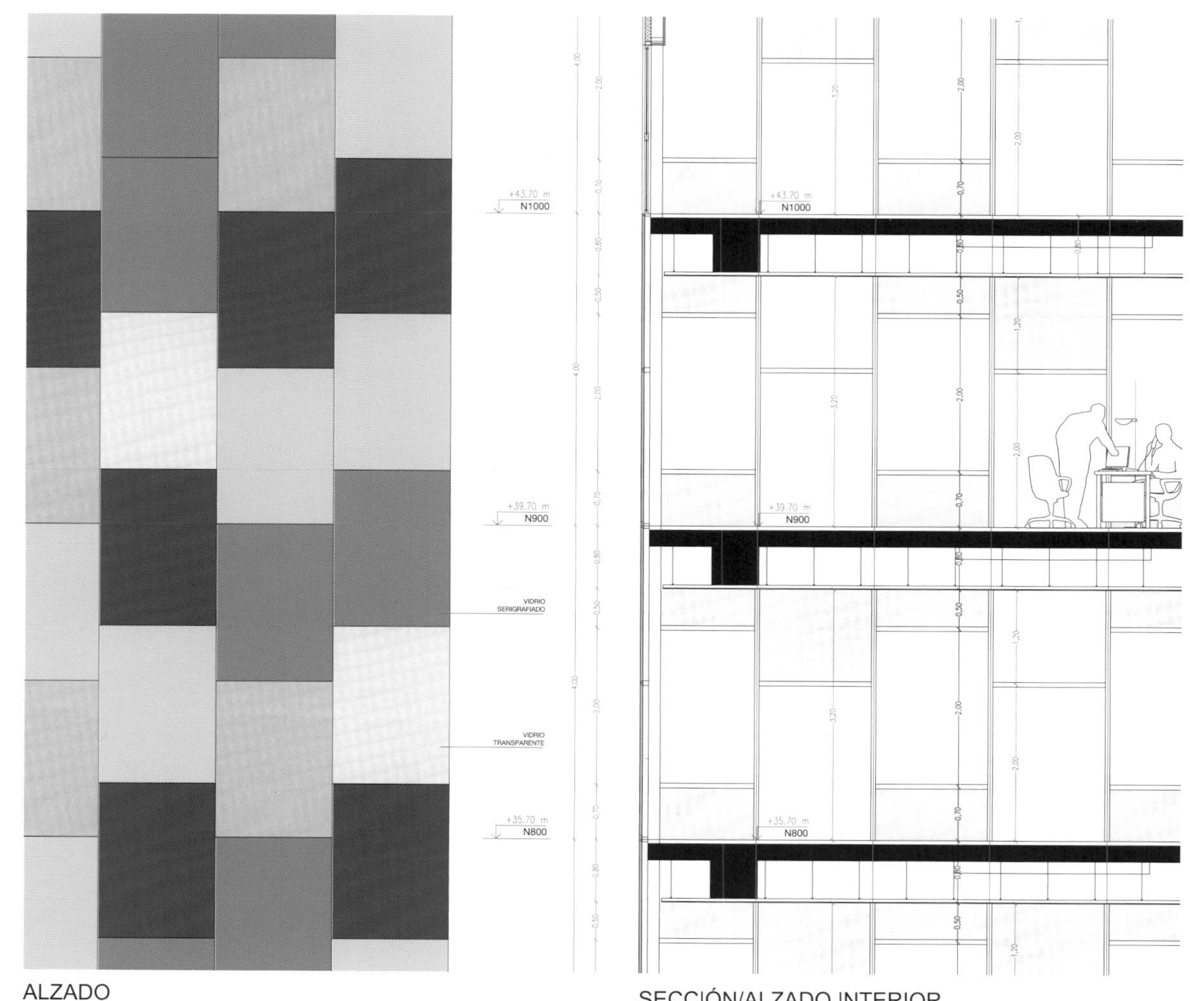

ALZADO

SECCIÓN/ALZADO INTERIOR

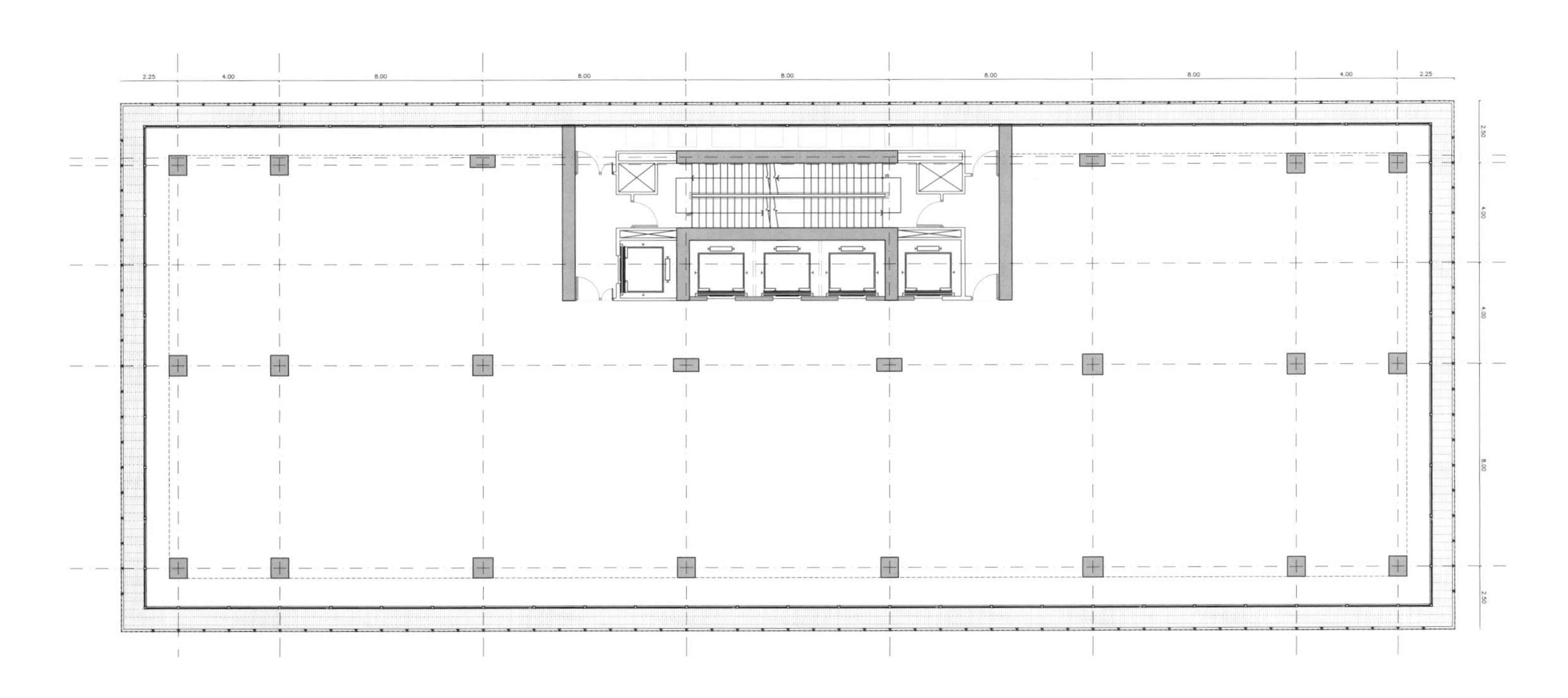

planta tipo

诺顿罗氏富布赖特大厦

NORTON ROSE FULBRIGHT TOWERS

ARCHITECT
Paragon Architects

LOCATION
Sandton Central, South Africa

AREA
24,000 m²

Located at the western edge of the bustling Sandton Central area, the glass and aluminium towers of 15 Alice Lane rise as statuesque architectural statements on the skyline of Johannesburg's north-western suburbs. This 24,000 square meters double tower structure is the new office space of well-known legal firm Norton Rose Fulbright. Set atop a six-storey basement the two towers rise 17 storeys into the Johannesburg sky.

The project employs cutting edge glass technology that has not been used in South Africa before. North and south facades are wrapped in a patterned "skin" of seemingly random planes of clear, dark grey and translucent glass. The east and west facades eliminate direct sunlight with sculptural hand-formed aluminium boxes set around deeply incised glass lines. These unique facades alter dynamically according to light and atmosphere changes due to their highly patterned and abstract surfaces.

The towers hug a dramatic canyon-like atrium space connected by walkways arranged in a fan-like pattern below its generous skylights. Paragon Interface, the space planning and interior fit-out business in the Paragon Group, undertook the sophisticated interior fit-out of the towers. Use of low-energy glass, good orientation and functional detailing contribute to minimising energy usage, which are imperative goals amongst responsible developers and tenants.

Commissioned and built entirely during the hardest economic recession that the world has seen, this building is a statement of faith in the future of Johannesburg and a measure of what can be achieved when well-integrated teams meet around singular challenges.

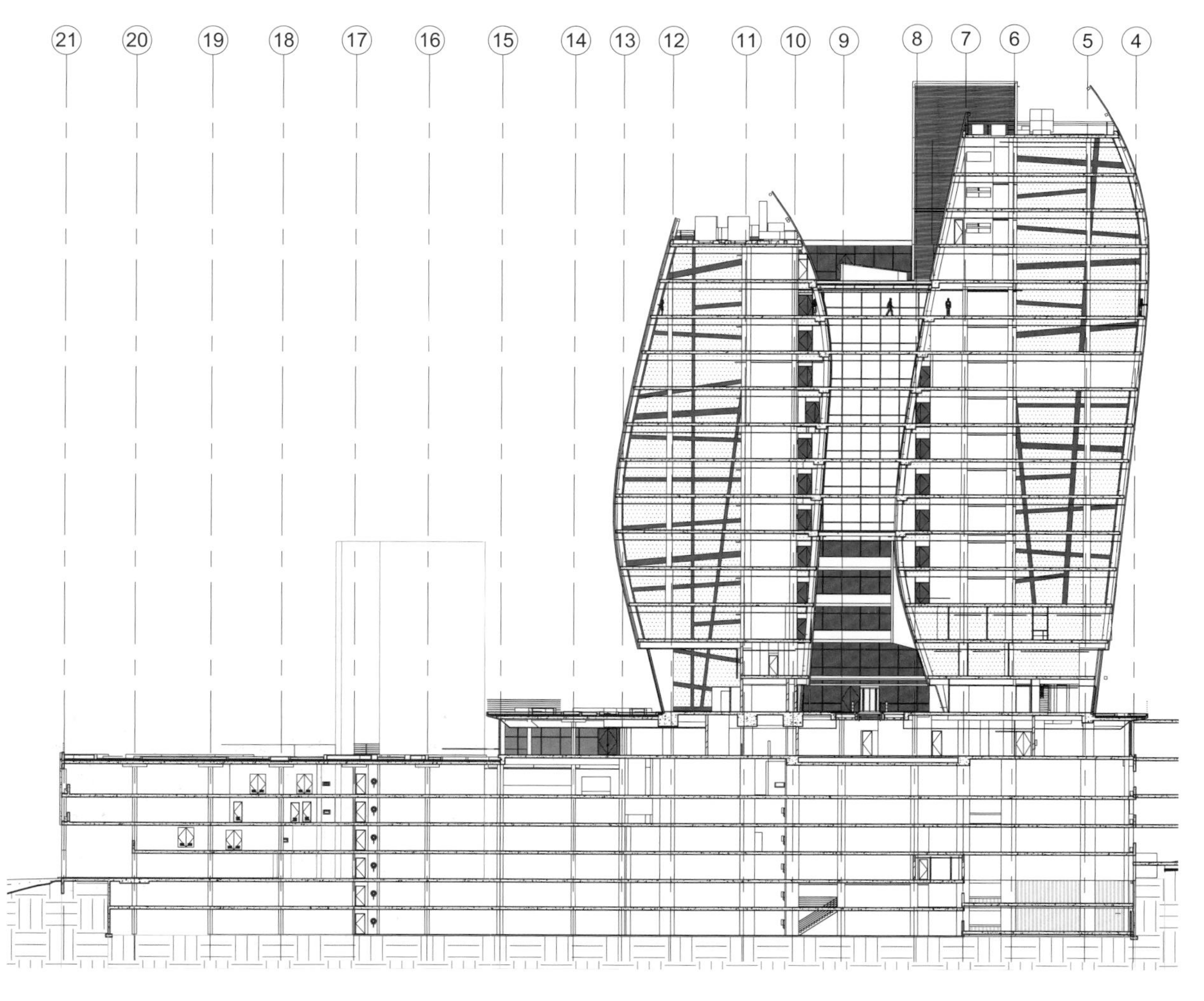

GA_Section A
1 : 500

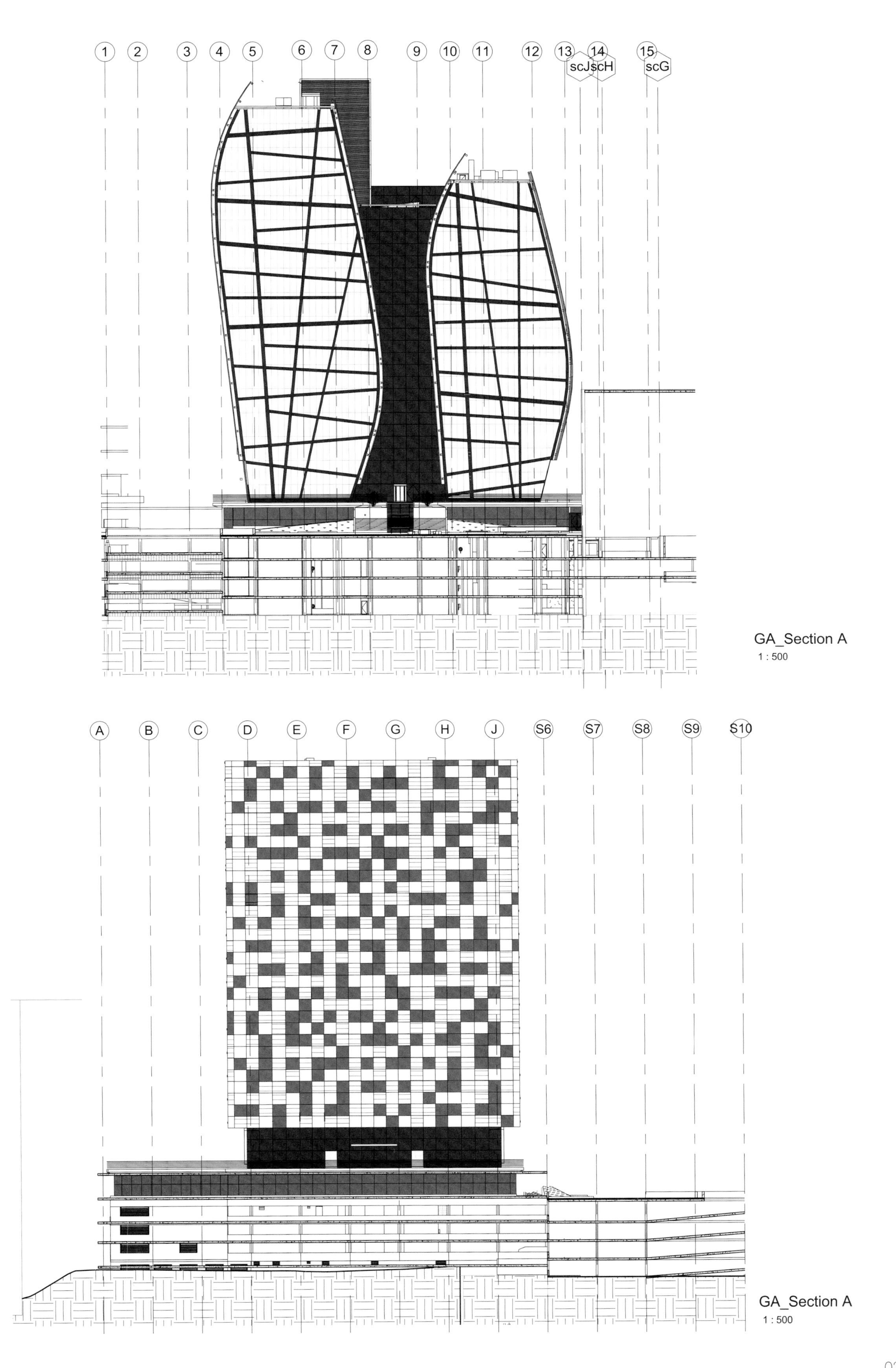
1
2
3
4
5
6
7
8
9
10
11
12
13
14
15
scJ
scH
scG
GA_Section A
1 : 500
A
B
C
D
E
F
G
H
J
S6
S7
S8
S9
S10
GA_Section A
1 : 500

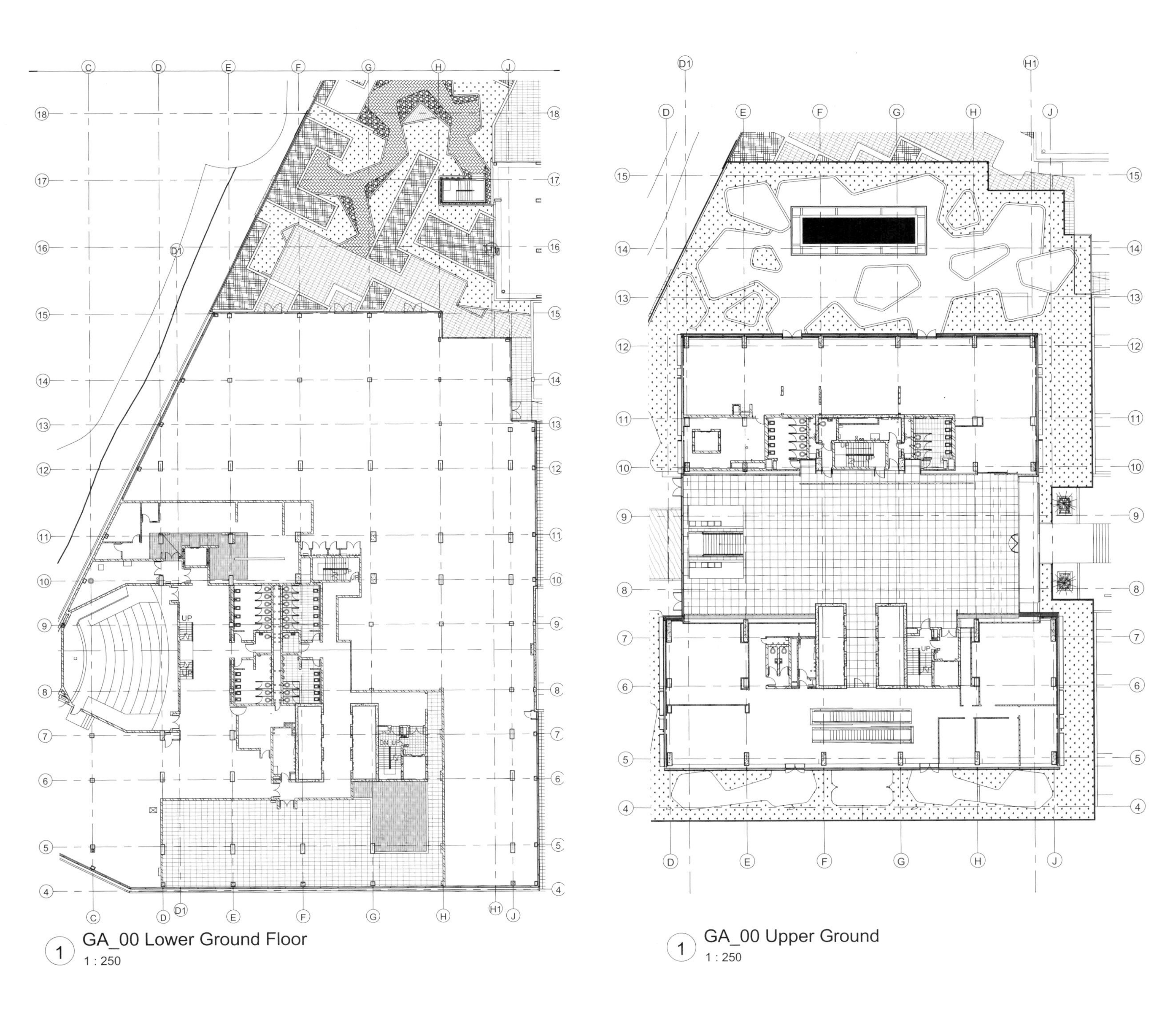

1 GA_00 Lower Ground Floor
1 : 250

1 GA_00 Upper Ground
1 : 250

在繁华的桑顿市中心区西部边缘，一座玻璃和铝合金材质的大厦在爱丽丝巷15号拔地而起，犹如雕塑般映衬在约翰内斯堡西北郊的天际线上。这座24 000平方米的双塔建筑就是知名的诺顿罗氏富布赖特律师事务所新的办公空间。该大厦有一个6层的地下室，地上17层楼高的双塔大厦高高地耸入约翰内斯堡的天空。

该项目采用前沿玻璃技术，这种技术之前在南非还没有使用过。南北立面均包裹在一个模板化的“皮肤”里，这一皮肤由一些随机排列的、清晰的、呈深灰色半透明的玻璃面板组成。东西立面将雕塑般的手形铝框深深嵌入玻璃面来消除阳光的直射。这些独特的立面由于高度模板化，在展示抽象艺术的同时，还可以根据光线的强弱和大气环境的变化不断地进行调节。

两座塔楼中间形成一个醒目的峡谷状中庭空间，高大的天窗下是呈扇形的通道。两座塔楼之间的完美对接工程及塔楼内部的豪华装修工程均由Paragon集团承接，这个集团主要开展空间规划和室内装修业务。低能玻璃的使用，良好的定位和功能细化都有助于将能源消耗降到最低，毕竟，节约能源是开发商和承租商共有的责任。

大厦虽然修建于世界经济大萧条的最艰难时期，却能成为世界一流的建筑，这不仅说明约翰内斯堡的未来是充满希望的，同时也表明一支技术过硬、团结协作的团队一定能够接受各种非凡的挑战，取得灿烂辉煌的成就。

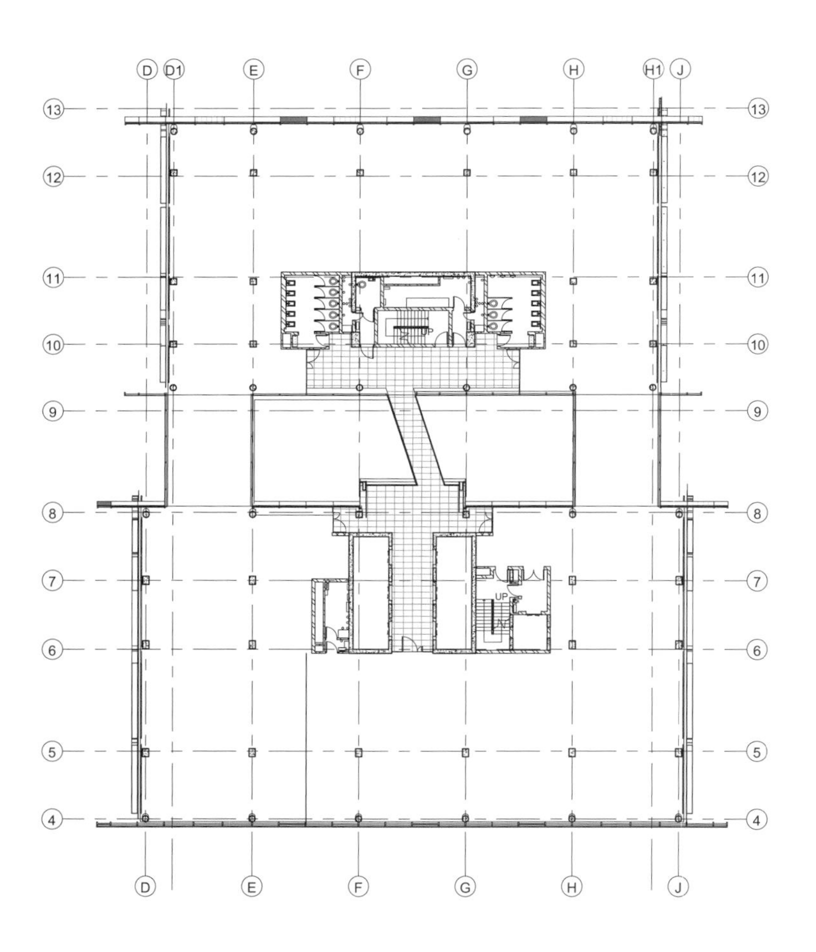

GA_9th Floor
1 : 250

GA_12th Floor
1 : 250

别具一格的双塔大厦

SIGNATURE TOWERS

ARCHITECT
Junglim Architecture Co., Ltd.

CLIENT
Doosan AMC, Ascendas Korea

STRUCTURE
RC, SRC

LOCATION
Seoul, Korea

AREA
6,933.9 m^2 (Site Area), 4,159.1 m^2 (Building Area), 99,992.33 m^2(Gross Floor Area)

Given the situation of the metropolitan city that connects the Nam Mountain and Inwang Mountain, the form of an opening gate between two towers was taken, and a method of reducing heat loss by using stone as a finishing material for the east and west elevations was incorporated. The ground level of the building was arranged to allow people to travel freely among shops, maximizing pedestrian convenience and access to retails. The two masses are connected up to the sixth floor, and the image of a bottom of the gate supporting the towers was achieved instead of the general form of a podium.

The biggest challenge was to add interest to the proportion due to the limited height of the 17-storey building. The inclined face of the outer section was designed to supplement the insufficiency of the proportion and was re-interpreted as a way to reinforce the image of the gate. As the location is seen when traveling from the entrance of Insa-dong via the Nam Mountain to Gangnam, the district south of the Han River and the peak of the building is visible from the starting point of the Cheonggyecheon, the building is valuable enough to stimulate a challenge in the spirit of the architecture. There is no doubt that the building will breathe new life into this area with all the innovative touches and enthusiasm the team members weaved into it.

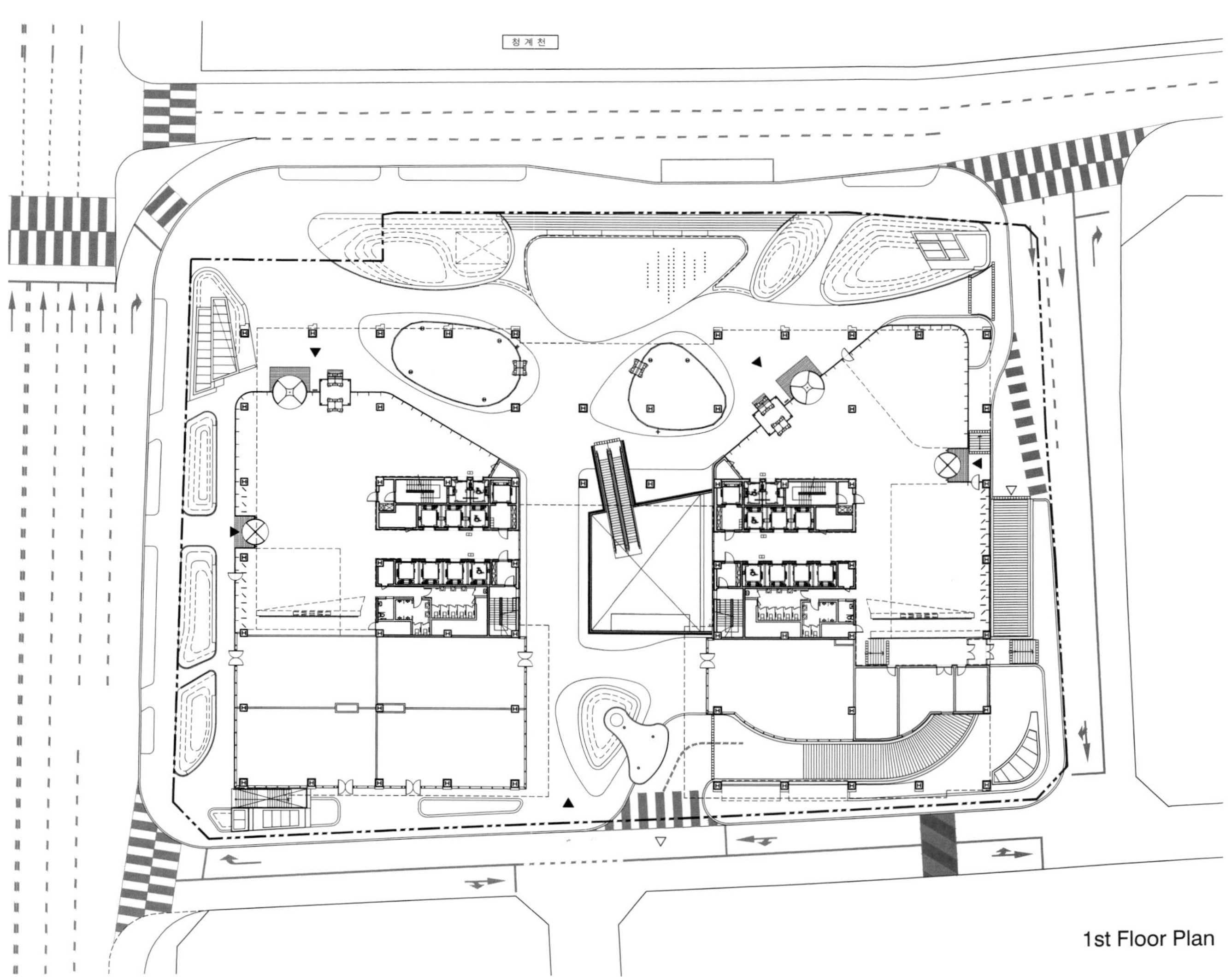

1st Floor Plan

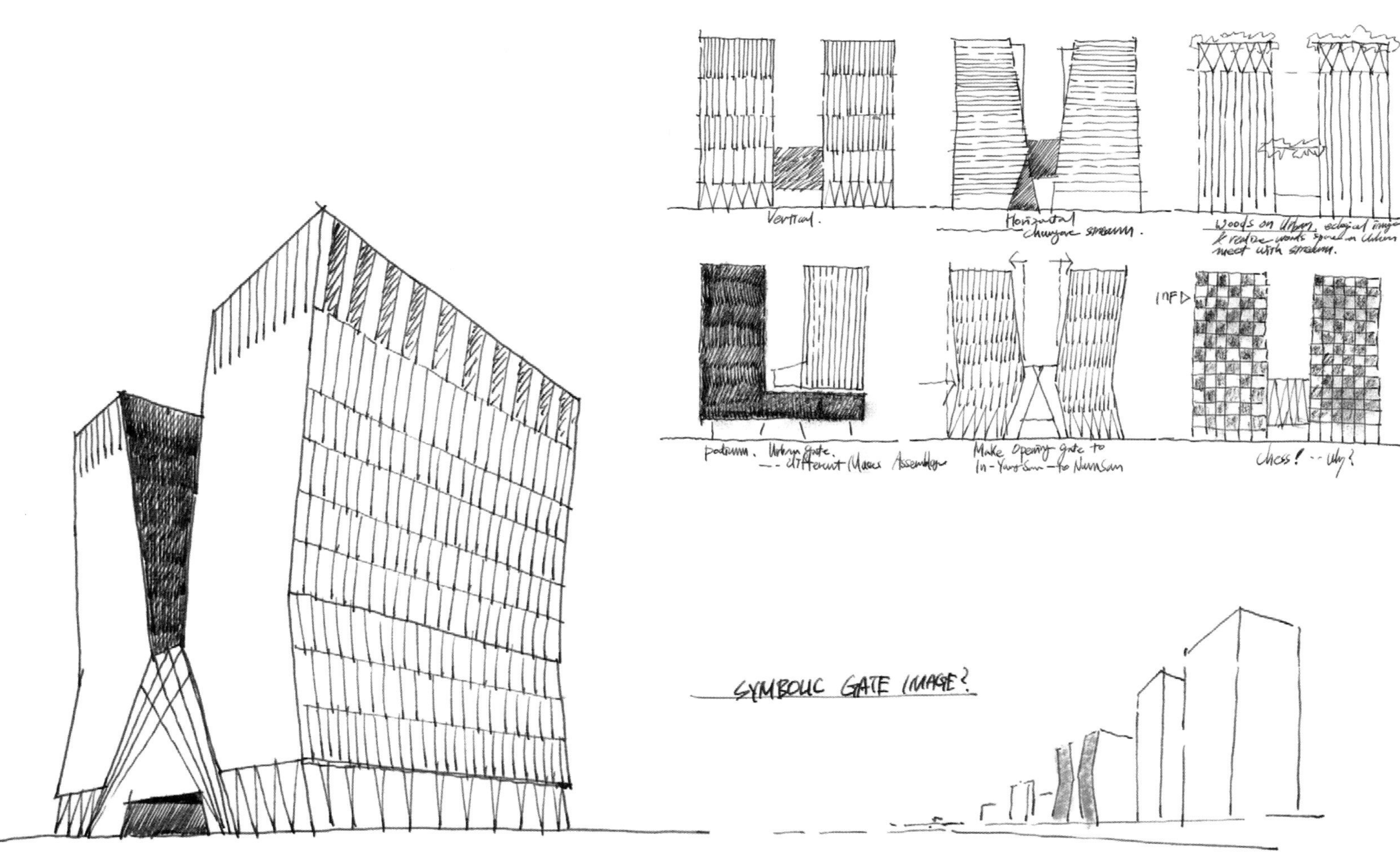
Vertical.
Chess! -- why?
SYMBOLIC GATE IMAGE?

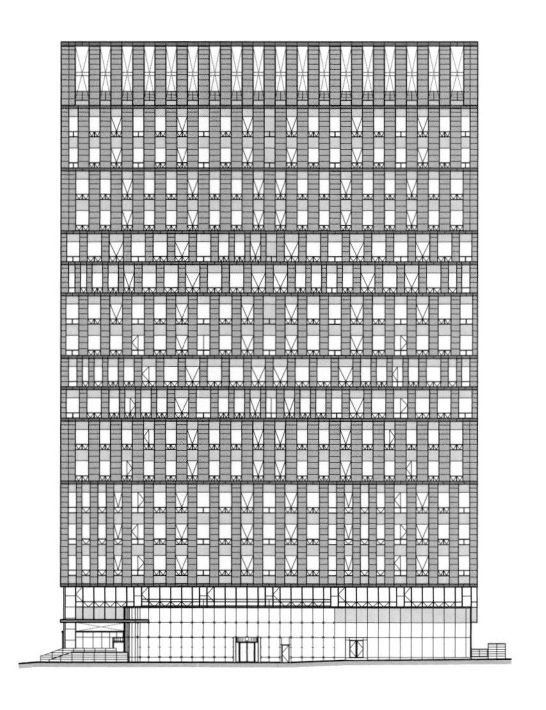

West Elevation

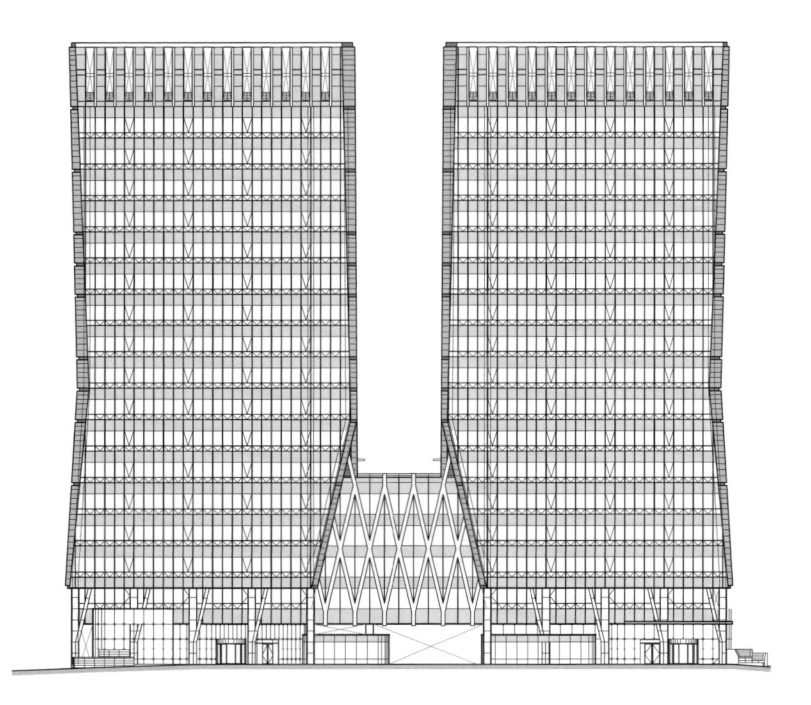

North Elevation

Section 1

Section 2

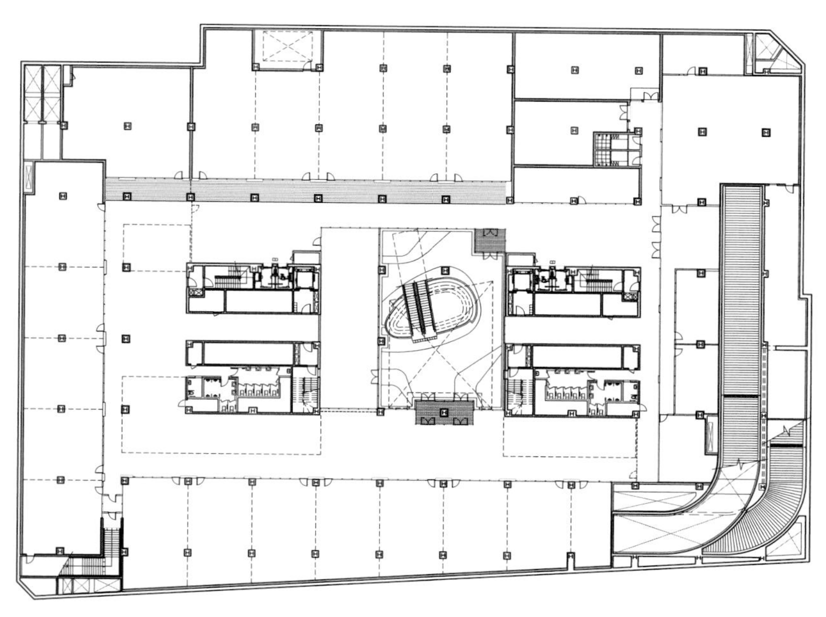

1st Basement Floor Plan

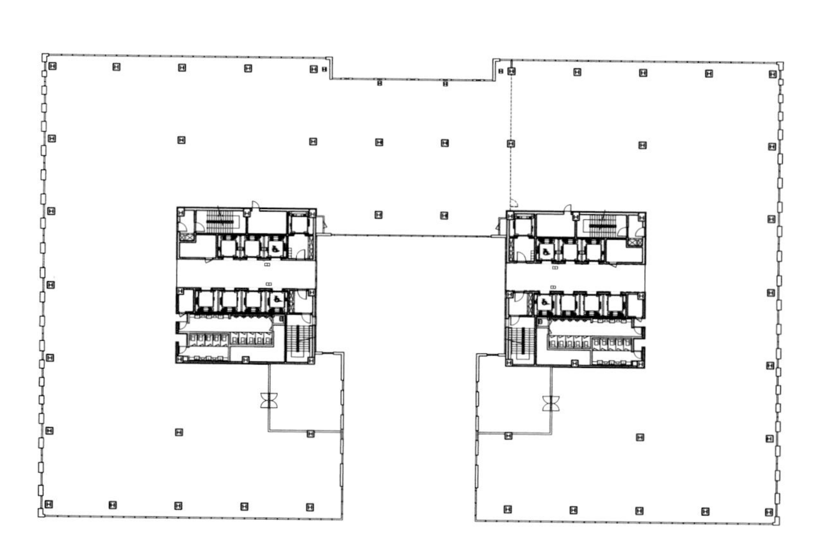

Typical Floor Plan

双塔大厦位于连接南山和仁王山的大都市，在两座塔楼之间使用了开放式的大门设计，并且通过使用石头作装饰材料来减少东西立面的热损耗。大厦地面层设计合理，可以让人们在独立商店自由穿行，给行人的通行及进入零售店提供了最便利的条件。两座塔楼在六层相连，看上去支撑着两座塔楼的好像是底部的大门而不是裙房。

由于大厦高度仅 17 层，所以该项目的最大挑战是在这个比例上添加令人感兴趣的元素。外倾斜面的设计既是为了弥补大楼比例的不足，也是为了突出大门的造型。游客在通过南山到江南区的仁寺洞入口时可以看到这个大门，从清溪川的起点位置也可以看见汉江以南的地区和大厦的顶端，因此，这座大厦足以和其他优秀的建筑相媲美。毫无疑问，这座建筑经过设计团队成员的创新设计，将给这个地区注入新的活力。

光州大都会市尚武区呼叫中心

GWANGJU METROPOLITAN CITY SANGMU DISTRICT CALL CENTER

ARCHITECT
SAMOO Architects & Engineers

CLIENT
Gwangju Metropolitan City Corporation

LOCATION
Gwangju, Korea

AREA
32,840 m^2

PHOTOGRAPHER
Yum Seung Hoon

The inspiration of form came from a stop motion of a Buddhist monk dance. The soft and dynamic curve of the clothes represents the culture of Gwangju, which is known as a city of art lovers. The space between the curves represents Gwangju's hope that is soaring with the spirit of the land also this signifies beginning of the ray of light. The light of hope reflects and spreads after meeting three layers of roof. The building is responding the existing street network and surrounding context. To secure openness and its view, an open space is positioned along the road. Each floor of call center is designed to accommodate more than 150 people, and a center core plan is introduced for efficiency. Considering the fact that majority of the staffs are female, more restrooms and refreshment areas for women were designed. The characteristic of the elevation is a result of combination of a curve and a straight line. While a curve expresses the image of light and calmness and sophistication is expressed by a straight line. By breaking down the mass and using set-backs, it expresses flexible and gentle image as well as avoiding the massiveness of the building. Dynamic and sophisticated outward appearance reflects creative and corporate citizenship in an urban setting.

광주광역시도시공사

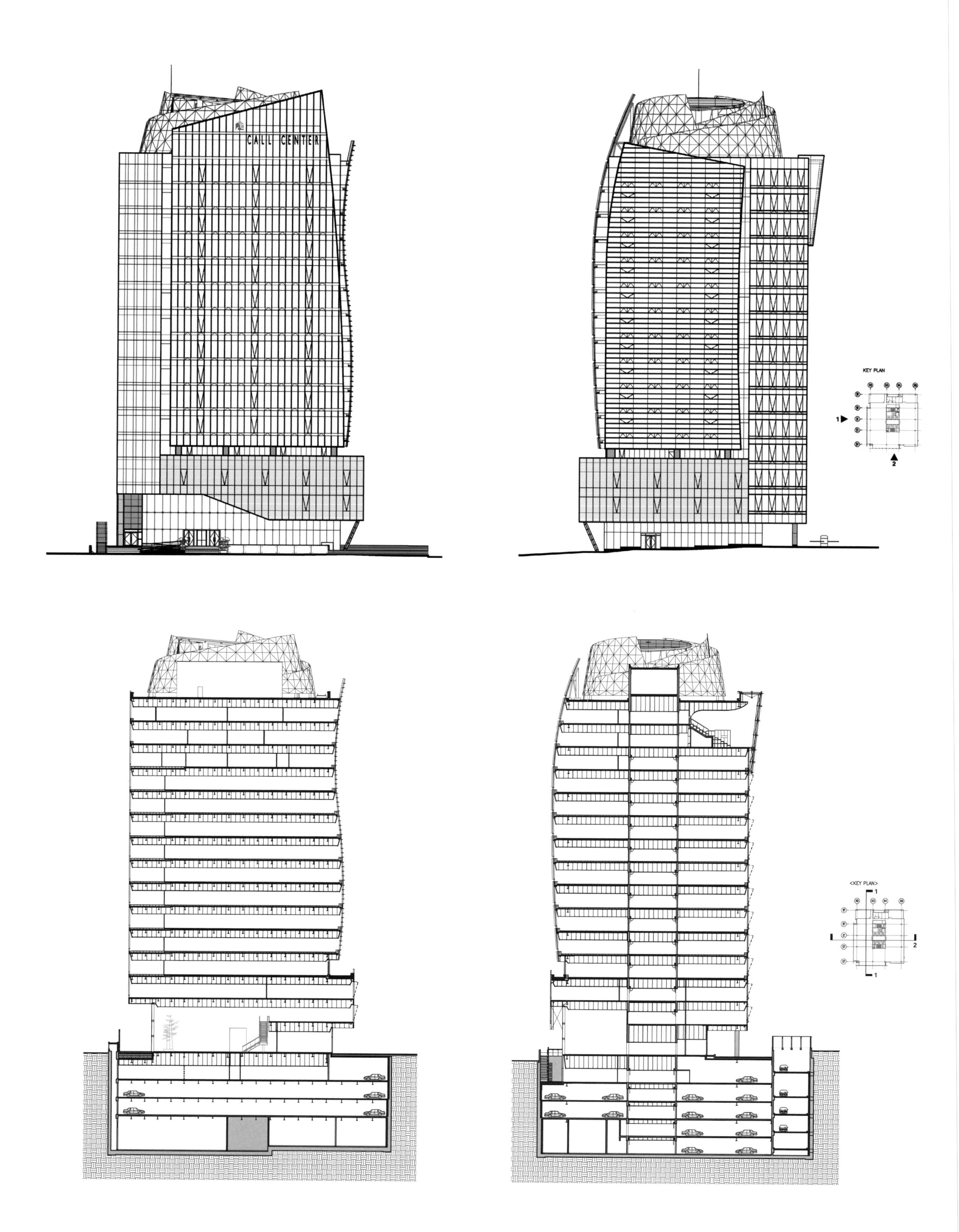
CALL CENTER
KEY PLAN
1
2
<KEY PLAN>
1
2
1

尚武区呼叫中心大厦的造型来源于佛教舞蹈中的一个动作，外形曲线像柔和而动感的衣服，代表着以爱好艺术而著称的光州的历史文化。曲线之间的空间代表着光州的希望，象征着光的起源，照耀着整个光州。象征着希望的光经过反射，汇聚于大厦顶部的三层后，继续向上延伸。大厦的造型与现存的街道网以及周边的环境相互协调。为了保证大厦的开阔视野和景色，设计师们在临街的一侧规划了一个开阔的户外空间。呼叫中心的每一层都可容纳超过 150 名工作人员，另设有一个中心区统筹规划，提高效率。考虑到呼叫中心大多数员工为女性，大厦规划了较多的卫生间以及休息区。大厦的整体造型由各种曲线和直线组合而成。曲线代表光明和平静，直线则表现出精明与简约。通过运用逆转和分解等手法，整栋大厦显得十分灵动与优雅，完全避免了常规办公楼宇的臃肿与庞大。动感与细致的外观反映了创新与联合的大都市公民精神。

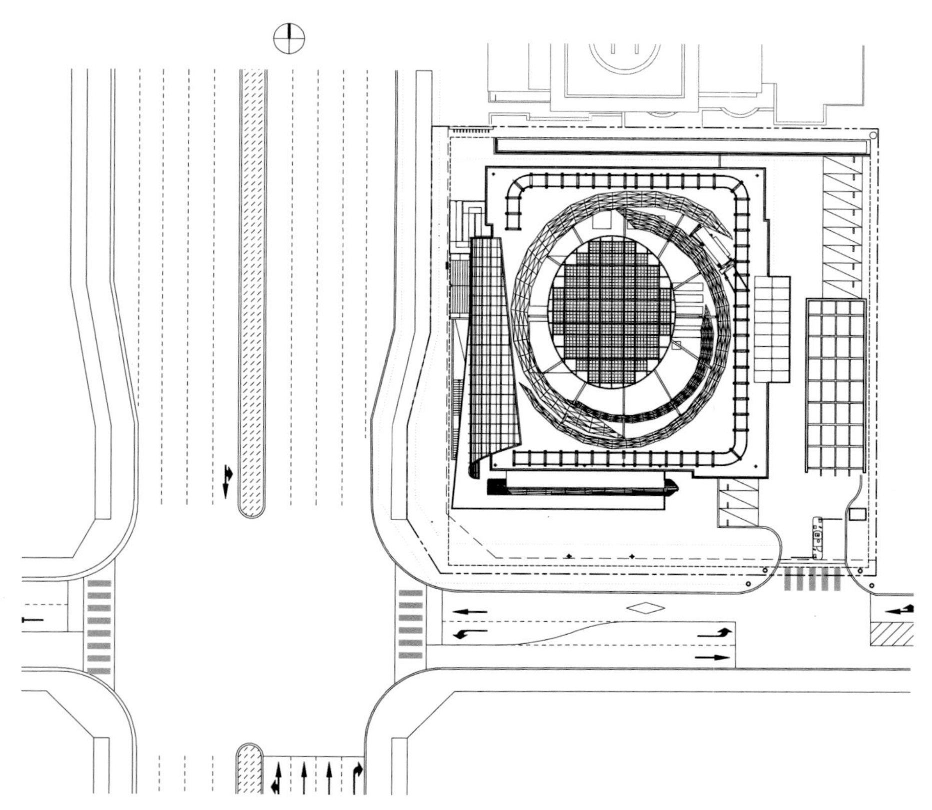

Site Plan

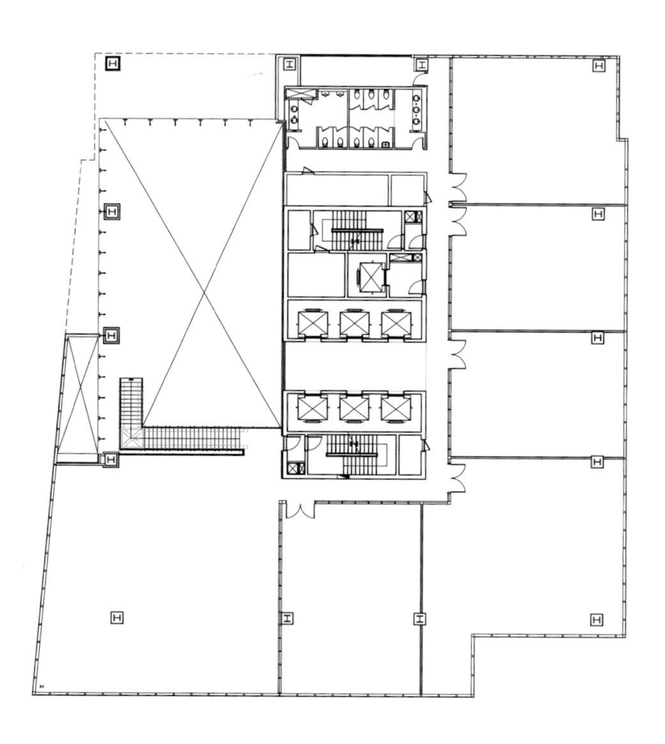

2F Plan

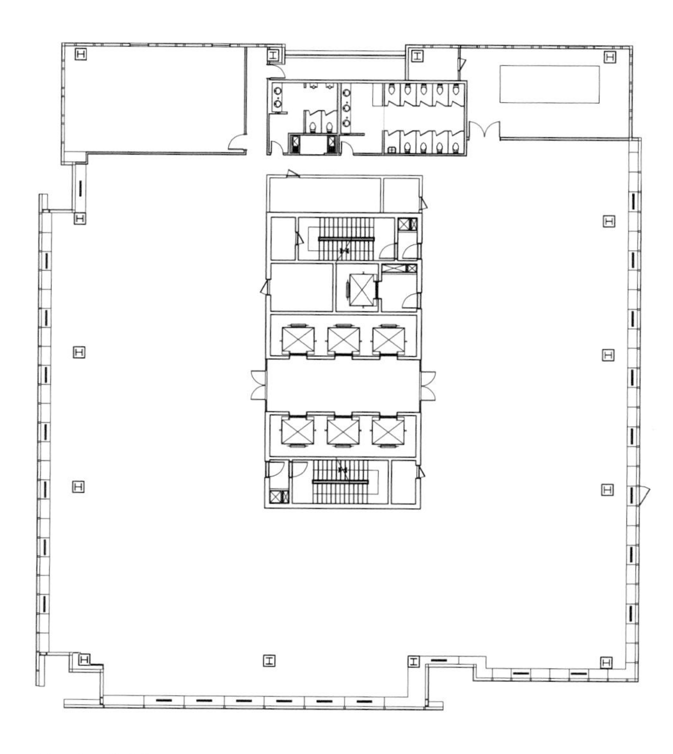

Typical Plan

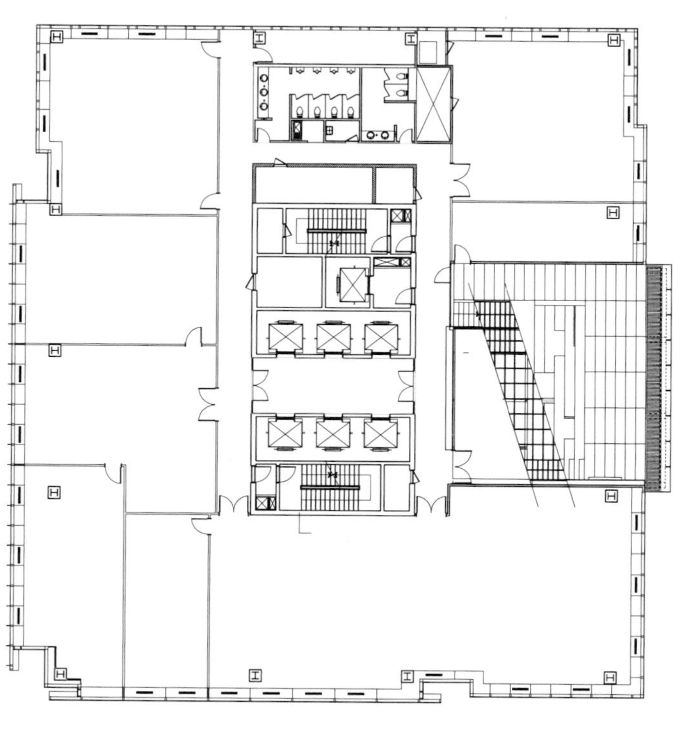

14F Plan

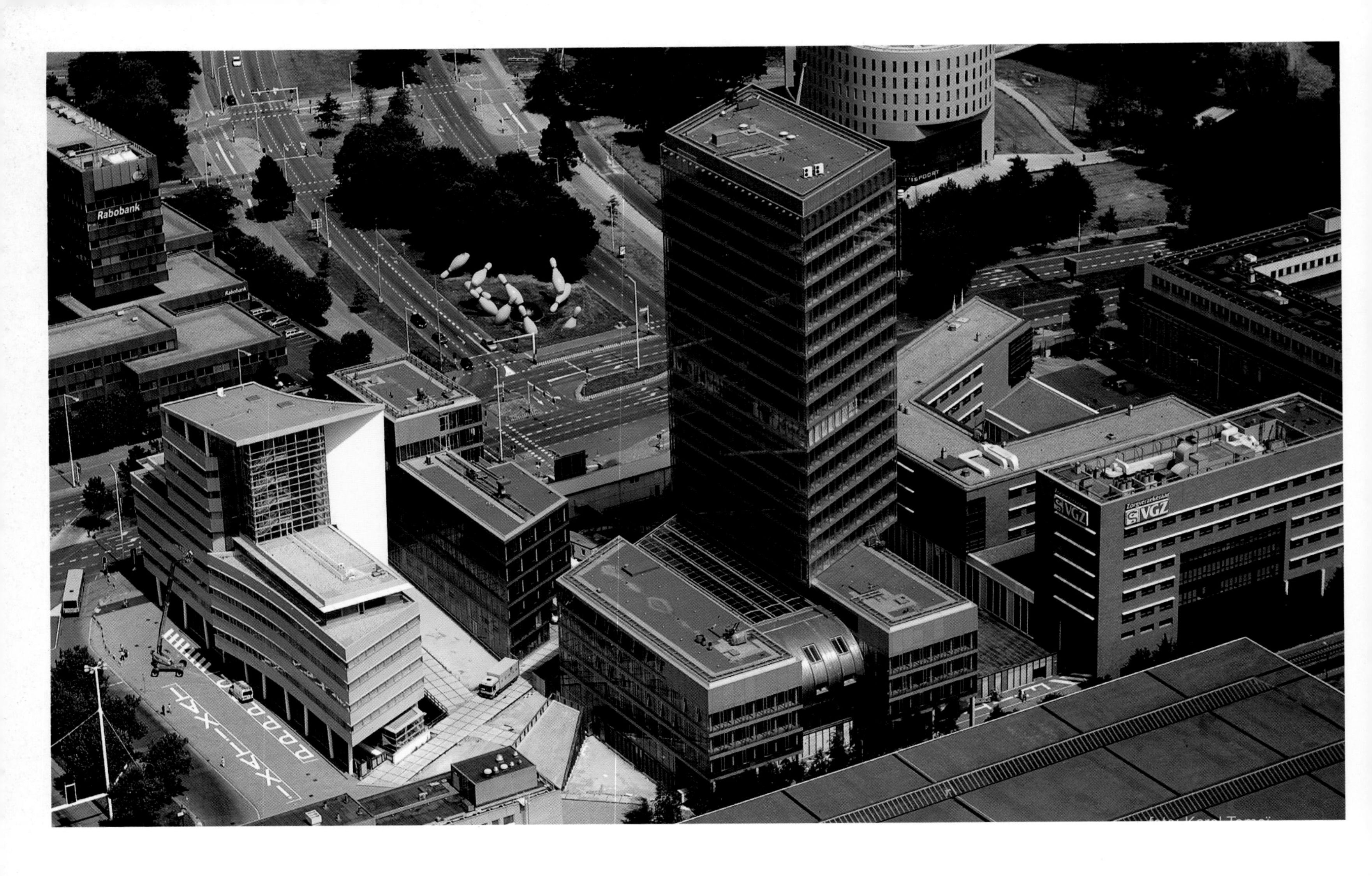

肯尼迪办公大楼

KENNEDY TOWER

ARCHITECT
Van Aken Architecten (VAA.)

PRINCIPAL
Bouwcombinatie (Hurks Bouw & Vastgoed / Heijmans IBC Vastgoed)

CONSTRUCTION CONSULTANT
Adviesbureau Tielemans Eindhoven

INSTALLATIONS CONSULTANT
Halmos Adviseurs

CONTRACTOR
Hurks Bouw & Vastgoed / Heijmans IBC Bouw

CONTRACTOR STEEL CONSTRUCTIONS
Hollandia

CONTRACTOR FACADES
Josef Gartner & Co

LOCATION
Eindhoven, the Netherlands

AREA
30,000 m^2

PHOTOGRAPHER
Van Aken Architecten

The Kennedy Tower is part of the Kennedy Business Center, a business area in the center of Eindhoven, north of the railway station. This model consists of five building zones perpendicular to the railway lines, assuring the transparent skyline to the inner city. The strips are cut by a diagonal opening to the railway station.

With it's entirely "double-skin facade" glass facades, the tower is an eye catcher in the area. On the one hand the transparency gives expression to the technological, innovative atmosphere, and on the other hand it forms a neutral transition between the architecturally different buildings of the Kennedy Business Center. The disposition of the tower in relation to the substructure is provocative and creates great excitement because of the corbeling. The decision to make a steel construction makes this possible and also eliminates the problem of a closed concrete core: the building becomes really transparent and offers great freedom in the layout of the office floors.

The structure was deliberately incorporated in the design with the choice of a Y-shaped structure, in which each oblique line of the Y ensures the stability of three levels. Between the steel hollow sections that cross through the floor an open zone of unconventional workspace is created that provides a large and very flexible lettable surface.

Because of the double skin facade the energy saving is considerable and the optimal daylight and natural ventilation provide maximum comfort. Also it reduces noise from the railway station.

hurks
HEIJMANS

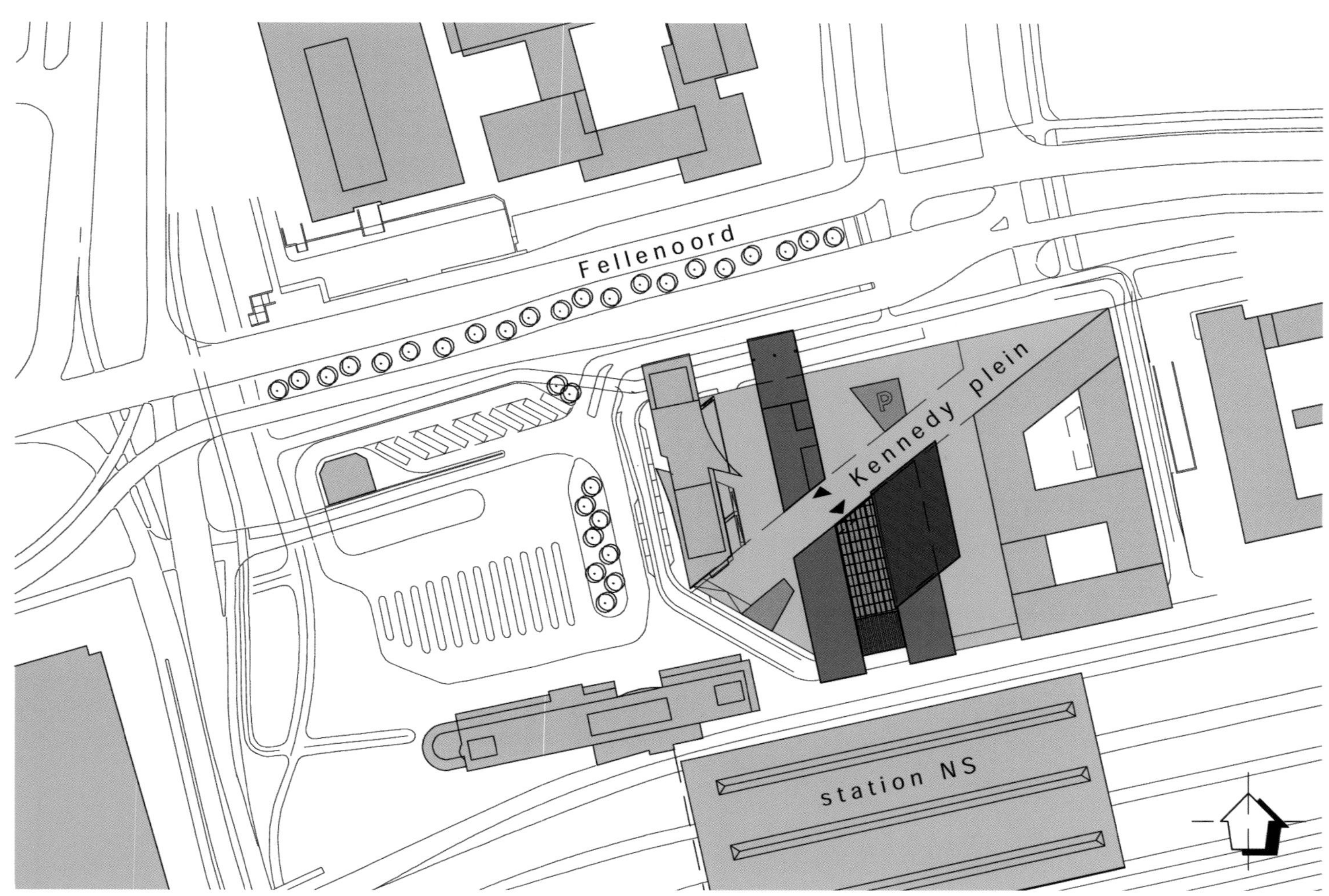

Kennedy Tower-Situatie

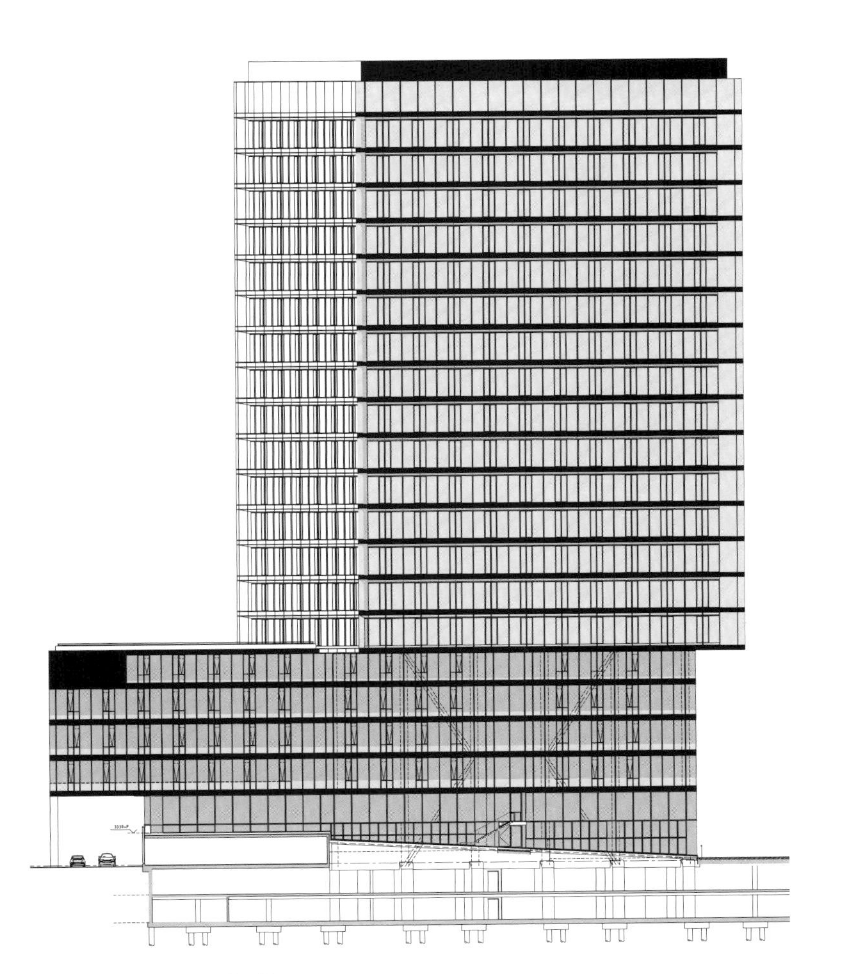
Kennedy Facade East

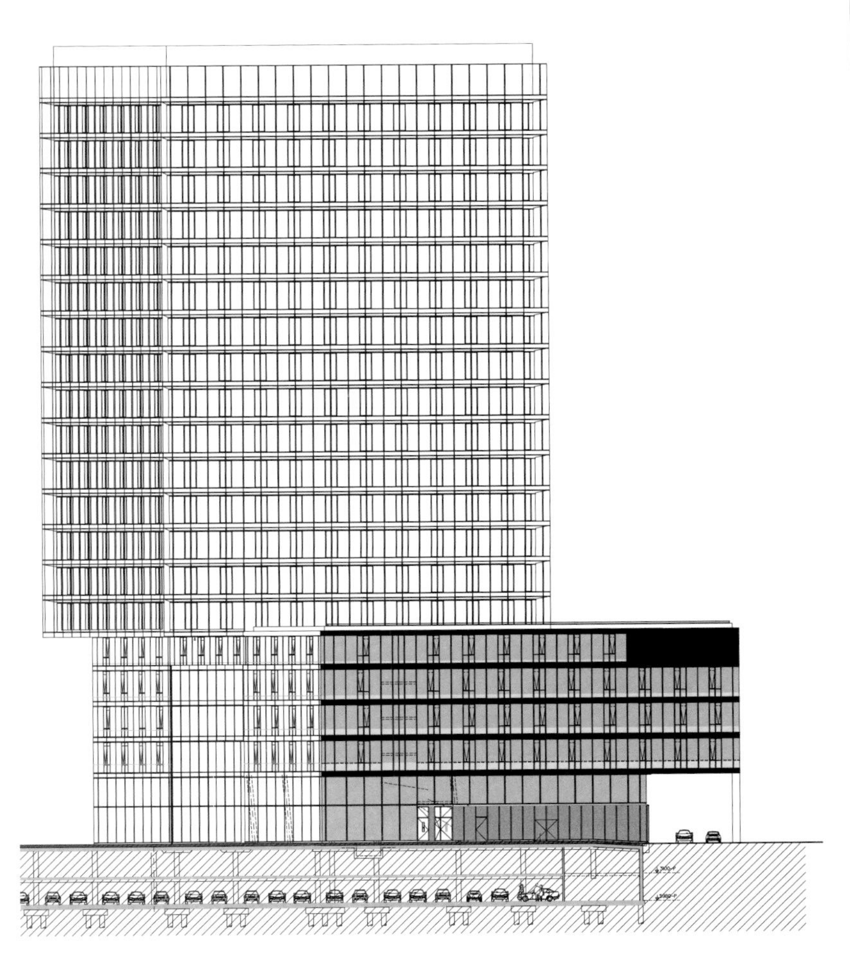
Kennedy Facade West

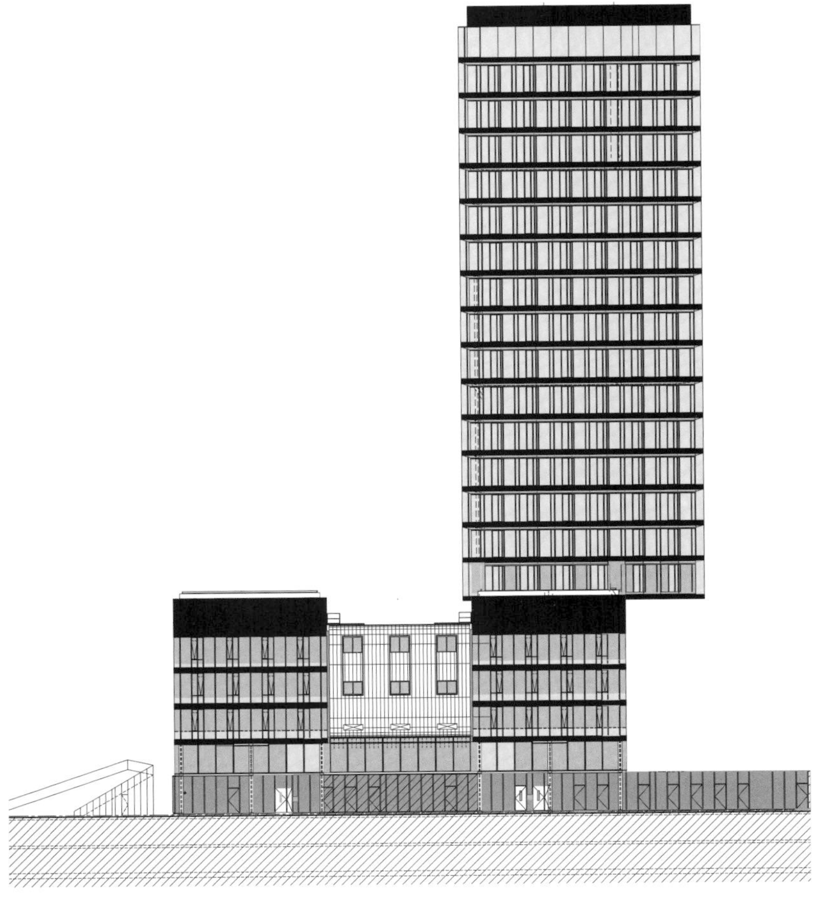
Kennedy Facade South

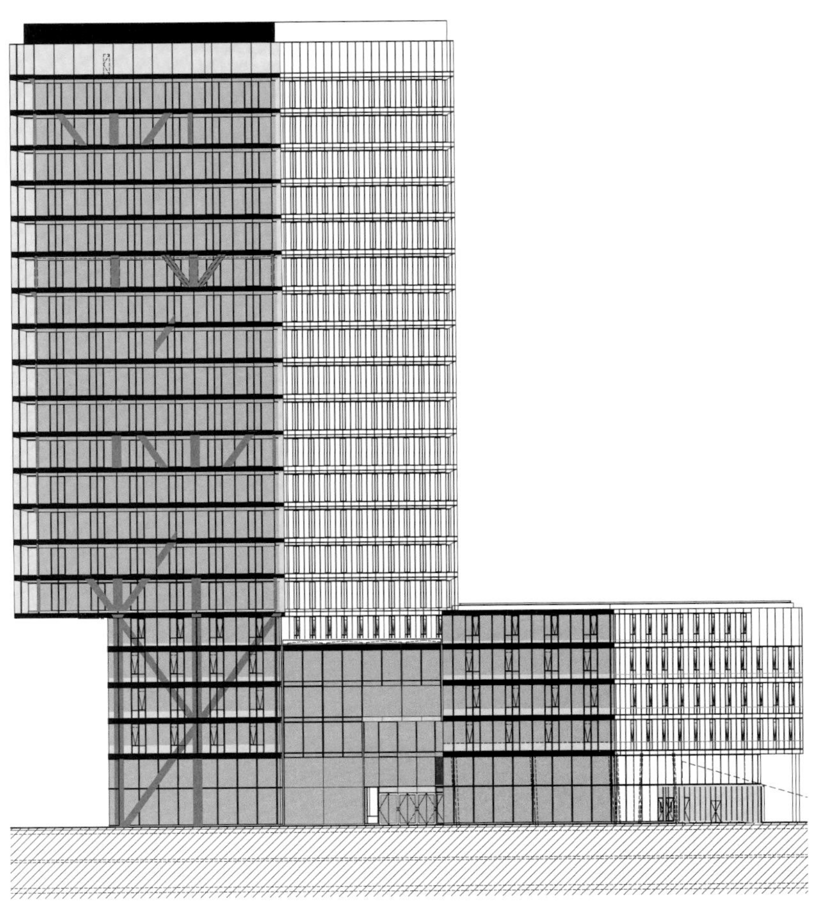
Kennedy Facade North

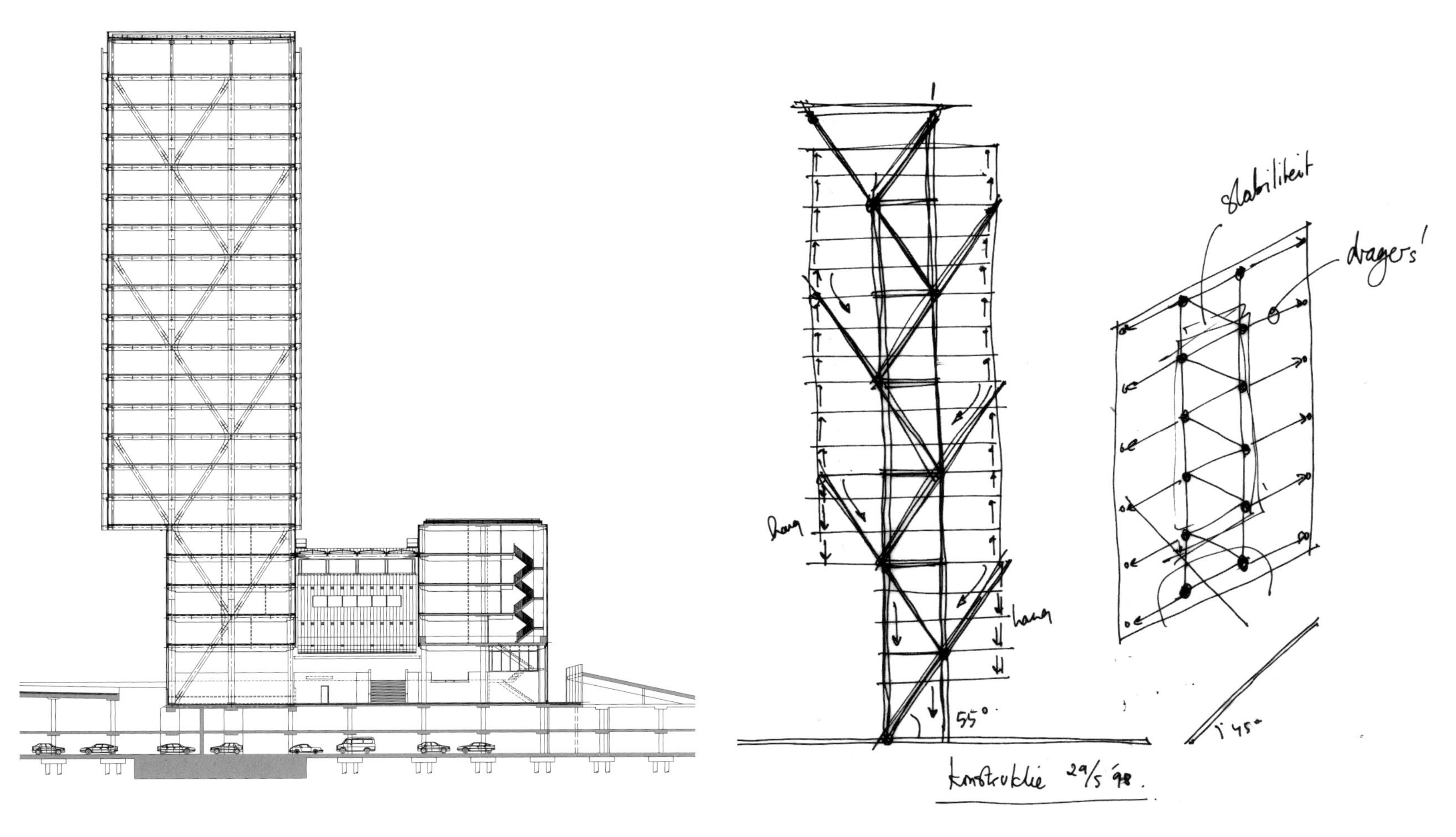

Kennedy Section

Kennedy sketch

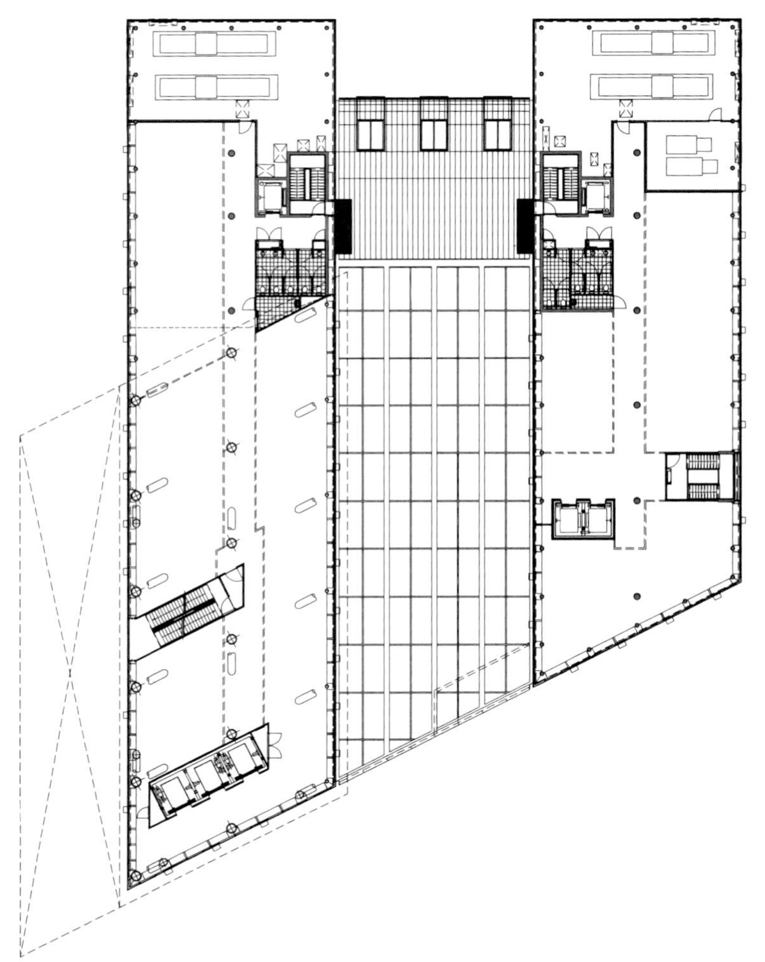

5th floor

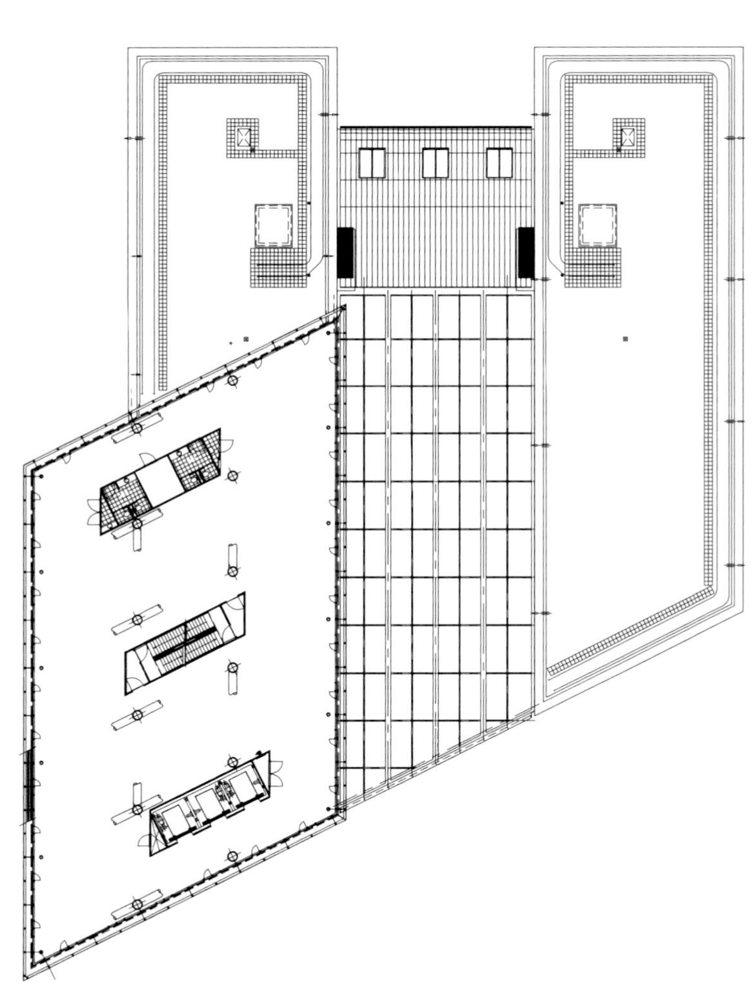

6th floor

肯尼迪办公大楼是肯尼迪商业中心的一部分。商业中心位于荷兰埃因霍温市火车站的北侧。这个经典的建筑包括五个垂直于铁道线的建筑区域，视野宽阔，远远望去，整个市中心尽收眼底。这些条形的建筑区域以斜线分隔，直接面向火车站。

肯尼迪办公大楼的外墙立面采用双层的玻璃结构，在当地十分醒目。一方面，透明的玻璃幕墙营造出了科技和创新的氛围；另一方面，肯尼迪办公大楼在商业中心众多不同风格和类型的建筑中保持了一种中立性的过渡。对肯尼迪办公大楼与其附属建筑的处理发人深思，建筑师们选用的钢枕梁结构产生了十分振奋人心的效果。所采用的钢结构同时解决了混凝土结构无法实现的效果：整栋建筑完全通透，为随后办公区域的规划设计提供了足够的发挥空间。

建筑师们特意将肯尼迪办公大楼的结构与Y形结构相结合，Y形结构的每一条斜线都为整个结构提供了支撑，使办公大楼的结构更稳定。在钢结构的中空区域横跨楼层的地方，设计了一个巨大的开放式区域，作为非常规功能区，从而获得了一个十分灵活的可控表面空间。

双层玻璃的外立面设计节约了大量能源，同时优化了采光和自然通风系统，还降低了来自火车站的噪音，为内部的工作人员提供了一个舒适的工作环境。

NET 公司办公大楼

NET CENTER

ARCHITECT
Aurelio Galfetti, Luciano Schiavon, Carola Barchi

FIRM
LVL Architettura

PROJECT TEAM
Riccardo Albertini, Giano Bernasconi, Nicola Bolpagni, Sergio Campana, Alessandro Corrò, Silvia Drago, Mirco Fiorati, Giulio Gennaio, Kim Kiuri, Florian Ludewig, Chiara Malfitano, Camilla Marchiori, Stefano Marconato, Linda Marini, Silvia Pinazza, Mariangela Riello Pera, Thomas Selmin, Lucia Talpes

CLIENT
Bruno Basso(Progetto Acciaio s.r.l.), Mauro Bertani(Net s.r.l.)

GENERAL CONTRACTOR
Edilbasso S.p.a.

LOCATION
Padua, Italy

AREA
32,000 m^2

PHOTOGRAPHER
Enrico Cano, Sergio Cancellieri, Paolo Frizzarin, Luigi Parise

The complex rests on a 16,000 square meters wide rectangular slab, covered in black slate, for pedestrian use only, while cars are parked underneath it on two levels. Four buildings rise from this slab: the red tower, 20 floors high; two 5-storey buildings, the eastern one with a 150 meters long gallery; and a steel and glass pavilion.

The design of the complex serves to anchor a new urban district, with the red and twisting tower creating a landmark for a larger surrounding territory. It is an attempt to give these suburban places the same quality of the historic centers of the main towns, through urban design and strong architectural expression.

Net Center has a sequential system of urban public spaces: the long gallery is connected to the big square and a smaller future one. Transparency and brightness are the main features of this architecture, the former allowing everyone inside feel like a part of a community, with other people working in the same building, or walking in the gallery or in the square outside, even in the nearby town.

The shading elements give primarily the buildings their architectural expression. In the low-rise buildings they are made of stretched aluminum sheets that operate automatically, following the movement of the sun. The twisting eastern and western facades of the tower are curved surfaces generated from straight lines made by horizontal red grids (three for every floor), while the glazing behind is actually flat. This envelope follows plans that change at every level: the lower floor plans are trapezoids with the shorter sides facing the street, rising up towards the top, the plans become opposite trapezoids with the shorter sides facing the public square.

B4 HOTELS

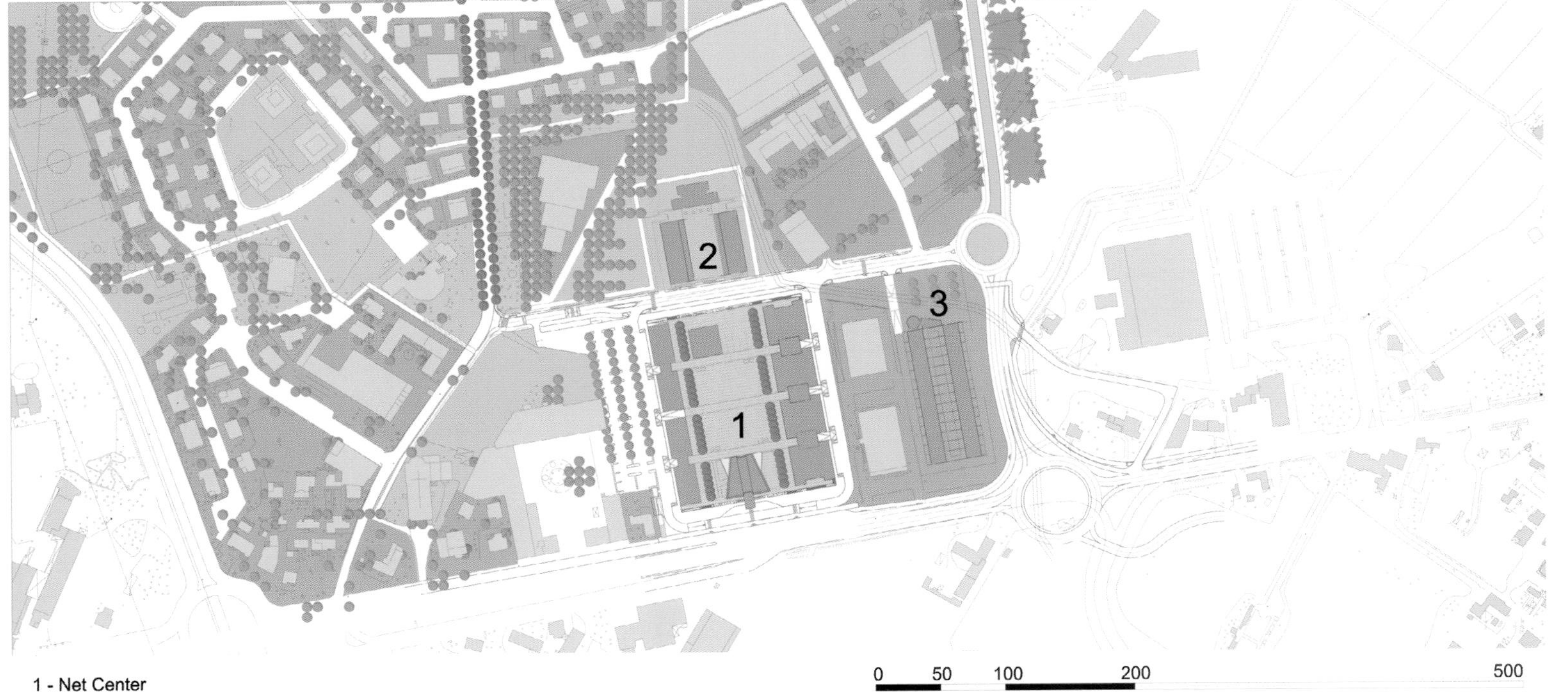

1 - Net Center

2 - Complesso di futura realizzazione

3 - Complesso di futura realizzazione

Net Center Siteplan

NET 公司办公大楼建在一块占地面积约 16 000 平方米的长方形地块上，上面铺设了黑色的石板作为人行步道，地下设有两层停车场。在这个地块上矗立着四栋建筑：一座 20 层高的红色塔楼；两栋 5 层高的大楼，其中东面大楼带有一条 150 米长的长廊；一栋钢结构和玻璃构成的阁楼式建筑。

NET 公司办公大楼的设计旨在打造一个全新的市区，使红色的、盘绕的塔楼成为周边更大范围地块中的地标性建筑。这实际上是对城郊的一次尝试性改造，通过都市化的建筑规划设计以及强有力的建筑表现形式，力图将这些城郊地区打造成为与主城区拥有同样的历史地位以及都市品质的区域。

NET 公司办公大楼具有典型的都市公共空间的建筑系统：长廊连接着大广场和一个即将建成的小广场。整栋建筑通透、明亮。这种通透感会给每个人带来一种归属感，与同一座大楼中的共事者，亦或是走在长廊、户外广场甚至是周边城镇的人们互动。

各种遮阳设施首先赋予了大楼强烈的建筑表现力。多层建筑上面利用自动拉伸的铝板，随着阳光角度的变换，不断改变遮蔽方向。盘绕塔楼的东西两侧则选择了红色的弧形表面，以每层三个格子状的外形构成一行直线，而背面则衬以平面玻璃。建筑表面的每一层都有不同的变化：低层使用梯形模块，短边面对着街道，随着楼层的增高，逐渐变为相反的方向，梯形短边朝向相反的公共广场。

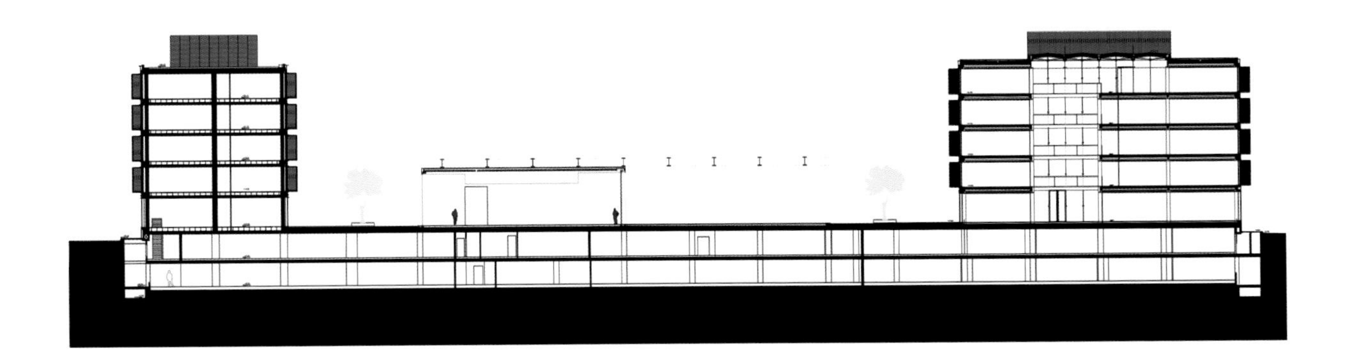

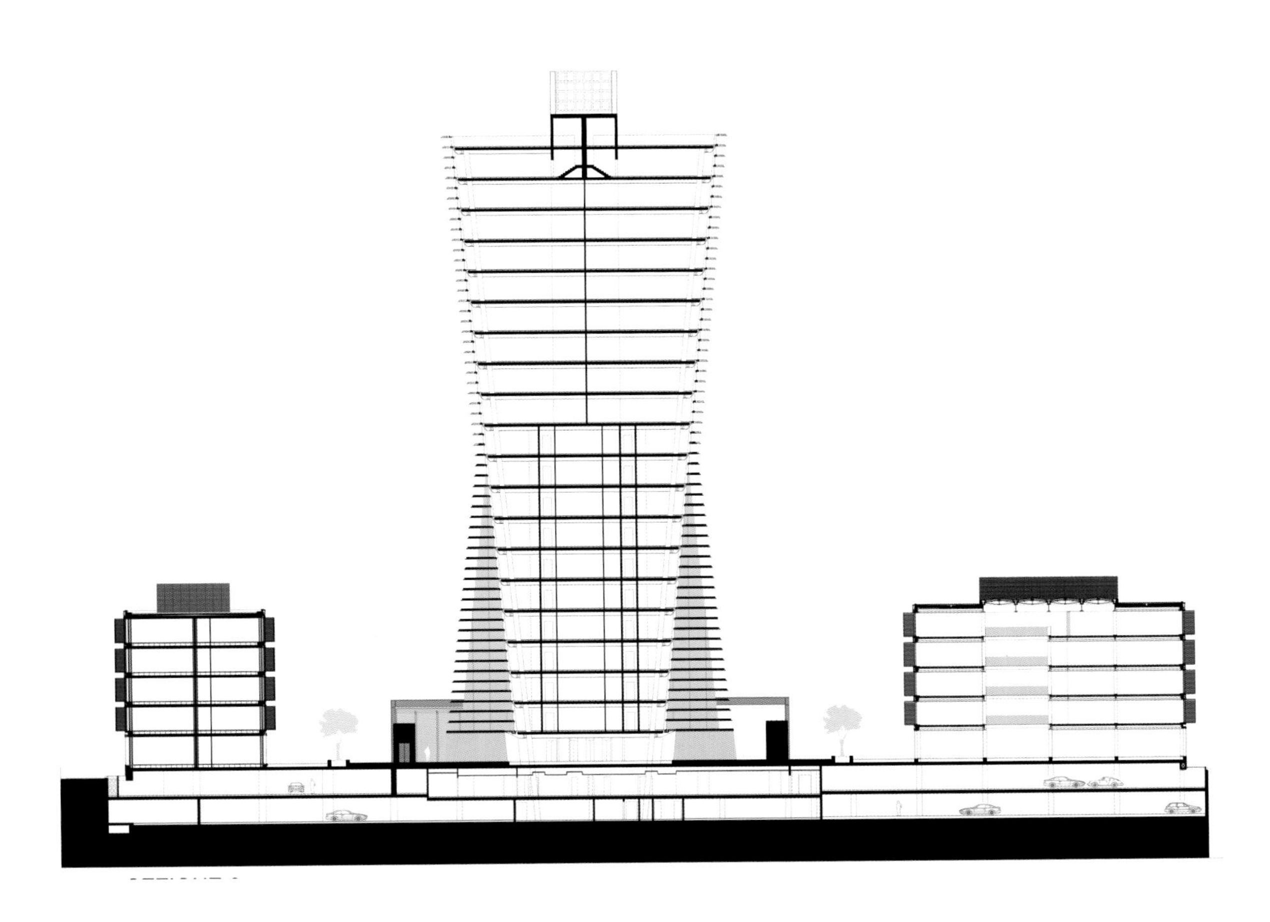

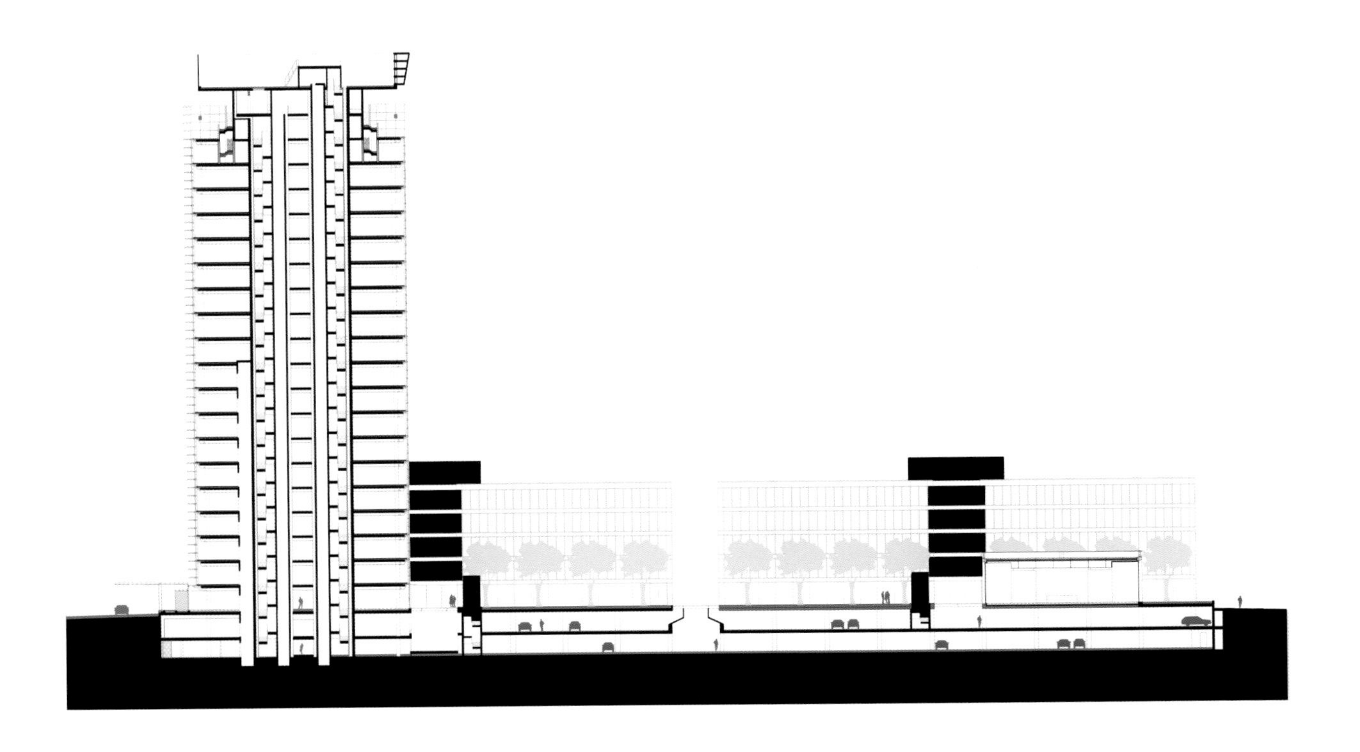

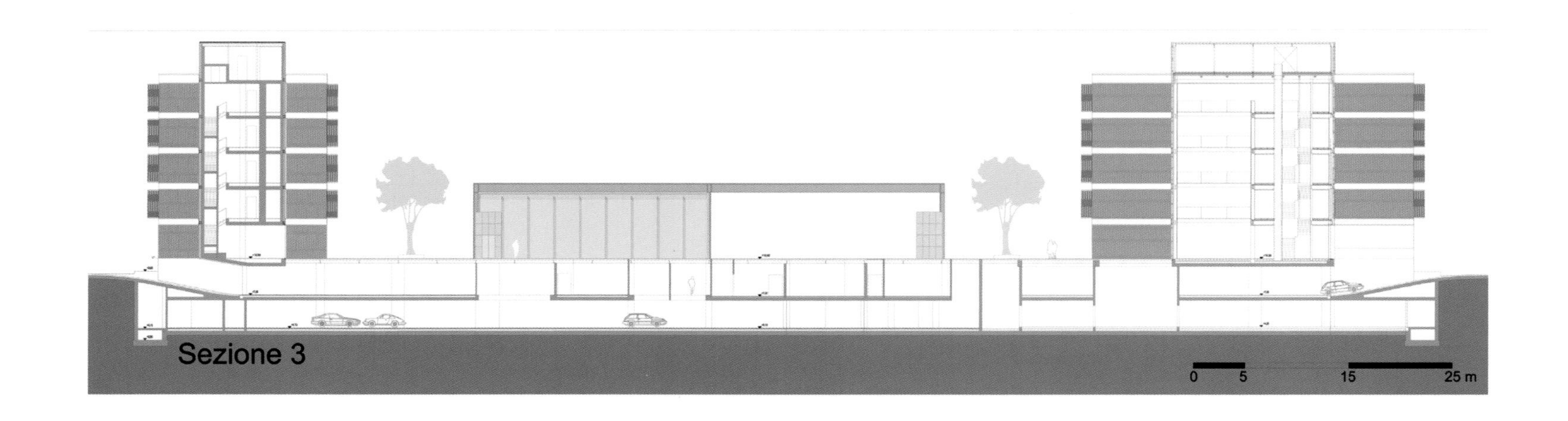

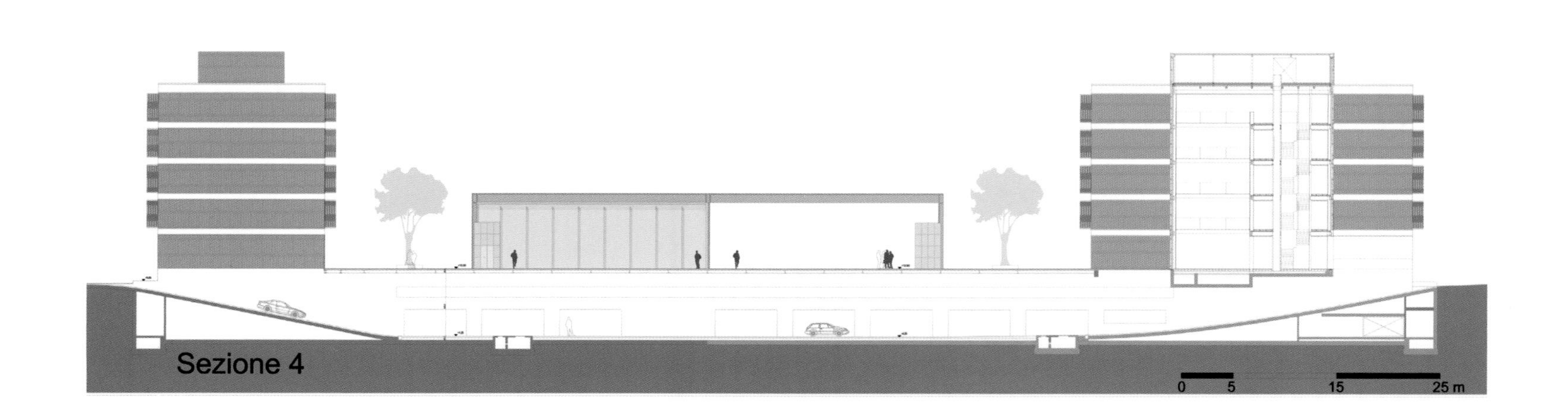

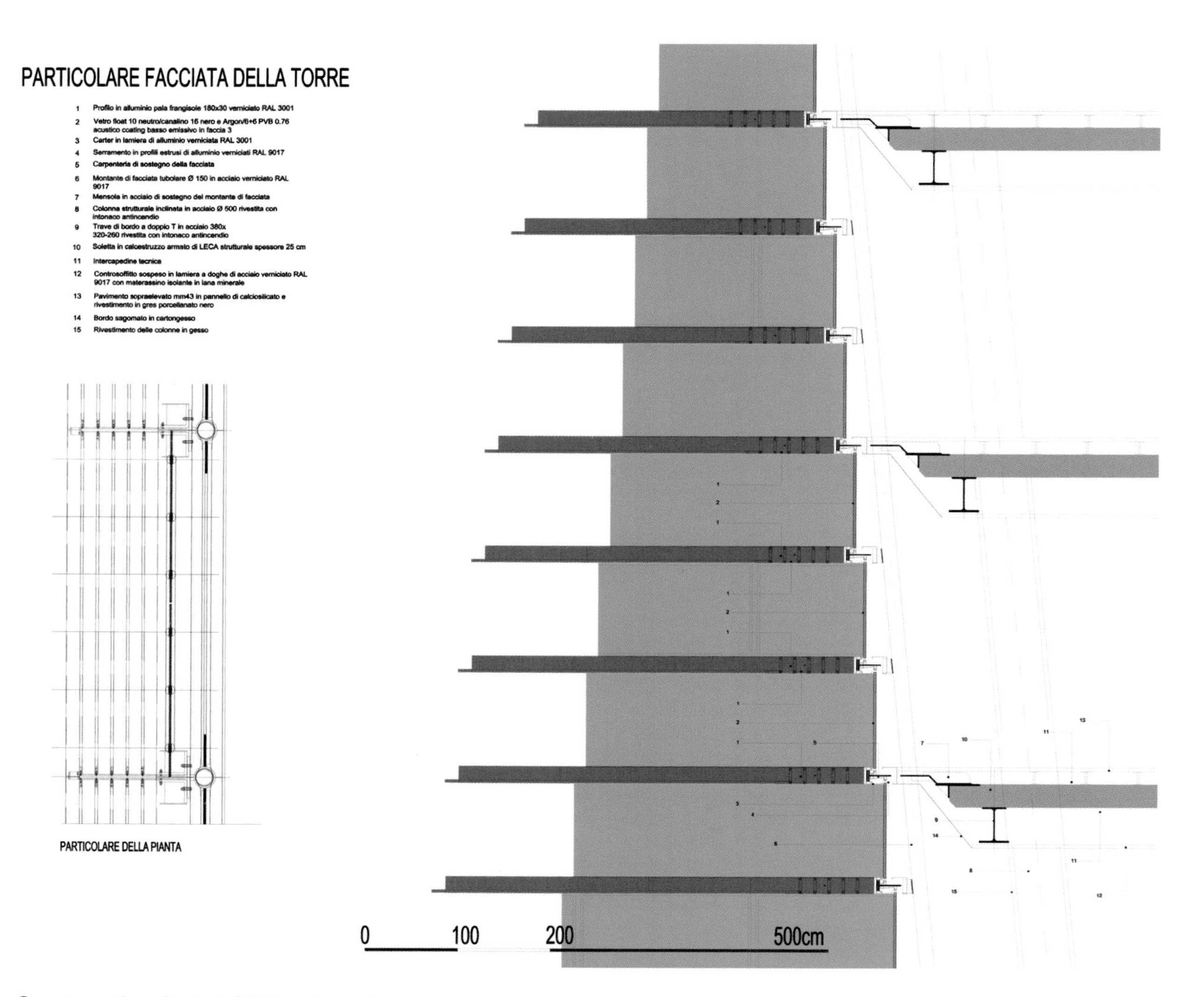

Construction Detail Of The Facade

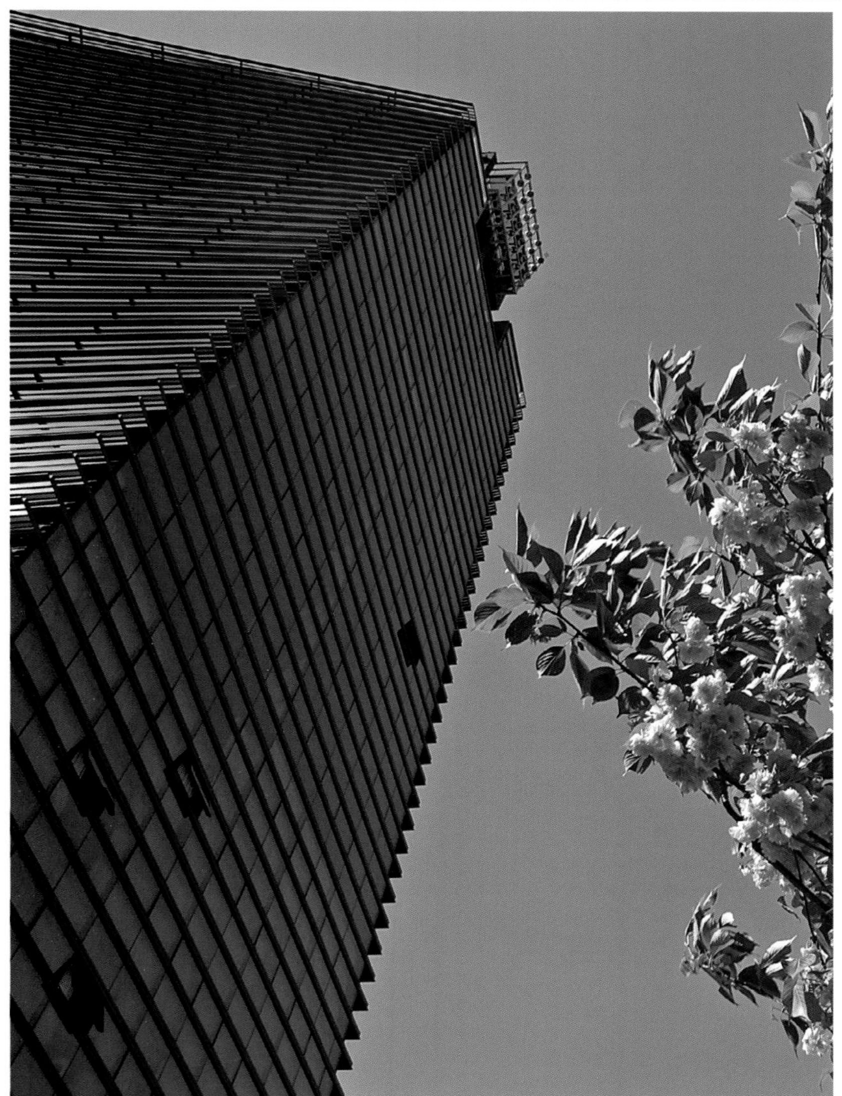

LV
ARCHITETTURA
LUCIANO SCHIAVON

BANG & OLUFSEN
PadovaStar

BANCA POPOLARE
ITALIANA

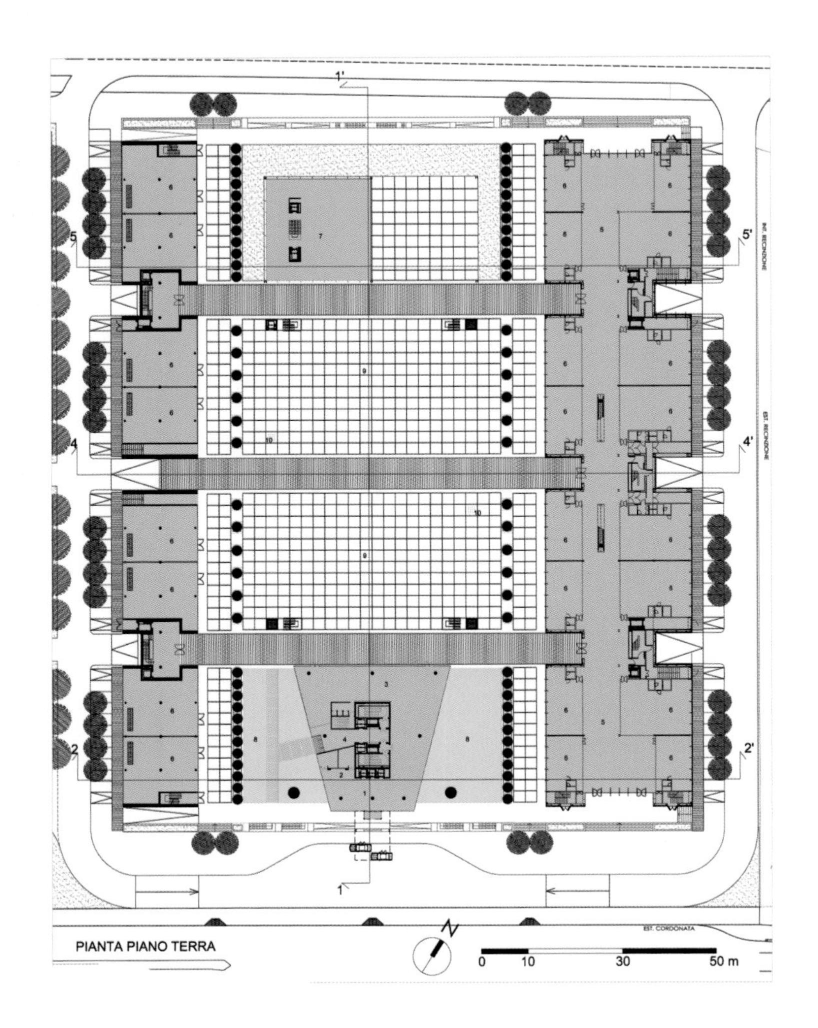
PIANTA PIANO TERRA
0 10 30 50 m
EST. CORDONATA

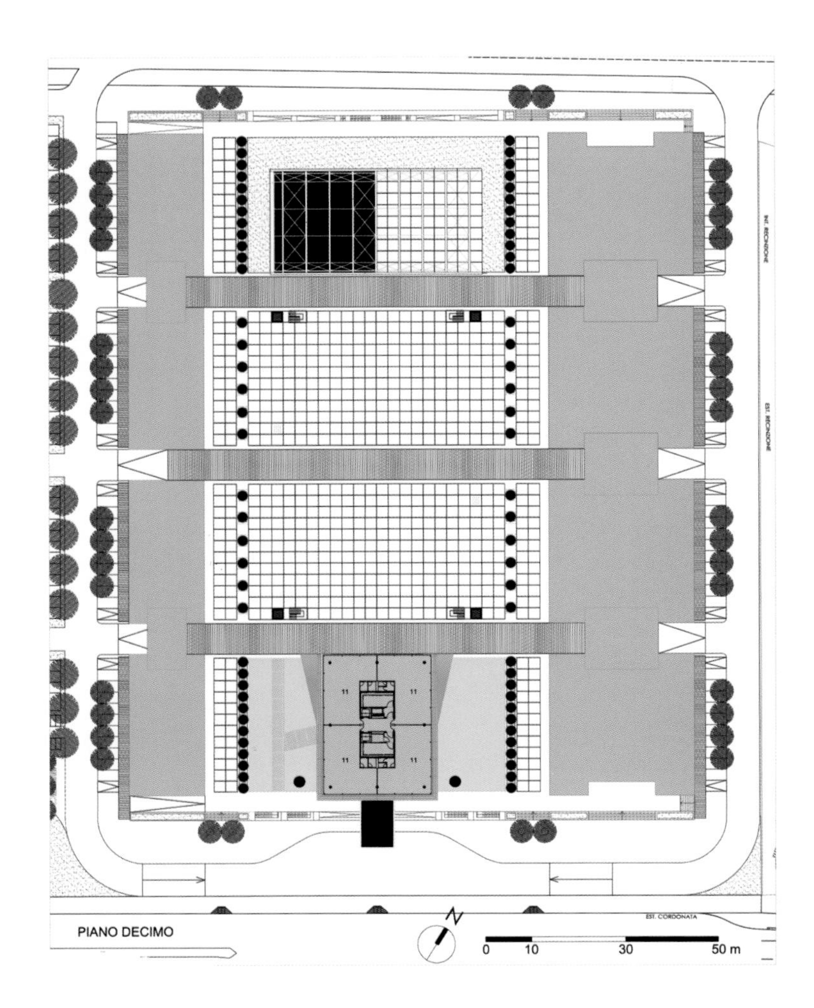
PIANO DECIMO
0 10 30 50 m
EST. CORDONATA

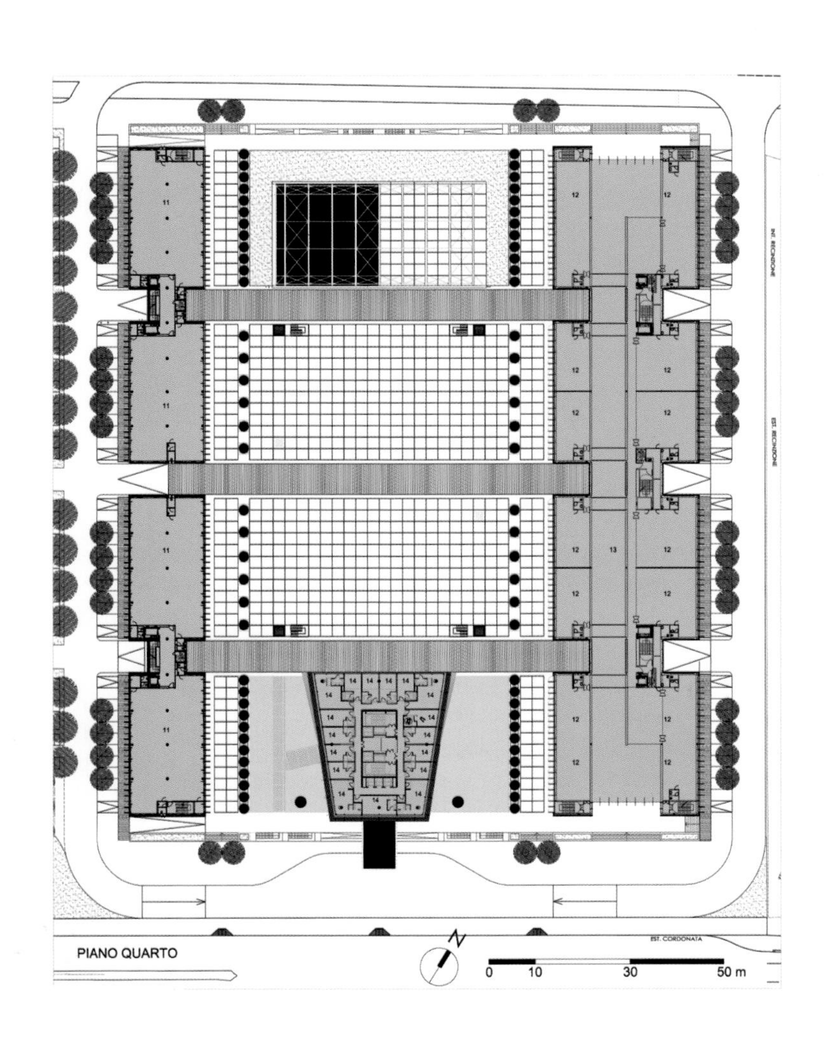
PIANO QUARTO
0 10 30 50 m
EST. CORDONATA

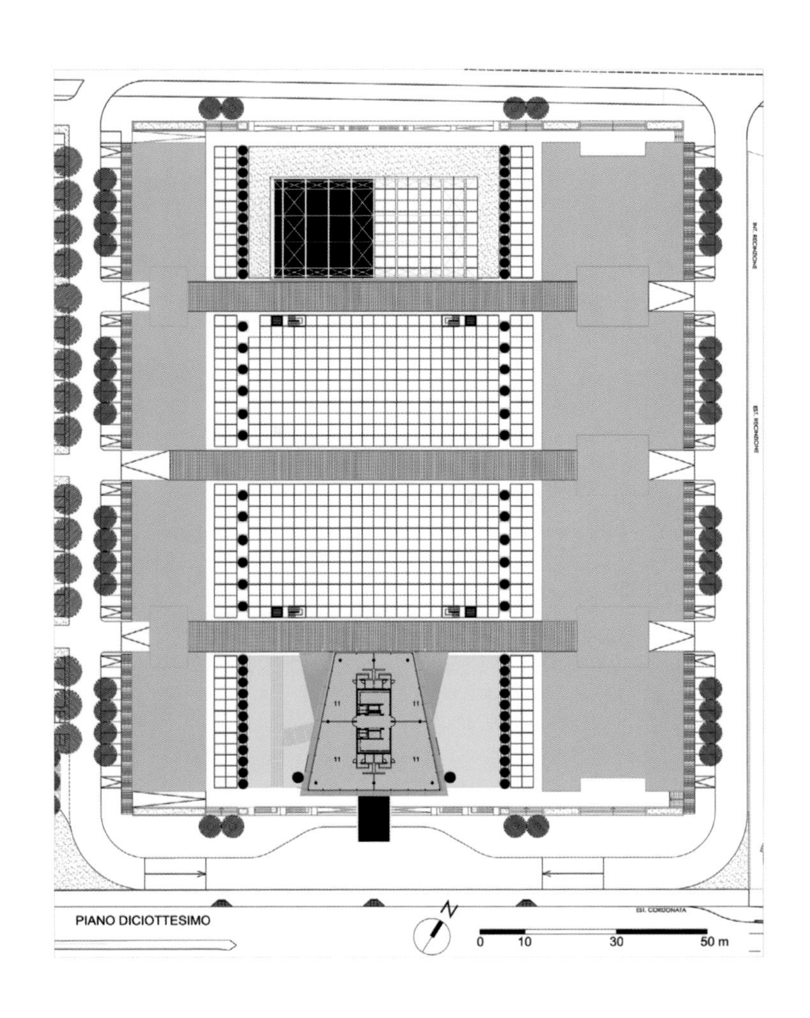
PIANO DICIOTTESIMO
0 10 30 50 m
EST. CORDONATA

埃里克・范・艾格利特办公大楼

ERICK VAN EGERAAT OFFICE TOWER

ARCHITECT
Erick van Egeraat

DRAWING
Erick van Egeraat

LOCATION
Amsterdam, the Netherlands

AREA
33,500 m^2

PHOTOGRAPHER
Christian Richters, J Collingridge

The Erick van Egeraat office tower, also known as The Rock, is part of an expressive high-rise urban development, south of Amsterdam, named Zuidas. Located in the proximity of the city center and with direct access to the urban network of public transport and highways, the project incorporates the potential of this unique location. The ambitious programming of a vibrant, high-density mixture of offices, housing, retail and public space, designed by nine international architects, all contribute to an exceptional development of metropolitan scale. The urban concept for this location as developed by De Architekten Cie. is based on a vertical layering structure with the anatomical analogy of legs, torso and head.

The Erick van Egeraat office tower challenges this masterplan further and proposes to create an explicit tactile and emotional experience out of the stacked block structure. Both an innovative composition of shifted volumes and a transformation from a light to heavy materialisation, it creates an expressive landmark which appears different from every angle. Each of the three sections of the building reveals its own character and material expression and offers space suited to potentially different tenant requirements. The lower part is transparent to allow light deep into the building and stimulate interaction with the direct surroundings. The upper part is characterised by natural stone layers composed in a facade pattern which creates a variety of openings and panoramic views. Both parts are connected by a combination of transparent and printed glass elements and aluminium panels, forming a subtle transition between the two.

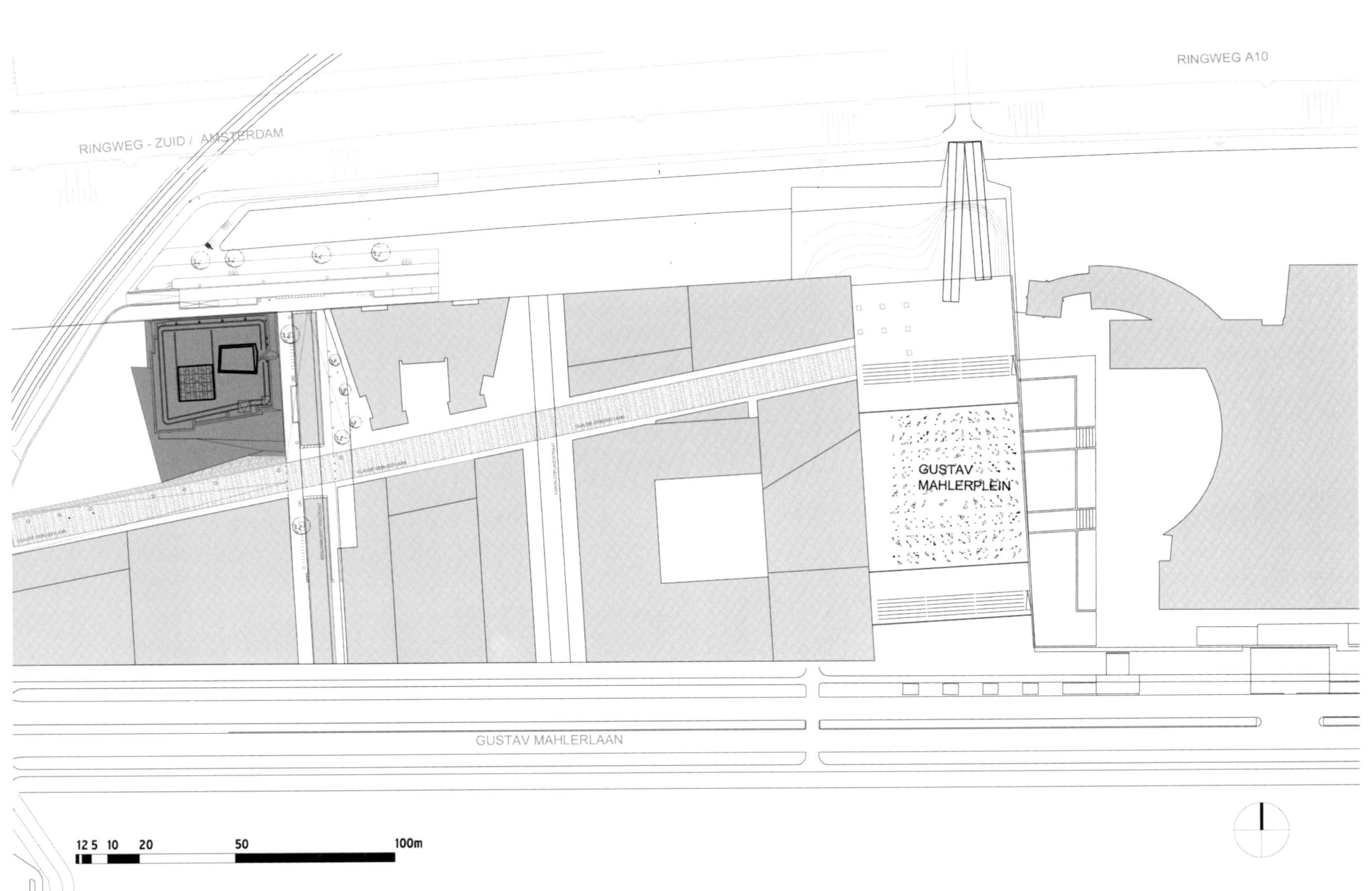
RINGWEG A10
RINGWEG - ZUID / AMSTERDAM
GUSTAV MAHLERPLEIN
GUSTAV MAHLERLAAN
12 5 10 20 50 100m

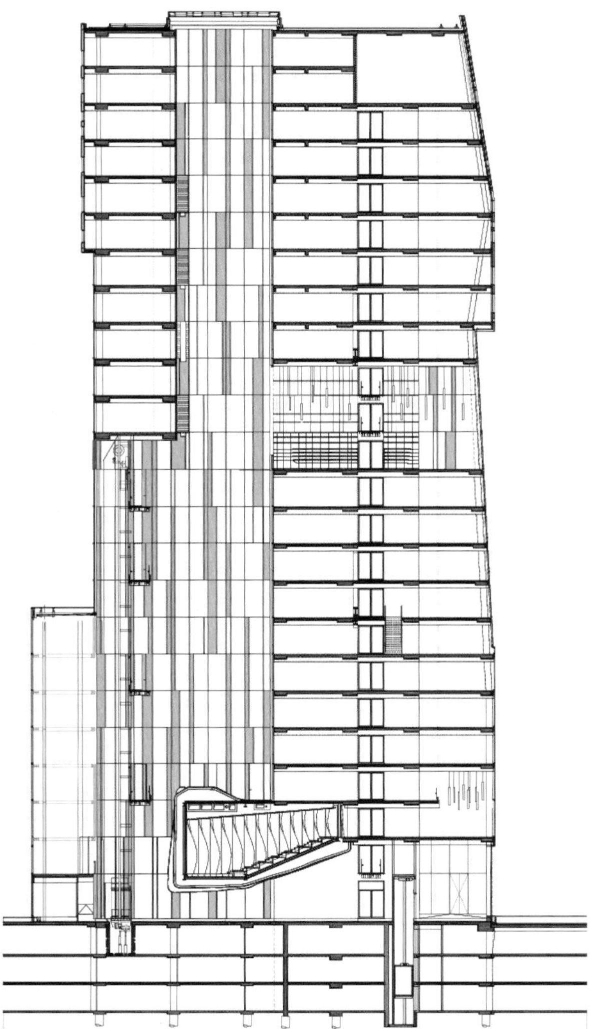

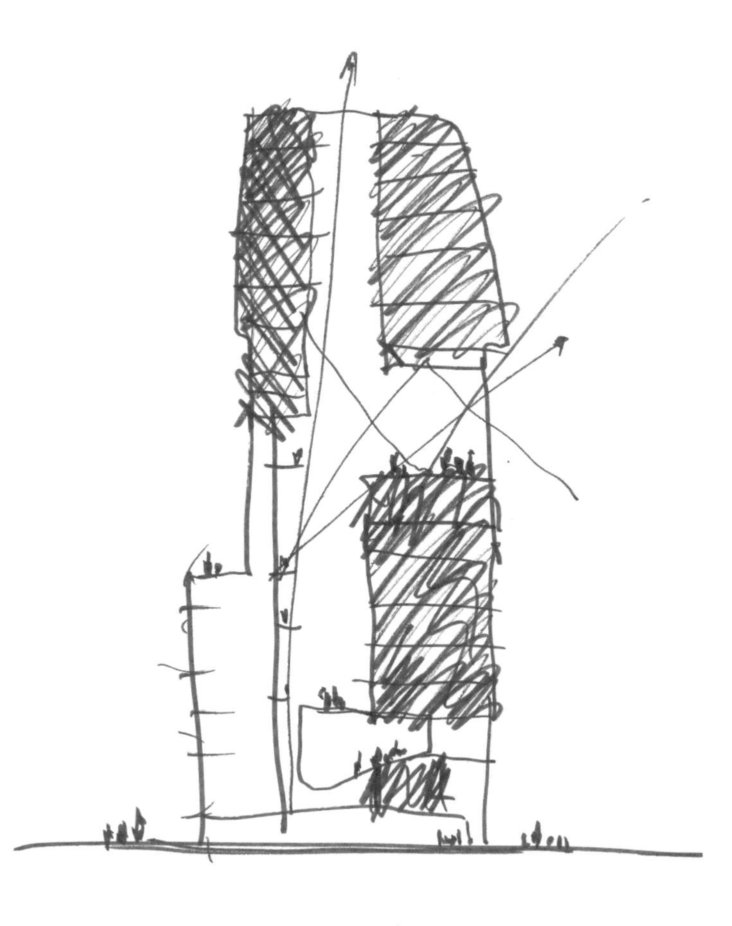
Mahler 4 Amsterdam

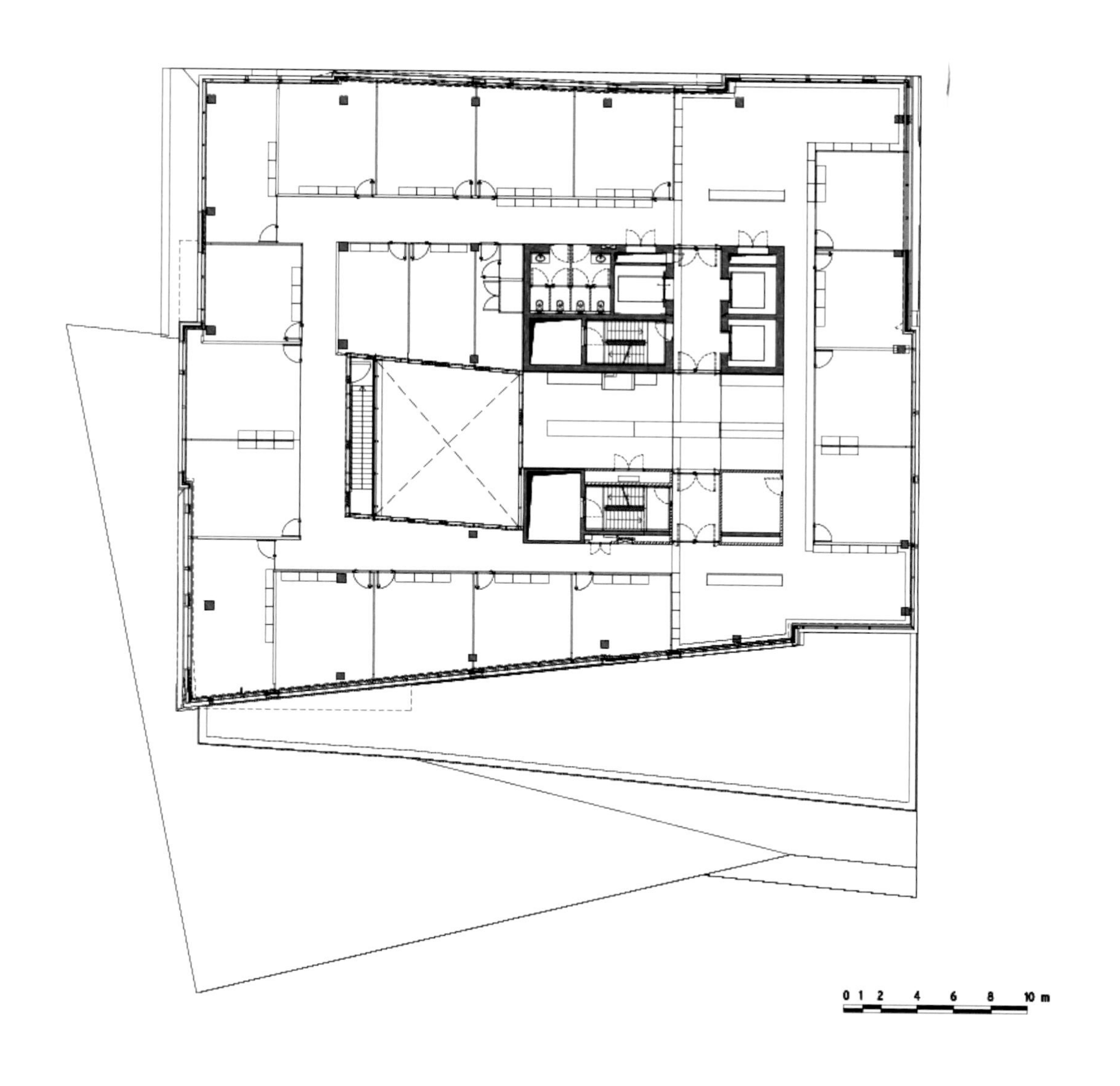
0 1 2 4 6 8 10 m

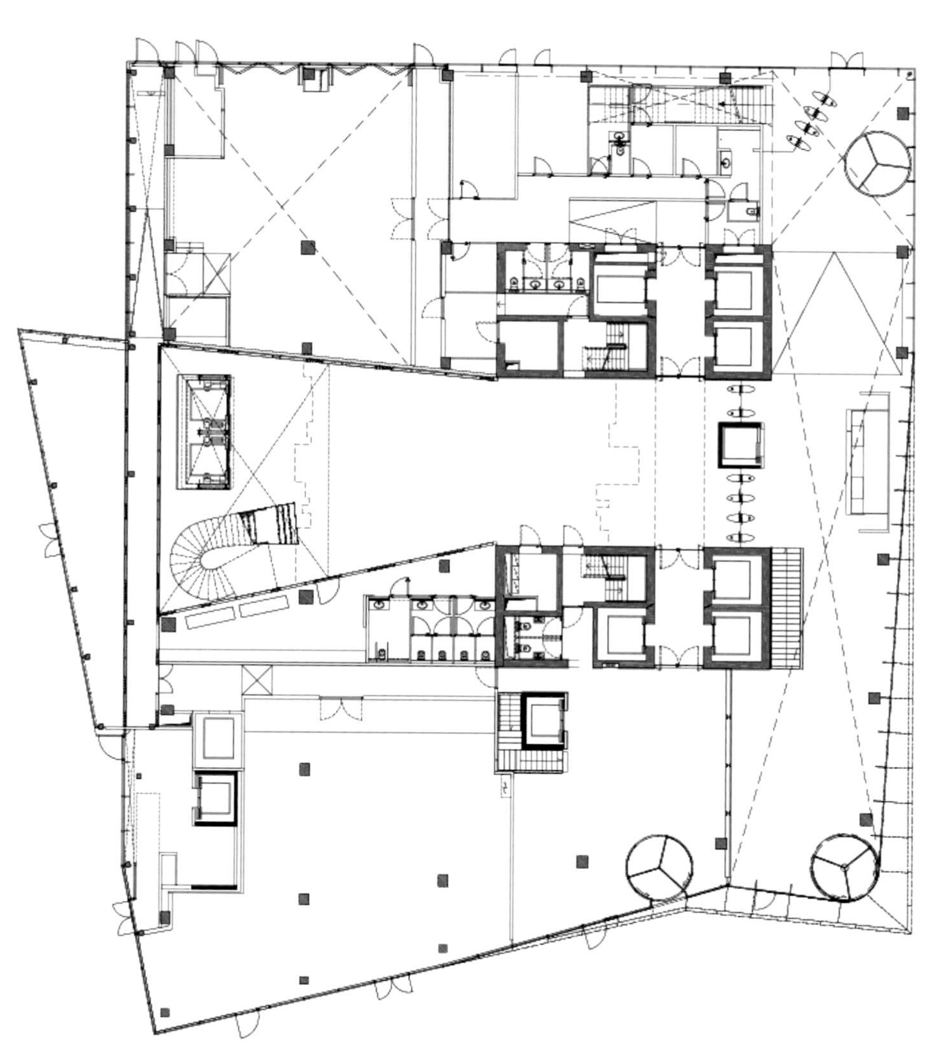

埃里克·范·艾格利特办公大楼又名岩石大楼，坐落于阿姆斯特丹南部高速发展的祖伊达斯城市开发区。岩石大楼临近城市中心区，并且坐落于城市交通与高速公路系统的枢纽位置，具有独一无二的地理优势。通过九名国际建筑师的通力合作，这个宏伟而杰出的大都市办公大楼内部规划了高密度的办公体系、仓储区域、零售区域以及相应的公共空间。通常，都市化办公大楼的规划设计理念来自于设计师阿史泰特·希伊，这是一种基于仿生结构的垂直层状结构设计，由上而下分别象征着人类的头部、躯干以及腿部。

埃里克·范·艾格利特办公大楼进一步挑战了这一总体规划，希望利用巧妙的设计方法，让人们在体积庞大的办公大楼体验一种清晰的触觉效果和强烈的情感体验。通过利用可移动式空间构成标新立异的空间结构，利用由轻到重的材料实现视觉的变换等，从而打造出一栋无论从任何角度观看，都有不同观感的地标性建筑。办公大楼的三个组成部分风格各异，分别使用不同的建筑材料，构成不同的办公空间，以满足潜在客户的不同需求。建筑的下半部分外墙完全透明，以保证自然光线直接射入大楼内部，同时这样的设计也可以让内部的办公人员能够自然而舒适地享受周边的景观视野。建筑的上半部分外墙利用天然石板进行装饰，呈现出集开放式和全景式为一体的多样景观。这两部分利用透明的彩绘玻璃以及铝板装饰的排列组合进行过渡，巧妙地将两者融合为一体。

莱姆斯商业中心

RAMS BUSINESS CENTER

ARCHITECT
Bogdan Stoica, George Mihalache

FIRM
ARHI GRUP

LOCATION
Bucharest, Romania

PHOTOGRAPHER
Andrei Margulescu

Rams Business Center is an example of A-class office building that escapes the common templates. The building stands in Bucharest's outskirts that have witnessed tremendous growth lately. The former industrial sites in the area are undergoing rapid urbanization and several residential and office complexes are being erected. The primary volume was some-what awkward, as it resulted from the planning regulations imposing different recesses for two parts of the same plot. The architectural response to such planning rules was a design that fragmented the building. The two resulting volumes obey the required heights and mellow the strong impact of the 9-floor bulky volume. Logically, the articulation between the two contains both the access and the main circulation core.

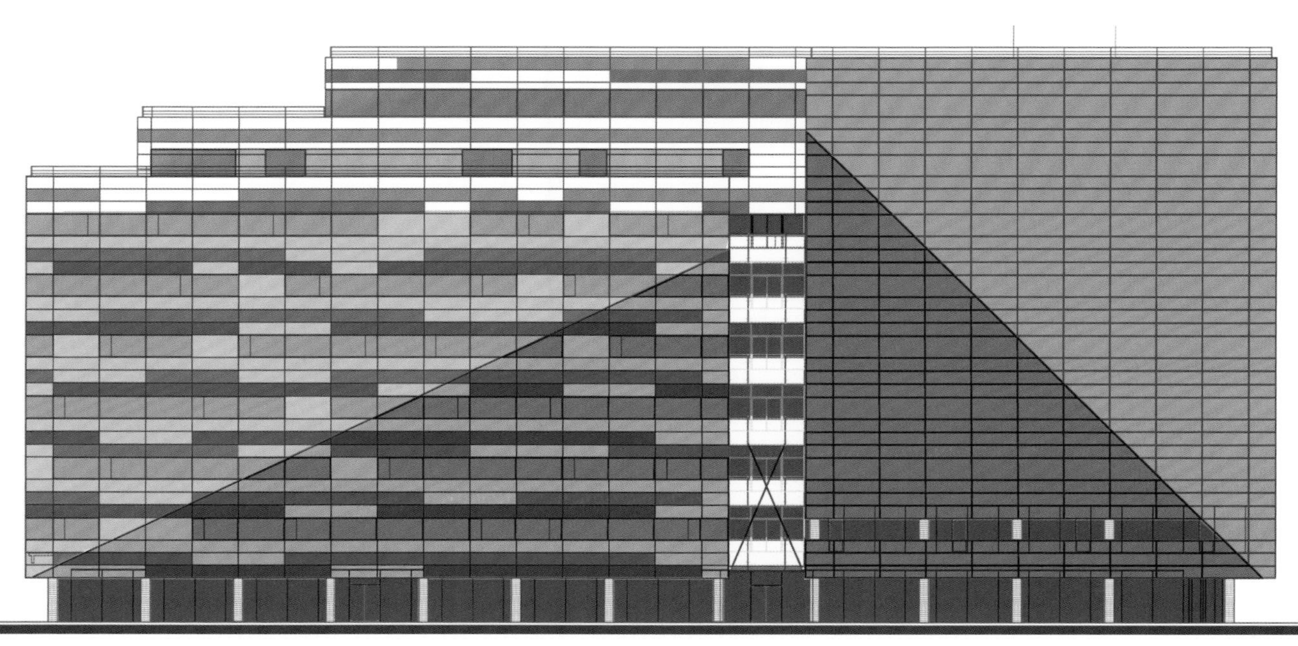

莱姆斯商业中心是一栋独出心裁的甲级办公大厦。它矗立在布加勒斯特市郊，近来见证了该地区的巨大发展。该地段原本是一个工业区，随着都市化的迅速推进，这里开始建造一些居住小区和办公楼宇。对于建筑师们来说，如何处理大厦的主体是一个棘手的问题。这是由于之前对同一地块的两部分采用不同的规划设计以及处理方法造成的，从建筑学角度来看有些支离破碎之感。最终建成的大厦主体的两个部分，高九层，完全遵照规划所要求的高度，并缓和了强烈的视觉冲击。在两个部分之间的衔接处，合理地设置了大厦的入口和主要的流通通道。

RAMS
CENTER
RAMS
CENTER

RAMS

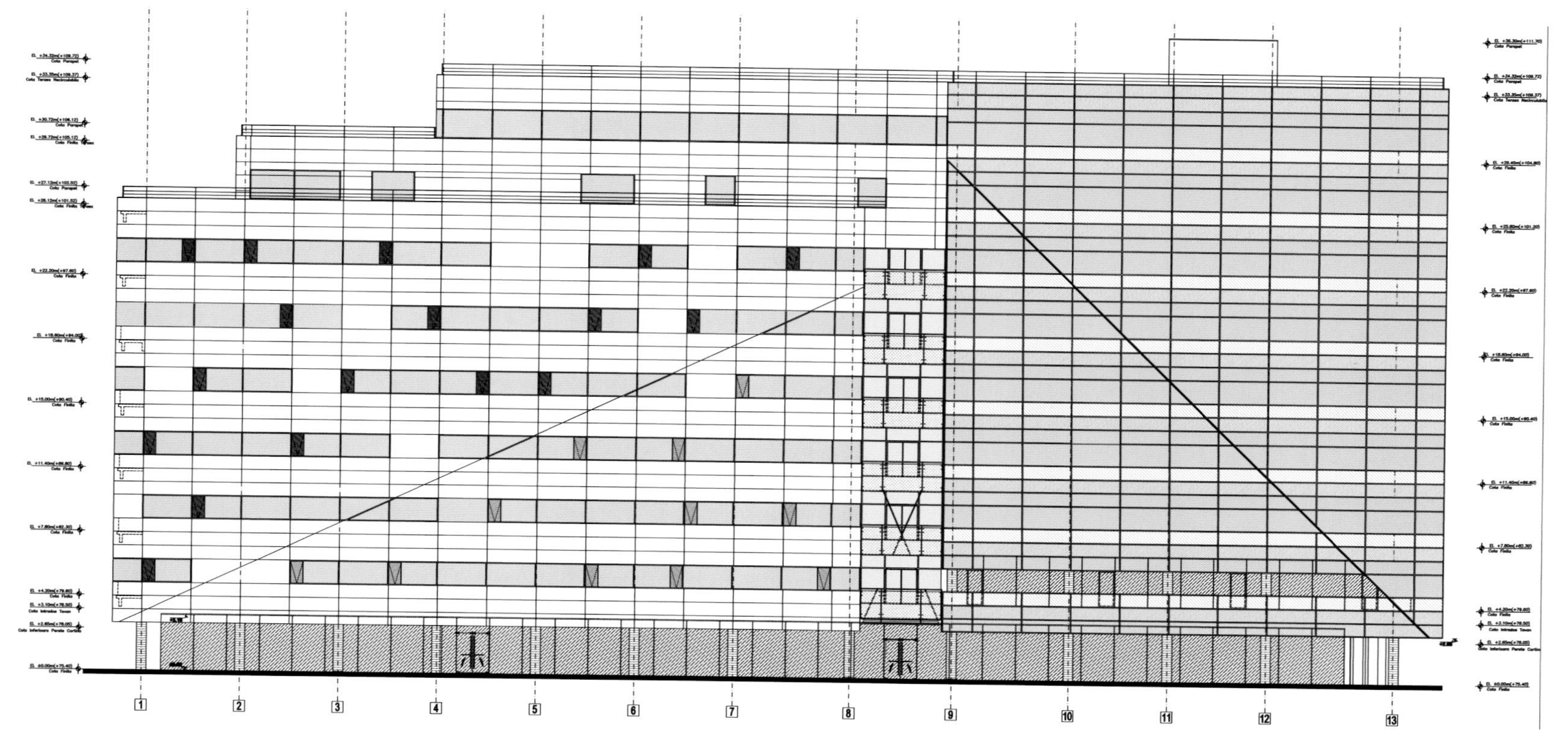

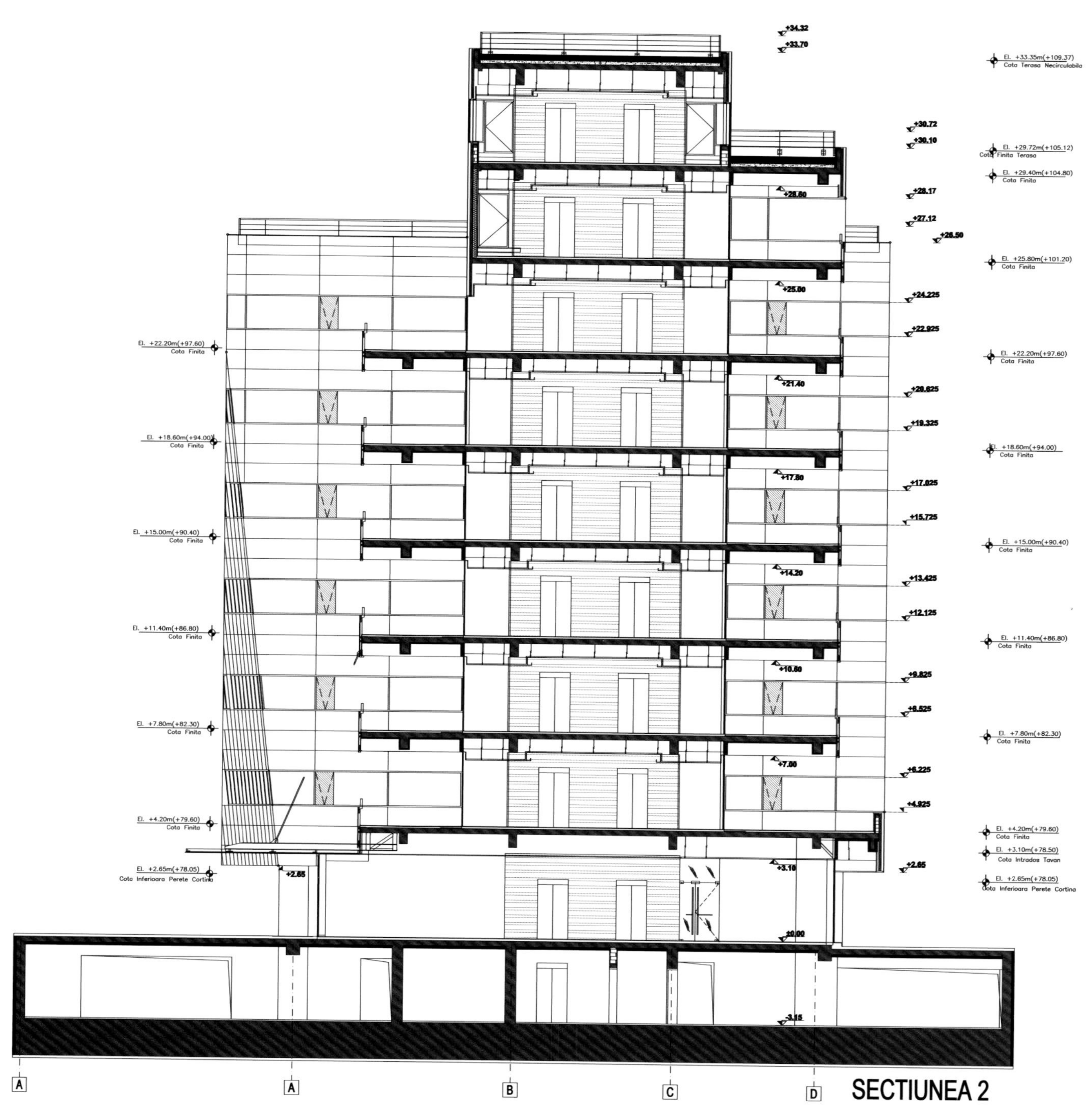

+34.32
+33.70
El. +33.35m(+109.37)
Coto Terasa Necirculabila
+30.72
+30.10
El. +29.72m(+105.12)
Cota Finita Terasa
El. +29.40m(+104.80)
Cota Finita
+28.17
+27.12
+26.50
El. +25.80m(+101.20)
Cota Finita
+24.225
+22.925
El. +22.20m(+97.60)
Cota Finita
+20.625
+19.325
El. +18.60m(+94.00)
Cota Finita
+17.625
+15.725
El. +15.00m(+90.40)
Cota Finita
+13.425
+12.125
El. +11.40m(+86.80)
Cota Finita
+9.825
+8.525
El. +7.80m(+82.30)
Cota Finita
+6.225
+4.925
El. +4.20m(+79.60)
Cota Finita
El. +3.10m(+78.50)
Cota Intrados Tavan
+2.65
El. +2.65m(+78.05)
Cota Inferioara Perete Cortina
+3.10
±0.00
-3.15
A
A
B
C
D
SECTIUNEA 2

naturalbond
NR.42

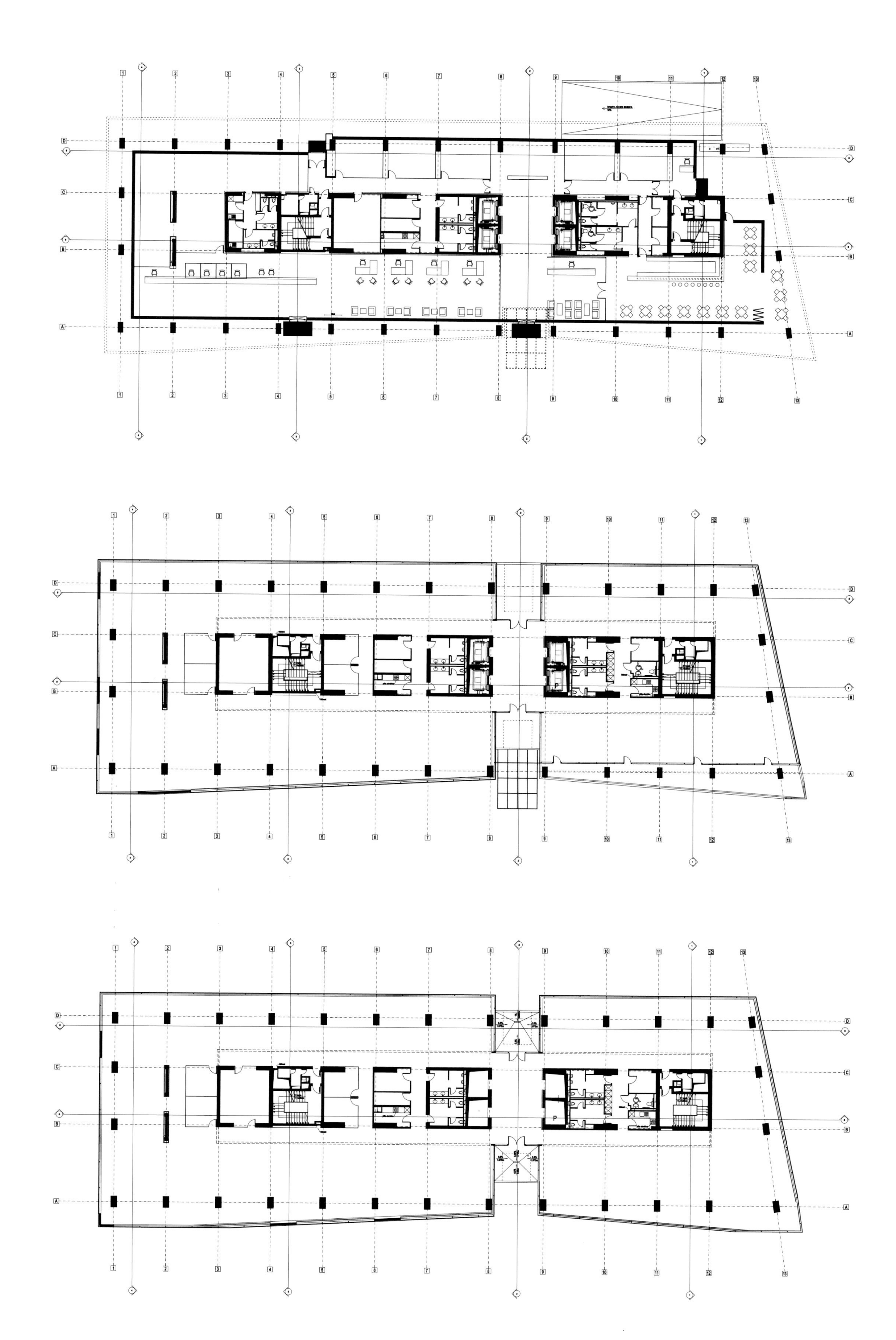

Quattro 商业园

QUATTRO BUSINESS PARK

ARCHITECT
Stefan Kuryłowicz, Marcin Goncikowski, Tomasz Bardadin, Krzysztof Pydo, Katarzyna Pielaszkiewicz, Małgorzata Kowalczyk

FIRM
Kurylowicz & Associates Architecture Studio

STRUCTURE
Project Service Biuro Inżnierskie

INVESTOR
BUMA INWESTOR Sp. z o.o.

CO-OPERATION AND DEVELOPMENT
Tomasz Kopeć, Grzegorz Szymański, Paulina Gutkowska, Krzysztof Popiel

AREA
12,193 m²

LOCATION
Krakow, Poland

The complex consists of an integral group of high 14-storey commercial and office buildings, designed on rectangular plan, grouped around an internal square-shaped atrium of 35x35 meters. In the southern part of the investment area, from the busy road side, a multi-storey free standing car parking facility and open car parking space are designed. The open car parking space is to be eventually built up with another, the fifth office building.

The inner courtyard is designed as a construction element connecting functional office and commercial buildings, being at the same time an attractive publicly available space. Single-storey, underground garage is planned under each building. Ultimately, the garages shall be connected by an internal communication system. The ground floors of the buildings are planned location of services that can operate independently, also beyond the office hours on higher storeys.

Windows visible on the facade are two-storey high; due to this solution the buildings seem slimmer. The facade is also determined by irregular surfaces with more extensive glazing. They provide more differentiation to the rhythm of checked facade and correspond with connectors, which also have more extensive glazing and with exposed glazing of the corners. Spatial composition, in spite of its large scale, seems to be dynamic and light. Next phases of implementation are distinguished by colors of their facade panels, presenting, however, the entire gamut of soft grey colors. The facade is made, in its most part, of glass and aluminium composite slabs, materials which are easy to clean and resistant to aging.

GRUPA
BUMA
Capgemini
QUATTRO B
APTEKA

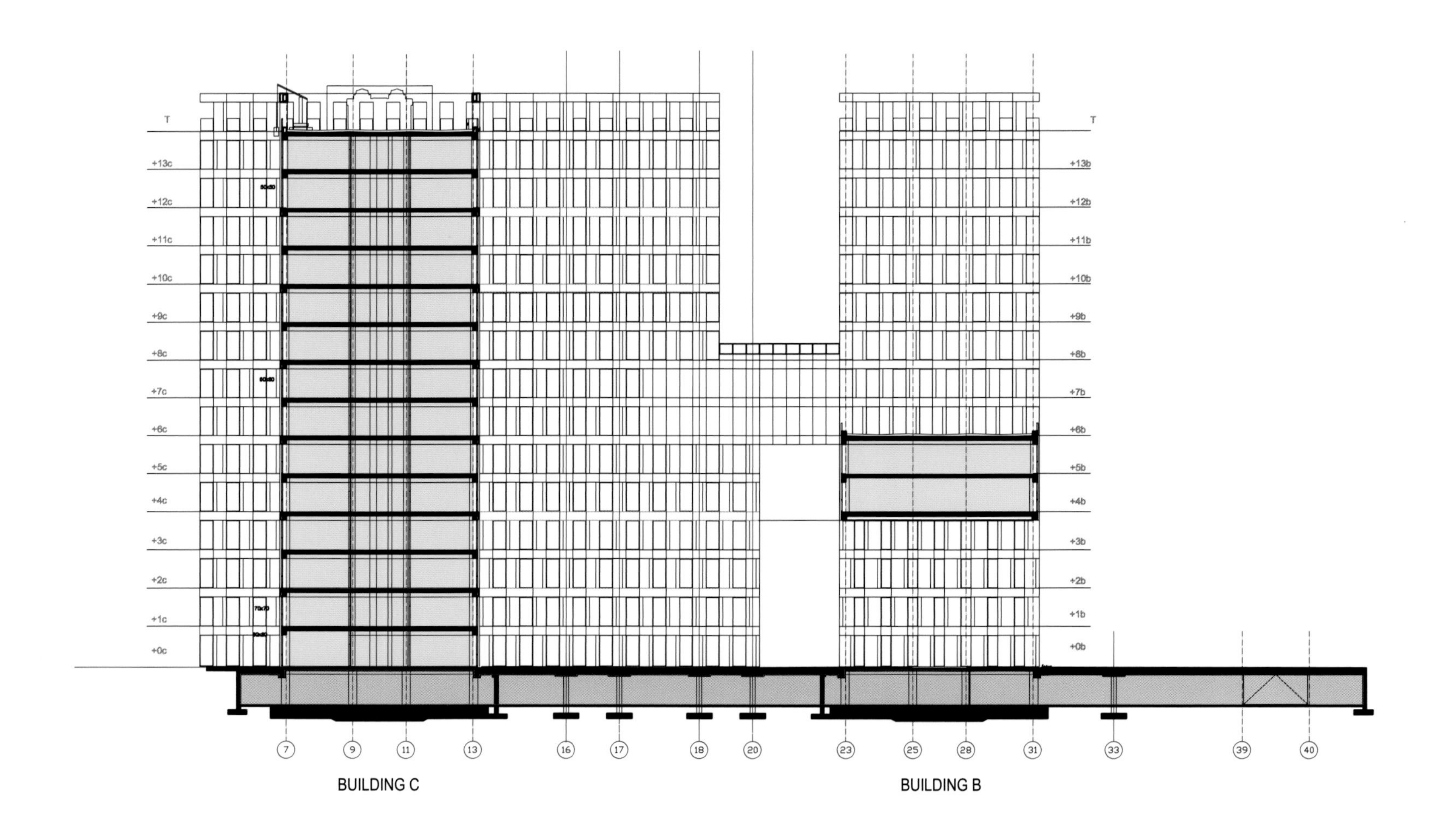
T
+13c
+12c
+11c
+10c
+9c
+8c
+7c
+6c
+5c
+4c
+3c
+2c
+1c
+0c
+13b
+12b
+11b
+10b
+9b
+8b
+7b
+6b
+5b
+4b
+3b
+2b
+1b
+0b
7 9 11 13 16 17 18 20 23 25 28 31 33 39 40
BUILDING C
BUILDING B

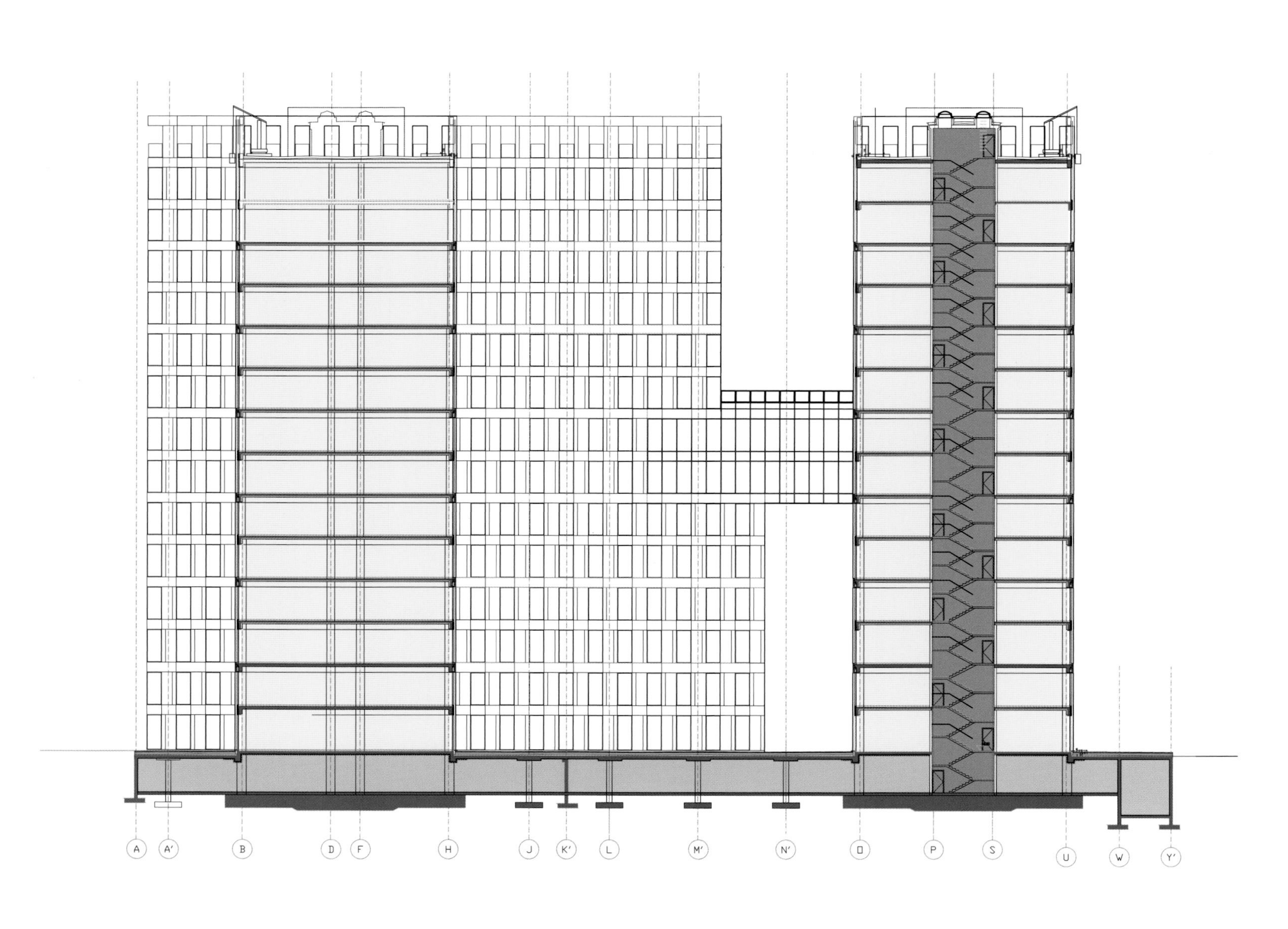
A A' B D F H J K' L M' N' O P S U W Y'

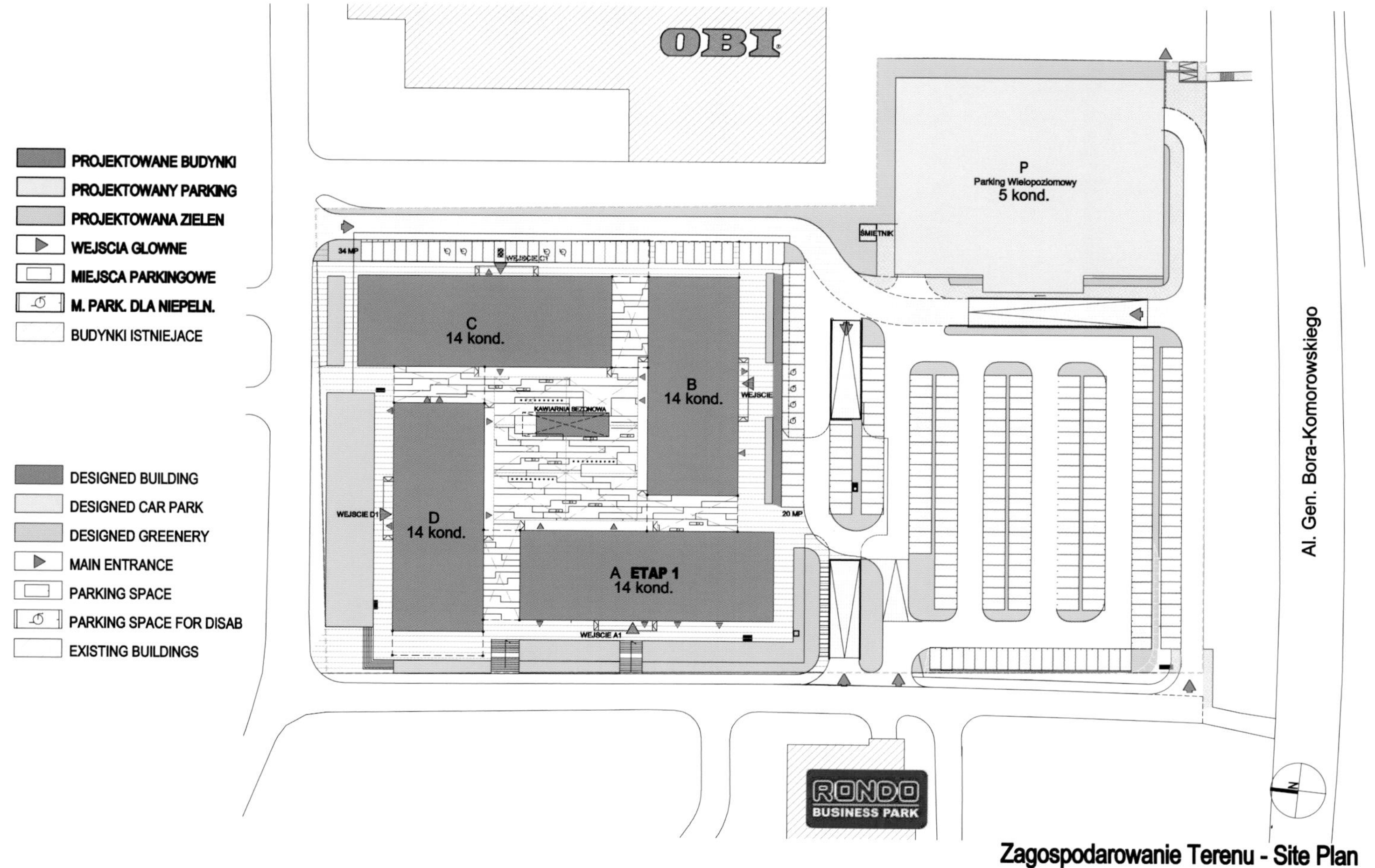

Zagospodarowanie Terenu - Site Plan

BERNER
BUMA

QUATTRO
BUSINESS PARK

Capgemini
GRUPA
BUMA
MOTOROLA
MOBILITY
QUATTRO B

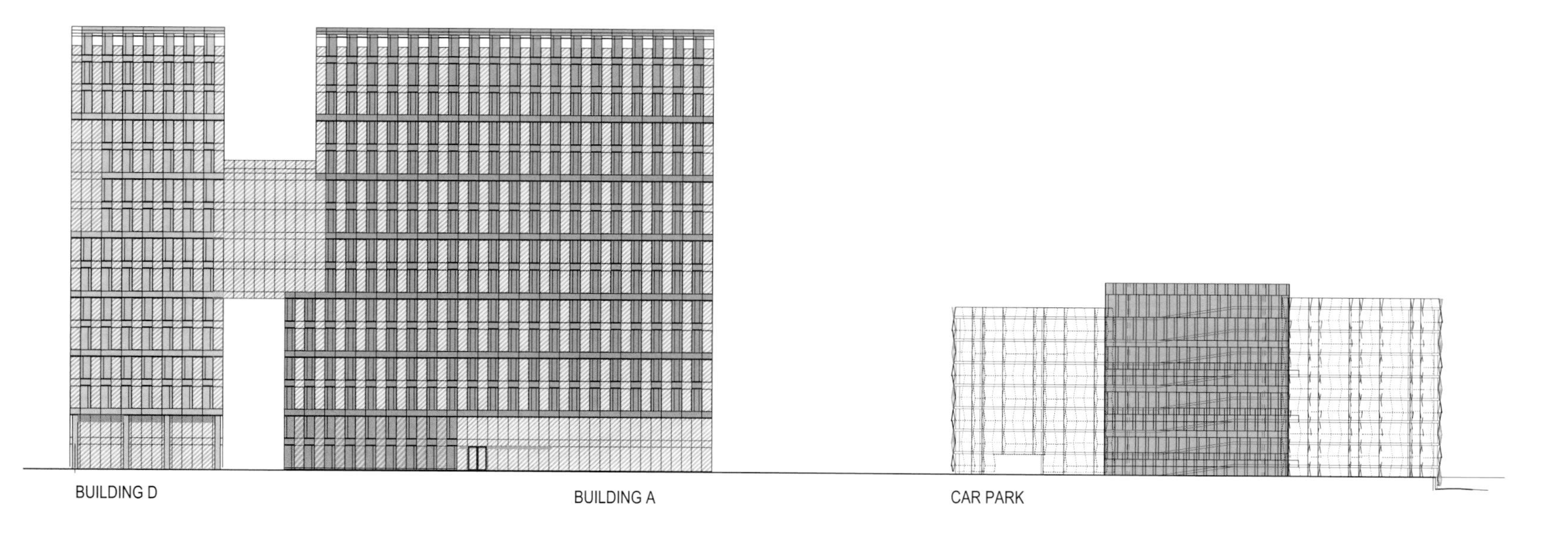

Quattro 商业园是一个建筑群，由整齐划一的 14 层高的商业大楼和办公大楼组成，办公楼均呈长方形，围绕一个 35 米见方的正方形内部中庭而建。在投资区域南部繁忙的公路边，还单独设计了一个多层停车场和一个室外停车场。室外停车场将和第五座办公大楼一起修建。

内部庭院将功能性的办公大楼和商业大楼巧妙地连接起来，同时也成为一个有吸引力的公共空间。每座大楼也设计有单层地下车库，所有车库最终将与内部通信系统相连。所有大楼的地面层都设计有可以独立运作的服务单元，便于在非办公时间也可以正常营业或工作，类似的服务单元在较高楼层中也有。

立面上的窗子有两层楼高，这样的设计可以使大楼看起来更修长。立面上也有大面积的由多块玻璃组成的不规则表面，这种不规则表面的使用增加了格子状立面的韵律感；楼与楼之间的连接廊桥以及东西大楼的转角也采用了这种不规则表面，这样，楼与楼之间的立面造型就可以相互呼应。在空间构成上，尽管大楼规模庞大，但从整体上看，轻盈而富有活力。在颜色上，立面板采用了柔和的灰色系列。在材料的选择上，立面大面积使用玻璃面板和铝复合面板，既易于清洁，又经久耐用。

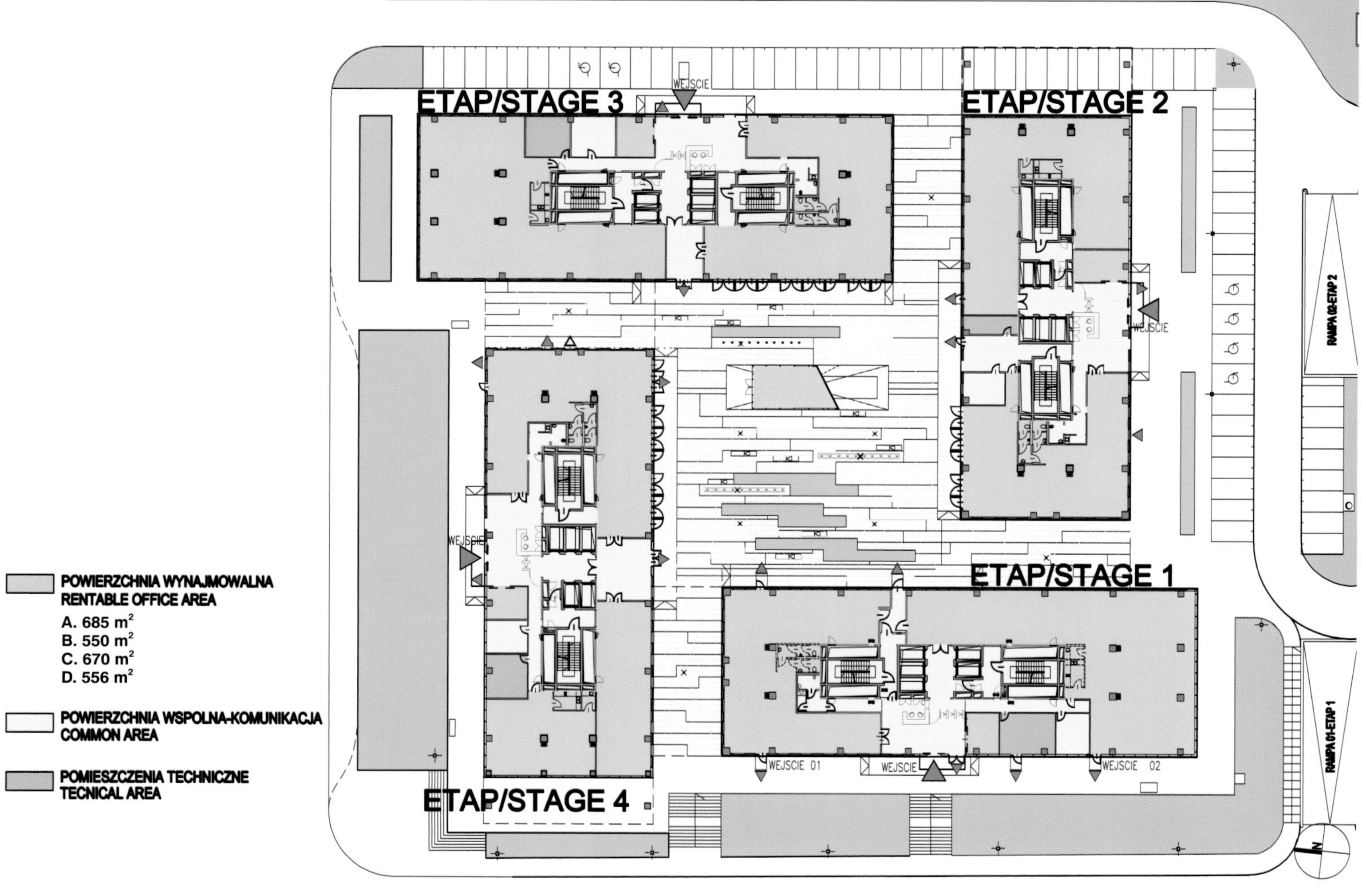

Rzut piętra - Floor 0

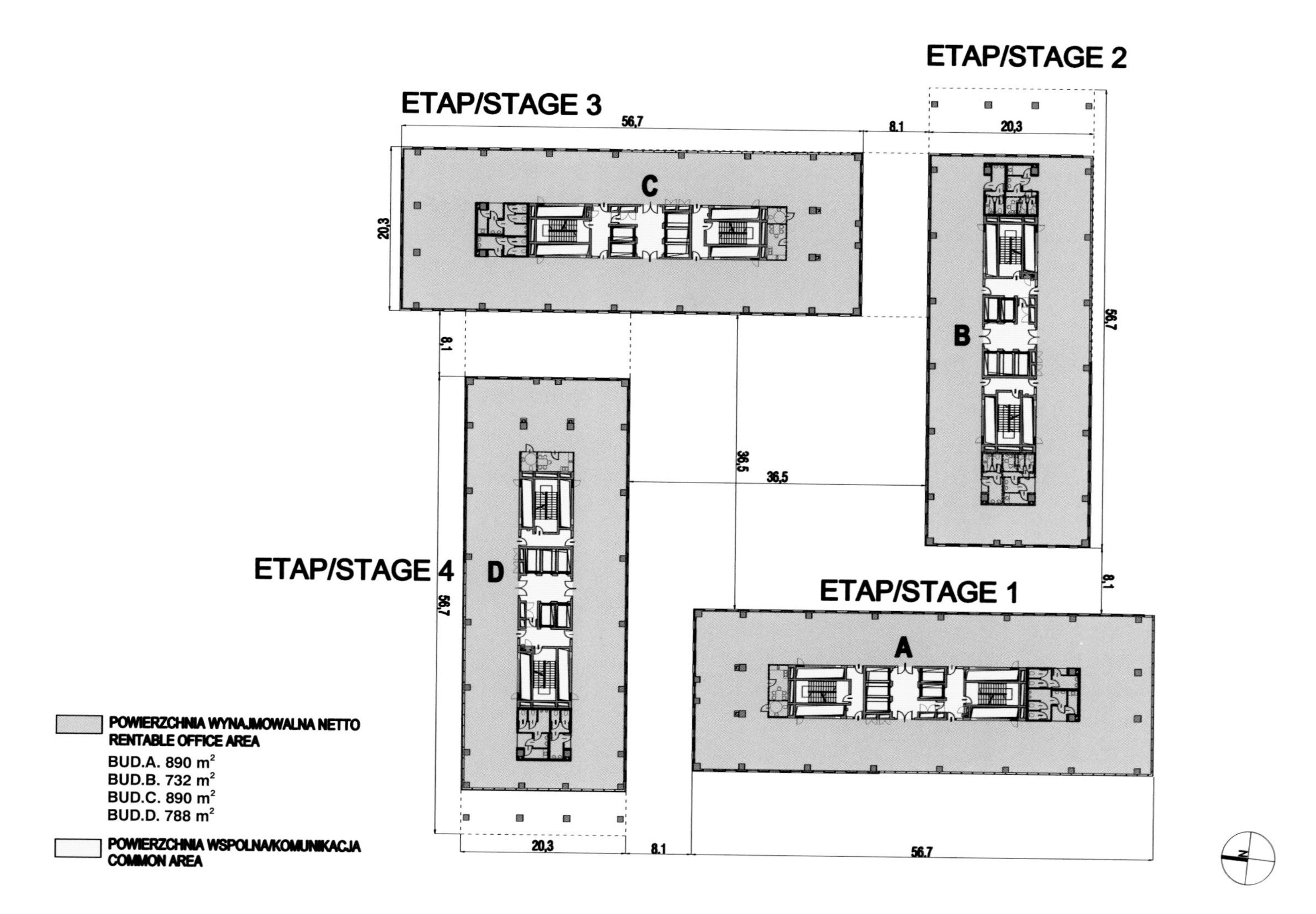
ETAP/STAGE 2
ETAP/STAGE 3
56,7
8.1
20,3
C
20,3
B
56.7
8.1
36.5
36.5
ETAP/STAGE 4
D
56.7
ETAP/STAGE 1
8.1
A
POWIERZCHNIA WYNAJMOWALNA NETTO
RENTABLE OFFICE AREA
BUD.A. 890 m2
BUD.B. 732 m2
BUD.C. 890 m2
BUD.D. 788 m2
POWIERZCHNIA WSPOLNA/KOMUNIKACJA
COMMON AREA
20,3
8.1
56.7
Rzut piętra - Floor 1/2/3

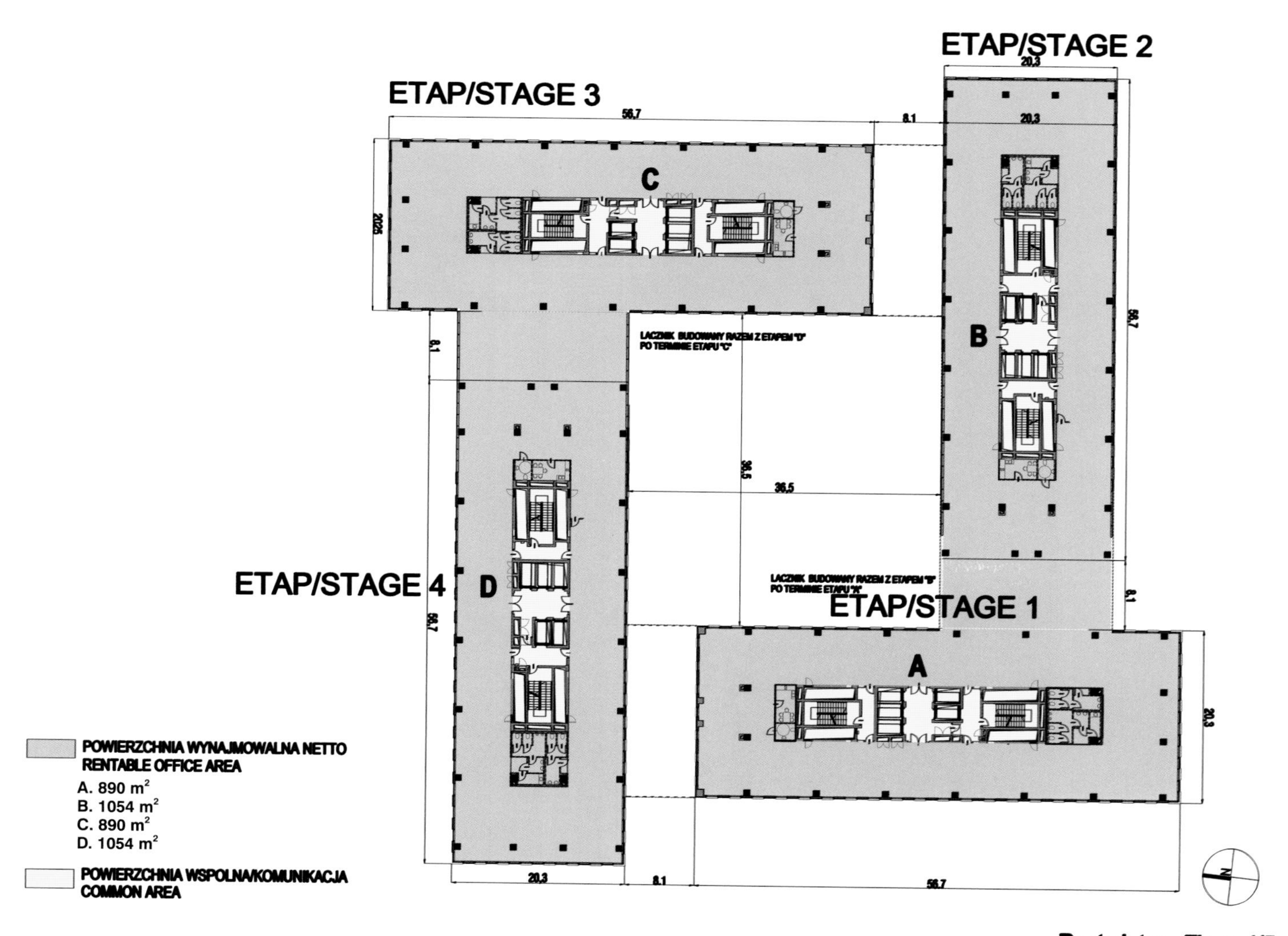
ETAP/STAGE 2
20,3
ETAP/STAGE 3
56,7
8.1
20,3
C
B
56.7
8.1
36.5
36.5
ETAP/STAGE 4
D
56.7
ETAP/STAGE 1
8.1
A
20.3
POWIERZCHNIA WYNAJMOWALNA NETTO
RENTABLE OFFICE AREA
A. 890 m2
B. 1054 m2
C. 890 m2
D. 1054 m2
POWIERZCHNIA WSPOLNA/KOMUNIKACJA
COMMON AREA
20,3
8.1
56.7
Rzut piętra - Floor 4/5

QUATTRO B
citi handlowy
XIII
MEDICOVER
ORACLE

MOTOROLA MOBILITY

TAX CARE
TAX CARE
EPSON
ABER GUYS!
HSE IN
ALL WE DO

STATOIL REGIONAL AND INTERNATIONAL OFFICES

ARCHITECT
A-lab

LANDSCAPE ARCHITECT
Østengen & Bergo

CLIENT
IT-Fornebu Eiendom

LOCATION
Fornebu, Bærum, Norway

AREA
117,000 m^2

PHOTOGRAPHER
Luis Fonseca, Trond Jølson, A-lab

Statoil Regional and International Offices in Oslo, is an ambitious and courageous project.

In the design the designers prioritized the synergy of the volume and the context. One of the main preconditions for the scheme is that the footprint has to fit inside the footprint of the existing multi-storey car-park. This is achieved by breaking the homogenous office program into five equally sized lamellas (A, B, C, D OG E see illustration), each is dimensioned in order to get the most flexible office plans. The stacking of these then creates sight-lines between and minimizes the visual impact of the height required to fit the program within the tight site area. The primacy of the park is ensured by allowing the office lamellas to cantilever beyond the basic footprint.

The in-between space created by the stacking of the lamellas, is transformed into a public covered "square" where all the activities came

across, is a monumental atrium, accommodating the public programs and the main circulation.

In the Facade, the further sub-division into prefabricated elements each in turn composed of 15 "pixels" introduces a human scale whilst simultaneously creating a pattern linked to the structure, legible as a "giant-order" from afar.

The office machine provides innumerable possibilities for configuring the workspace, both at the individual level and as a whole within the organization. The arrangement of the social cores within the overall framework of the circulation promotes positive interaction between employees, teams and departments. Optimization of the facades ensures the visual connection with the surrounding landscape, fjord and city in the distance.

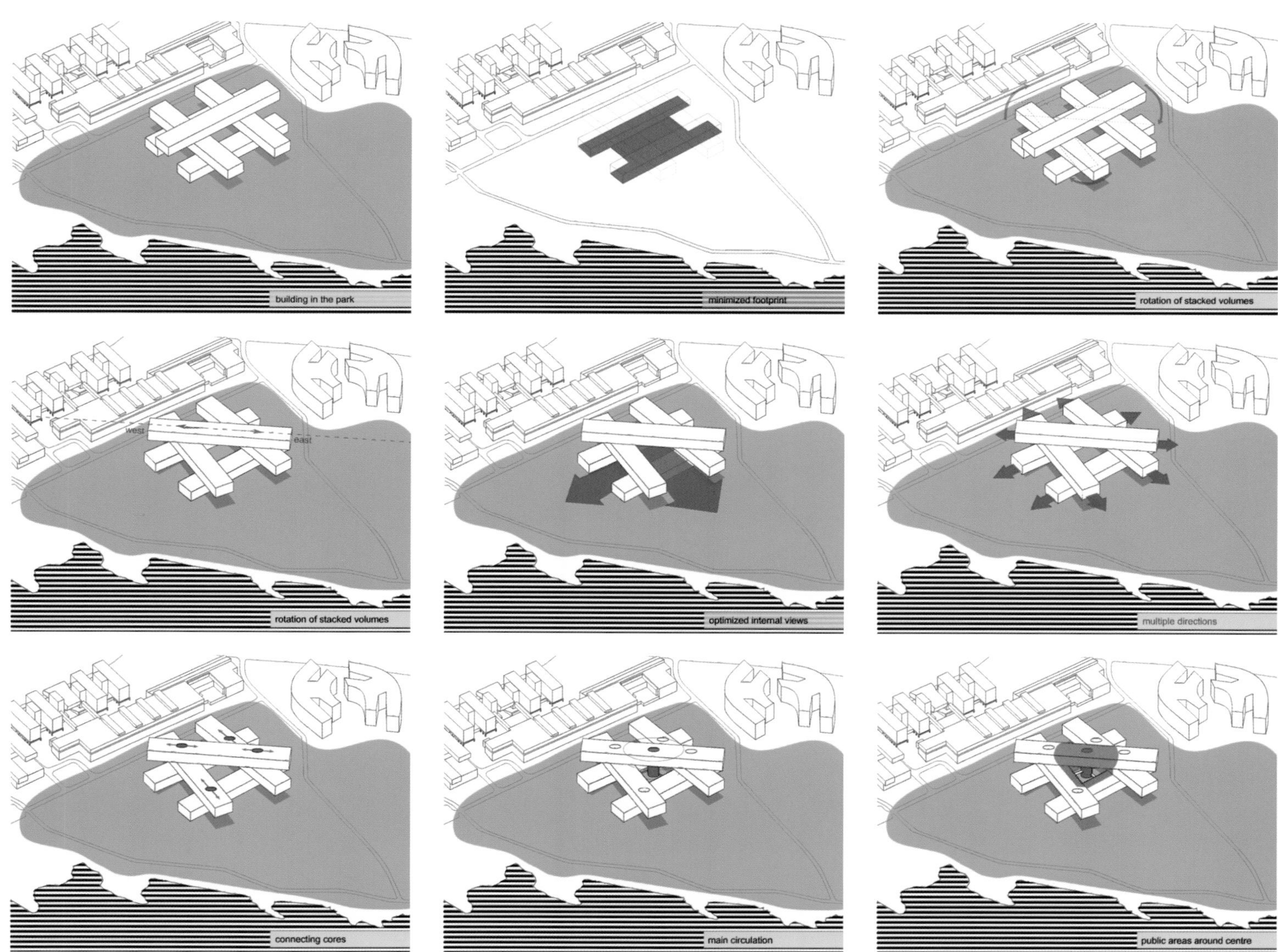

Concept Diagram

Situation Plan

Statoil 能源公司办公大楼位于奥斯陆，是一栋有着宏伟规划设计的壮观建筑。

在规划方案中，设计师们优先考虑的是空间与周围环境的协调一致。由于在规划中办公大楼要坐落于原有的多层停车场范围内，因此设计师们将整个办公大楼分为五个大小相同、风格相近的条块式模块（示意图中表示为A、B、C、D和E），每个模块都尽可能更多地获得有效的内部空间。这样五个模块叠加组合在一起，构成了一幅独特的画面，大大缓和了楼体高度所带来的视觉冲击，更好并更合理地利用了所在地块的紧凑地形。地面上利用钢悬臂结构树立在条块式模块四周，使得原有停车场分外醒目。

在条块式模块叠加的位置中央，设计师们设计了一个室内的公共广场，作为模块间的一种过渡，同时也是体现整栋办公大楼独特风格的天井。这里是主要的人流通行区域，也可以举办各种大型活动，为公众社交提供活动场所。

在大楼的正面，进一步划分出的每个预制组件依次由15“像素点”组成，引入了人性化设计，同时也与大楼主体结构相连，远远望去，像巨大的柱式建筑，清晰可见。

办公大楼为内部的办公人员提供了各式各样的功能空间，同时这种空间设计也不失为一个整体。在整个大楼的框架结构内，将各式各样的办公群体合理地整合在一起，同时提高并促进员工、团体以及各个部门相互之间的沟通。办公大楼的外观设计优美而不失壮观，与周边的风景及远处的峡谷和城市融为一体。

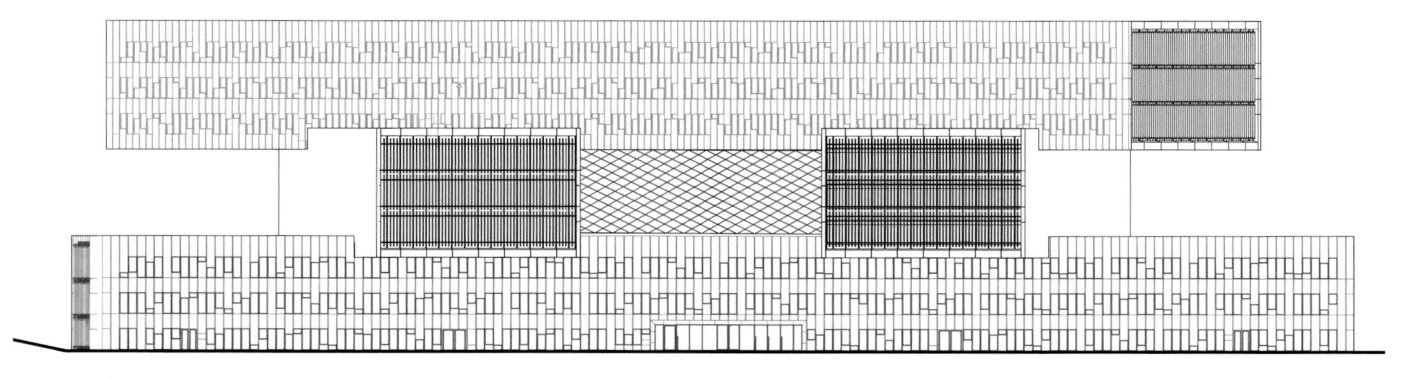

section A-A'

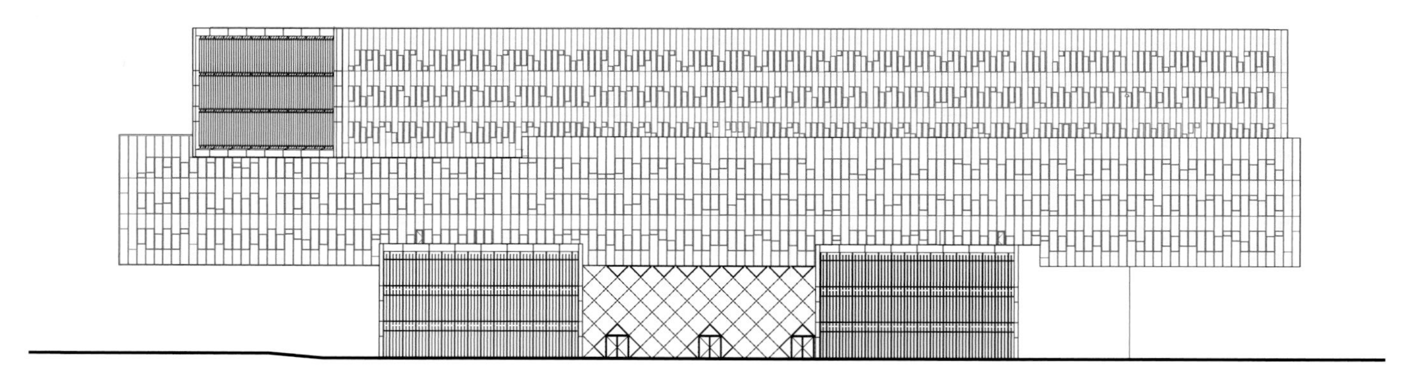

North West

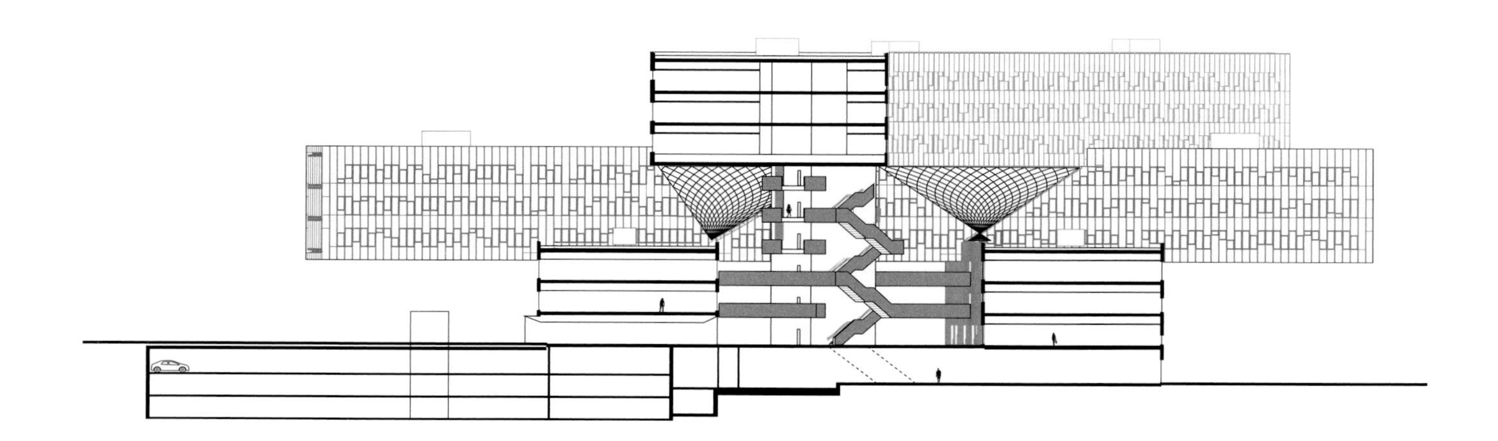

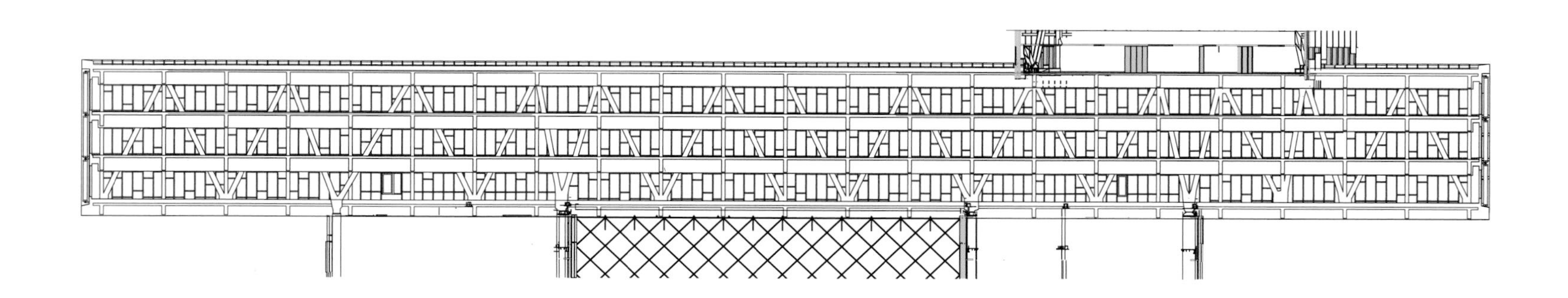

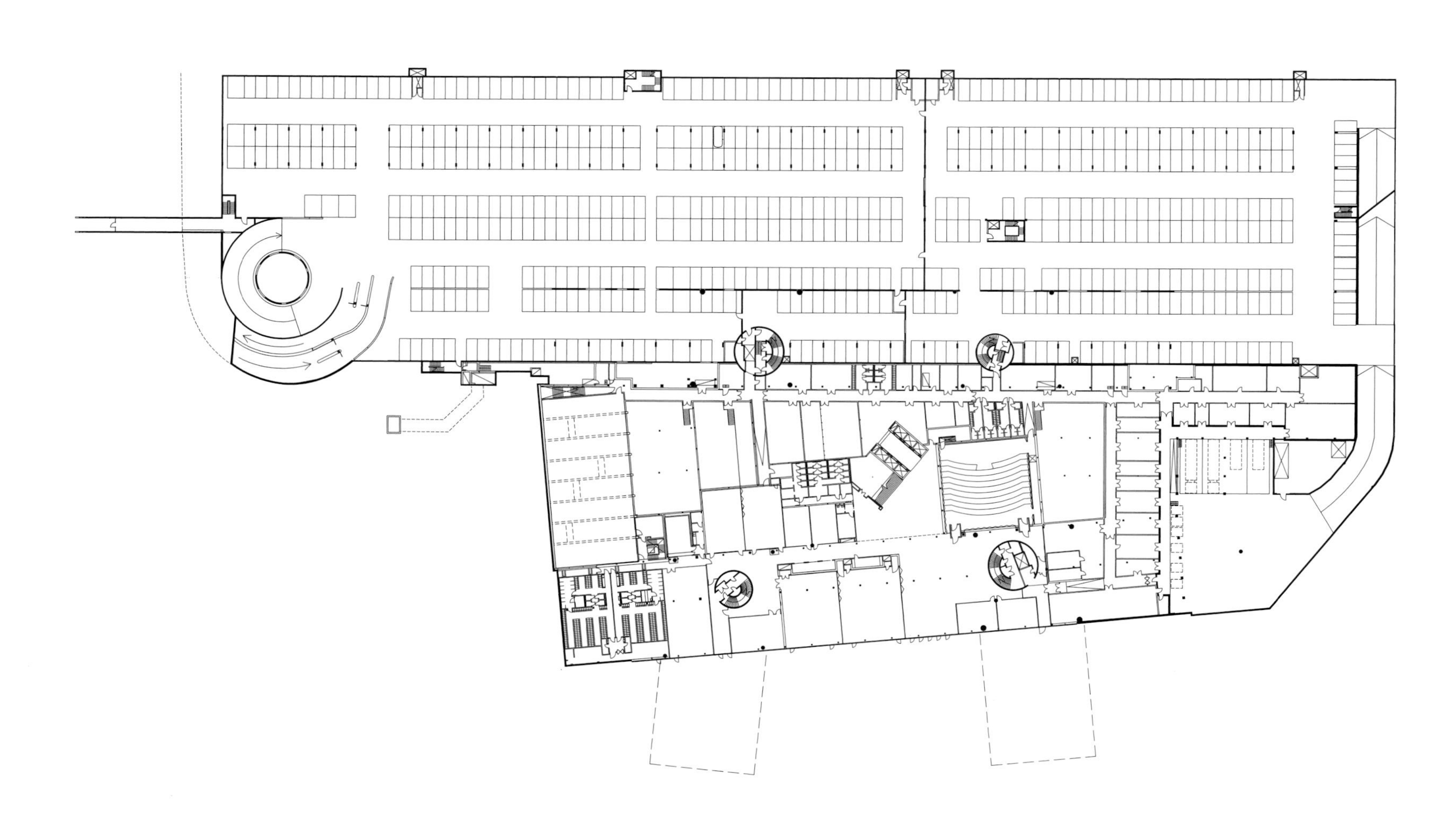

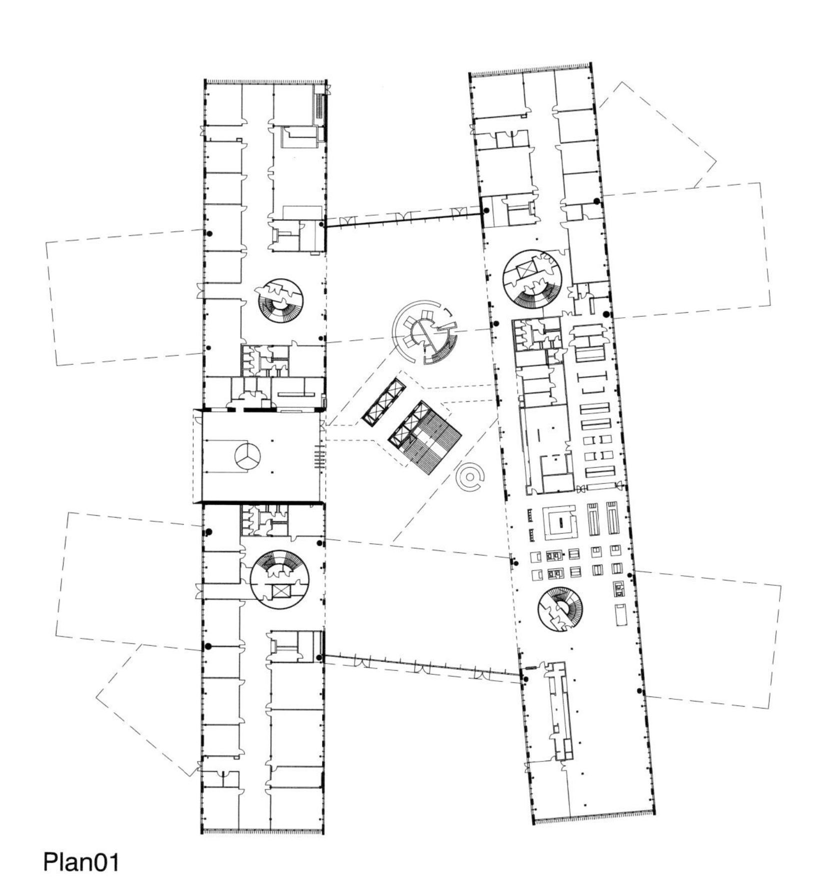

Plan01

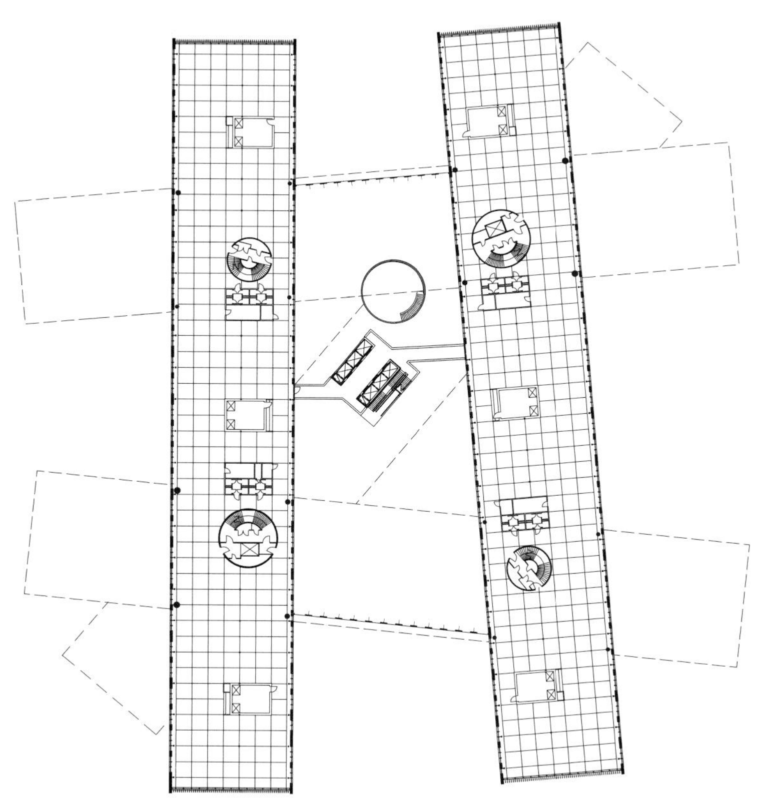

Plan02

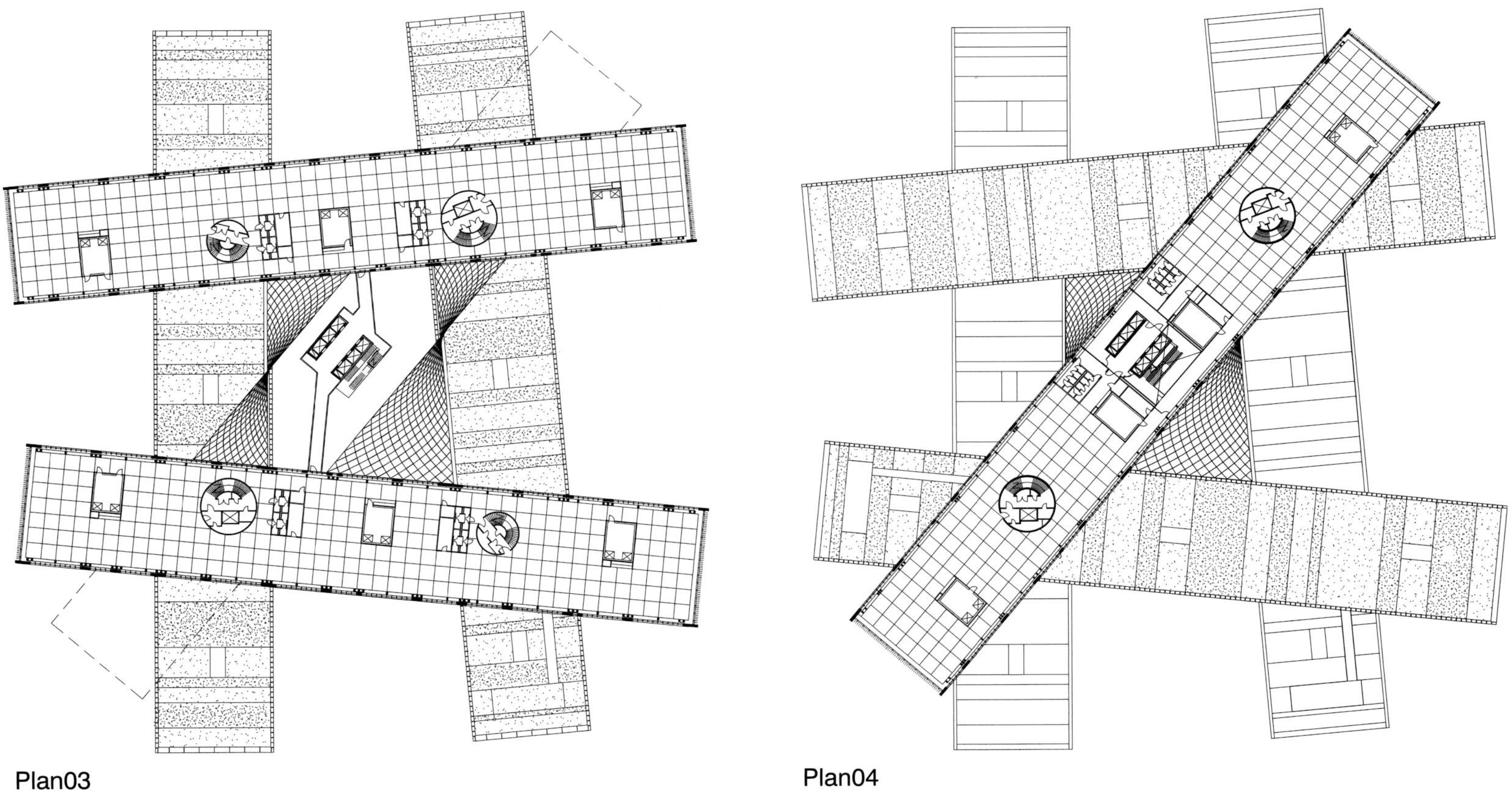

Plan03

Plan04

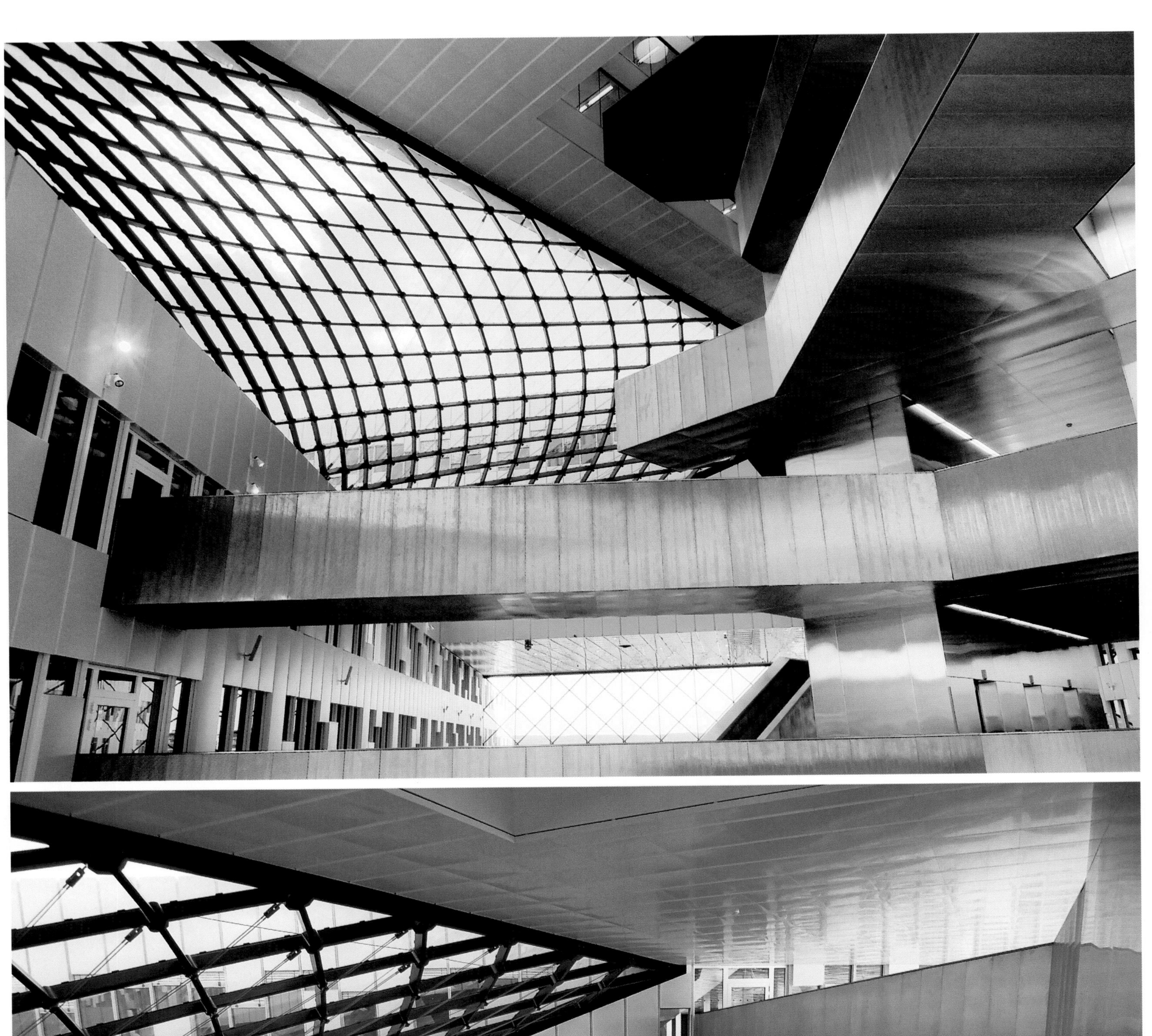

B2

帕拉莱罗办公大厦

PARALLELO OFFICE BUILDING

ARCHITECT
Mario Cucinella, David Hirsch, Julissa Gutarra

FIRM
Mario Cucinella Architects

CLIENT
DUEMME SGR SpA

LOCATION
Milan, Italy

AREA
12,048 m^2

PHOTOGRAPHER
Daniele Domenicali

This project is an A-Class office building in Milan. 2,500 square meters of photovoltaic panels on the roof produce enough renewable energy to completely satisfy the building's cooling needs. The building is on the outskirts of the city and is designed to give a new identity to the Famagosta area. It is raised 13 meters above ground level with piazzas, pathways and green areas below. The building is compact and extends horizontally over most of the site with a series of courtyards that allow diversification of function and maximum use of natural light and ventilation. The single building has three distinct parts that house different functions as required by the brief. The facades are glazed with high technology selective glass and the treatment of the glass varies depending on the orientation of the facade.

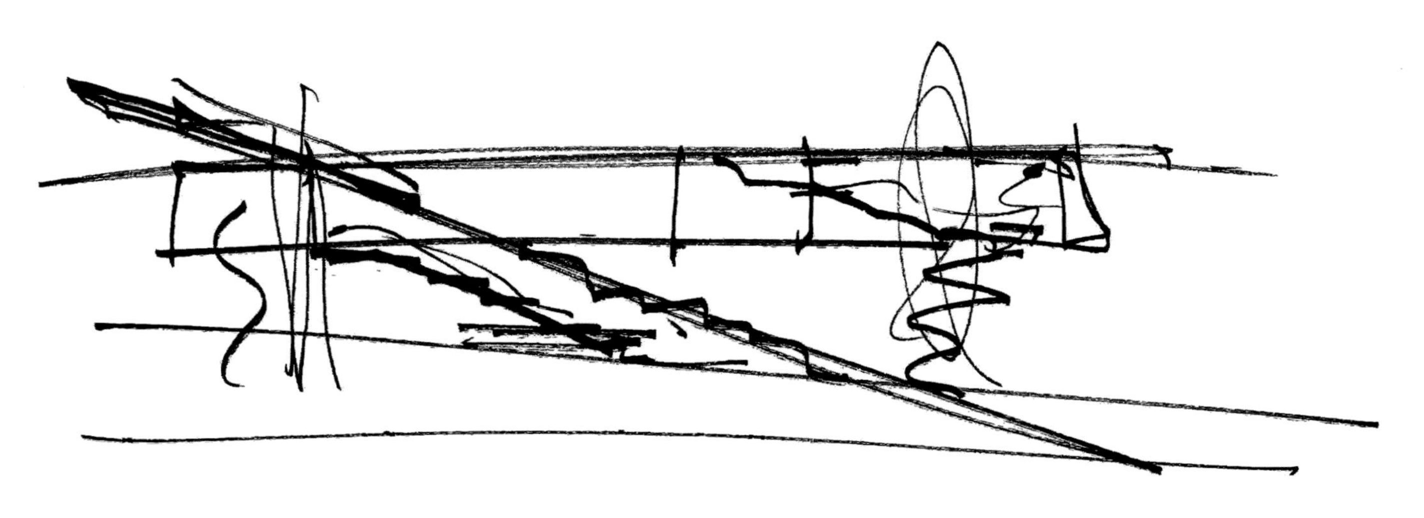

Schizzo

帕拉莱罗办公大厦是米兰市顶级的办公大厦。大厦顶部安放了约 2500 平方米的光电池板，利用太阳能生成可持续电能，完全满足了整栋大厦冷却系统所需的电力消耗。该办公大厦位于米兰市郊，是法马格斯塔地区新的地标性建筑之一。设计师们将地面抬高 13 米，广场、通道和绿地都分布在大厦下方。大厦主体十分紧凑，利用一系列庭院向周边地块进行横向延伸，使大厦的功能趋向多样化，并最大限度地利用了自然光和自然通风。大厦内部规划简明而规范，根据不同的功能分为三个独立的部分。大厦的外立面都由精心选择的高科技玻璃产品构成，而且不同墙体位置的玻璃都根据具体的情况，进行了不同的处理。

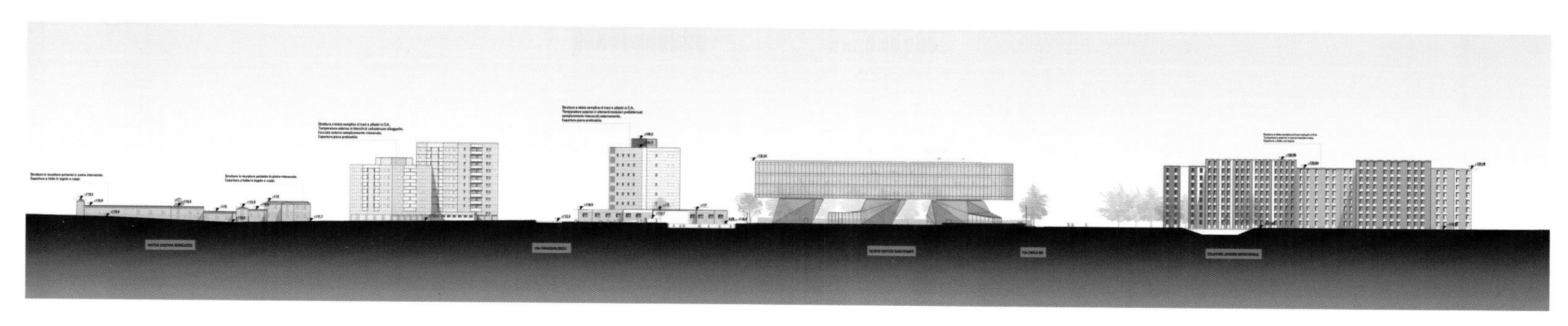

_Nuovo Edificio Santander_Milano
_prospetto ambientale nord Stato di Progetto

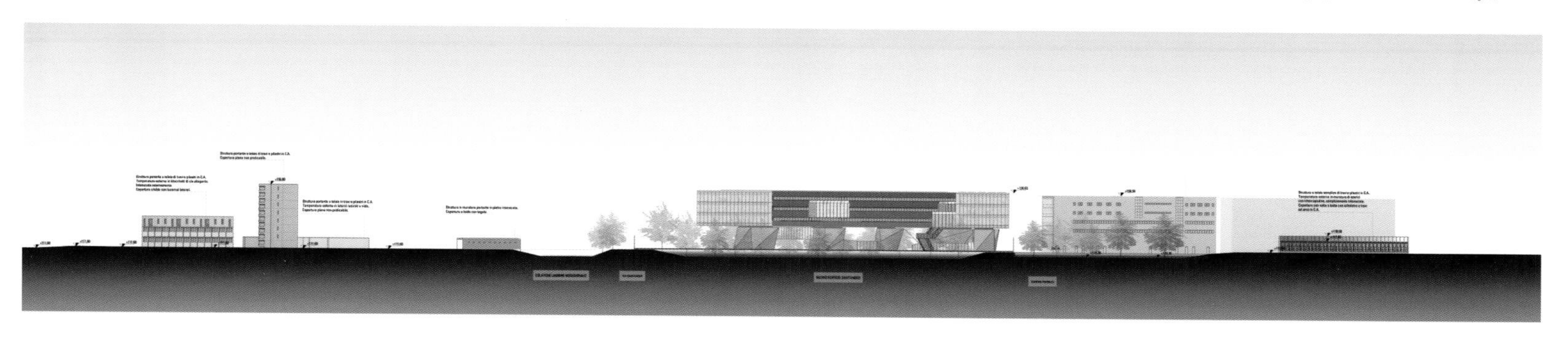

_Nuovo Edificio Santander_Milano
_prospetto ambientale est Stato di Progetto

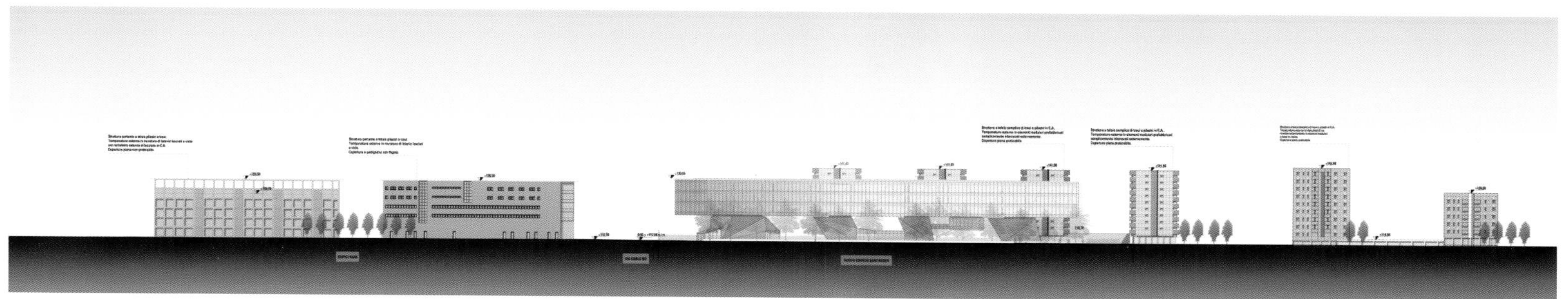

_Nuovo Edificio Santander_Milano
_prospetto ambientale via Santander Stato di Progetto

Prospetti Ambientali 2

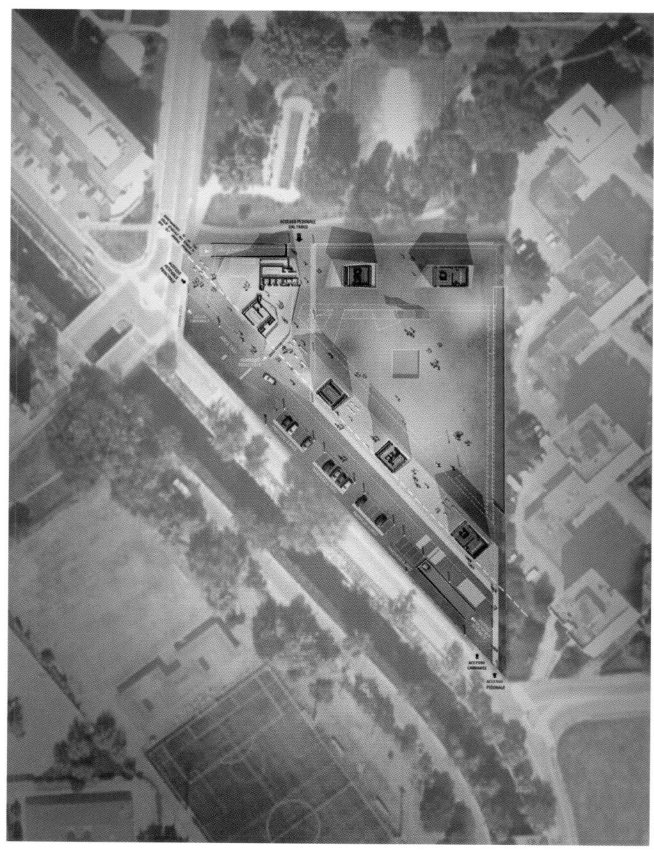

Fly View

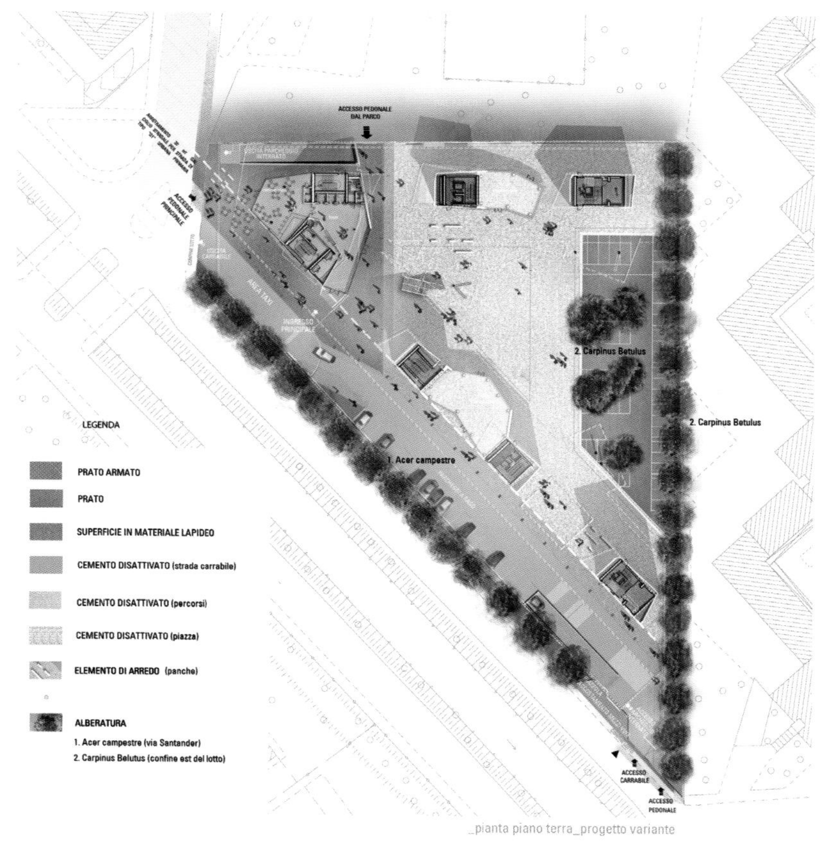

Sistemazioni Estene

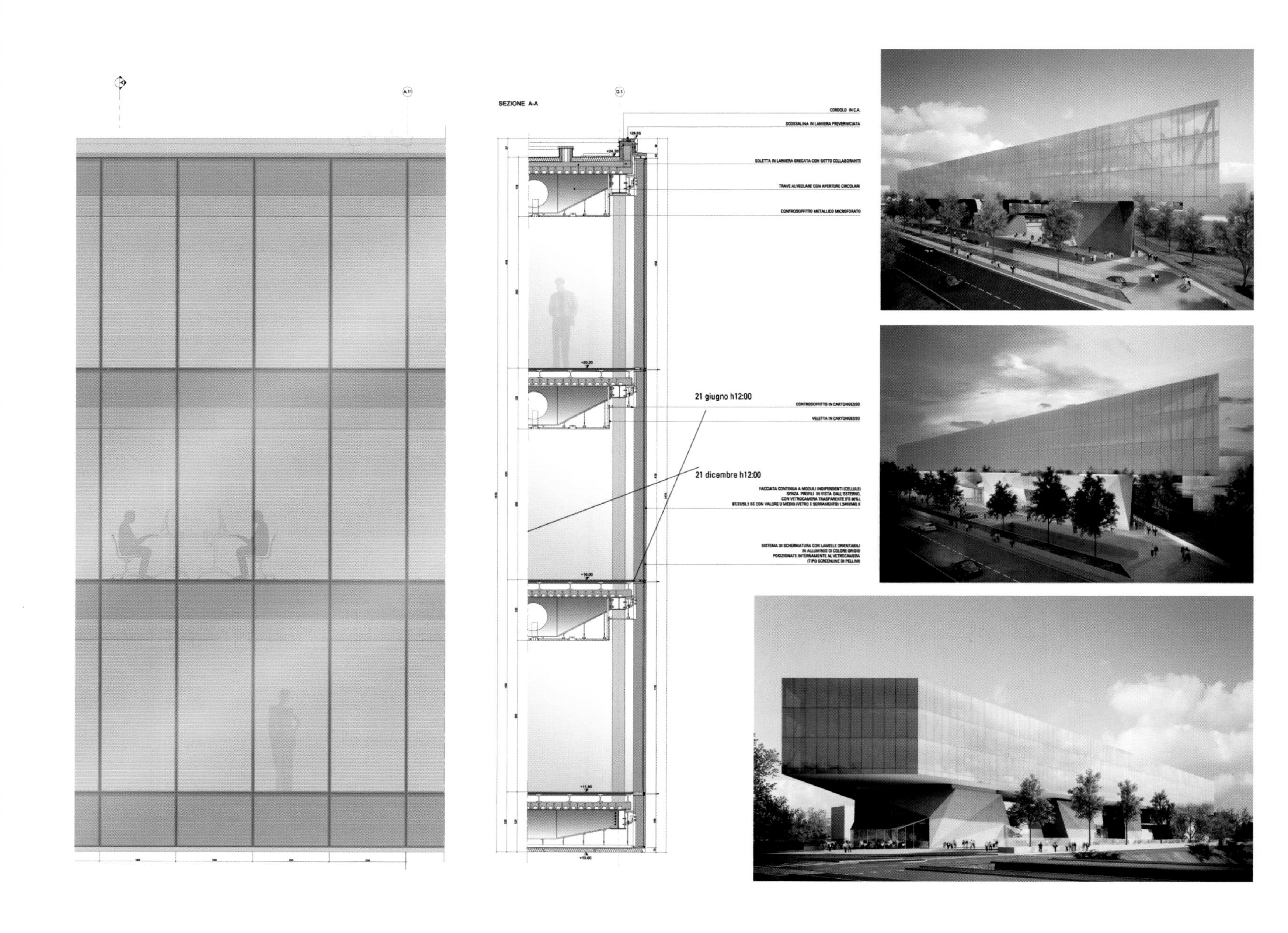

climatizzazione a 0 emissioni di CO2

l'impianto fotovoltaico è stato dimensionato per soddisfare interamente la domanda di energia per la climatizzazione dell'edificio

classe energetica A

certificato con software Cened, in conformità alle normative vigenti in materia nazionali e regionali

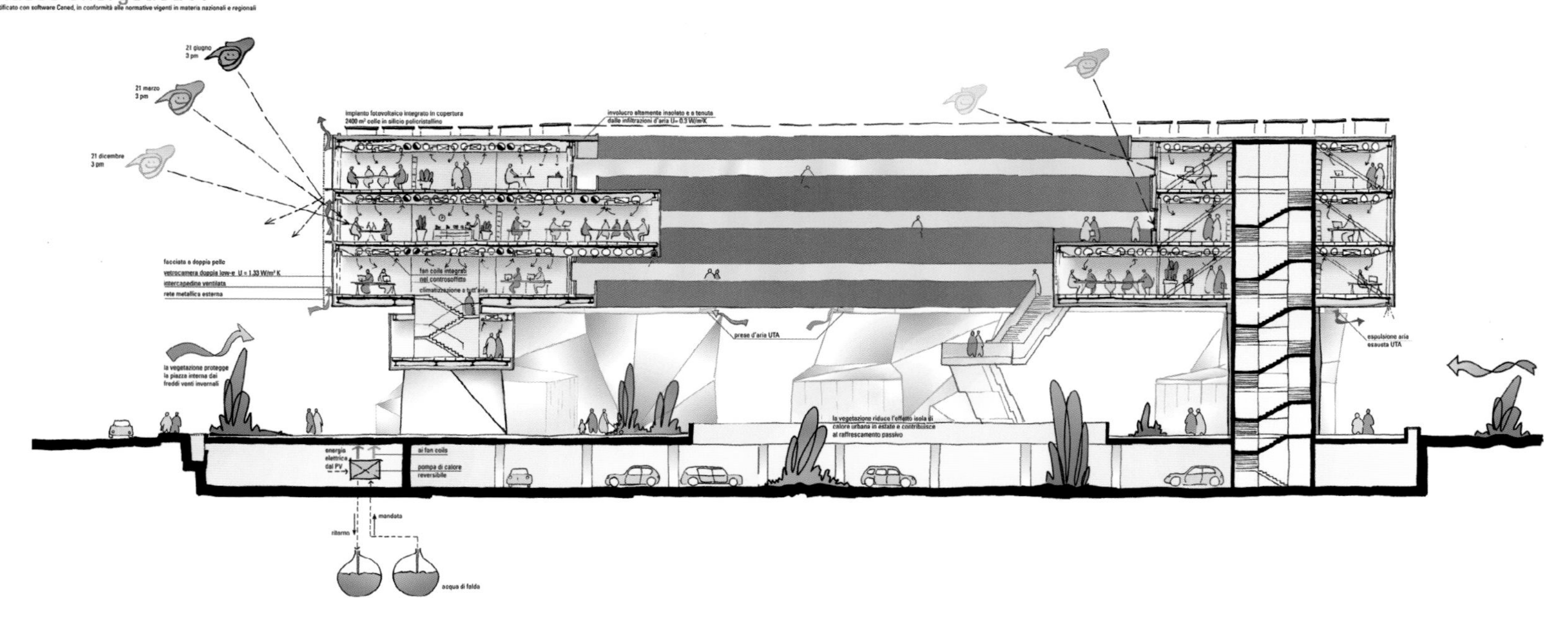

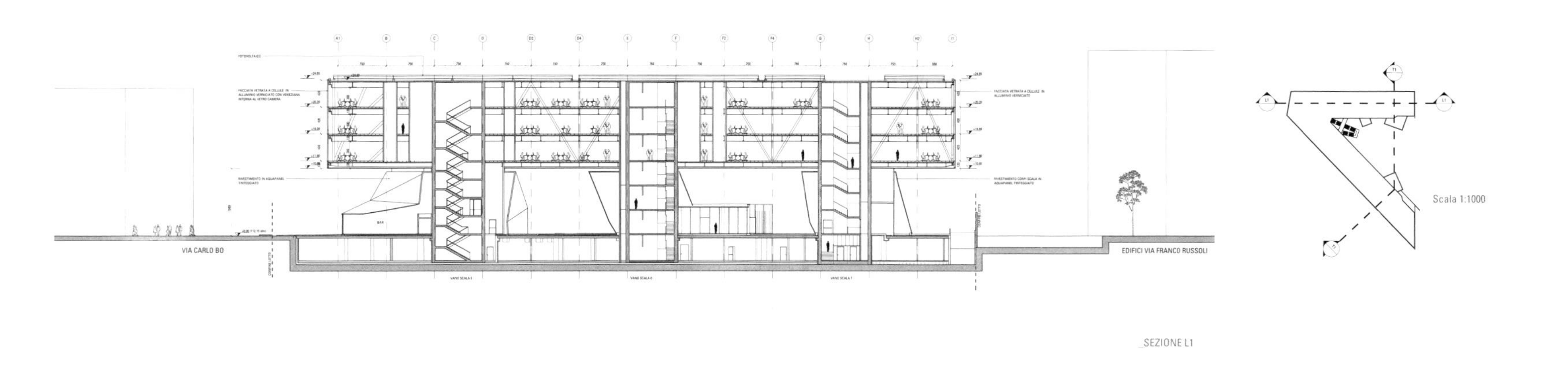

_SEZIONE L1

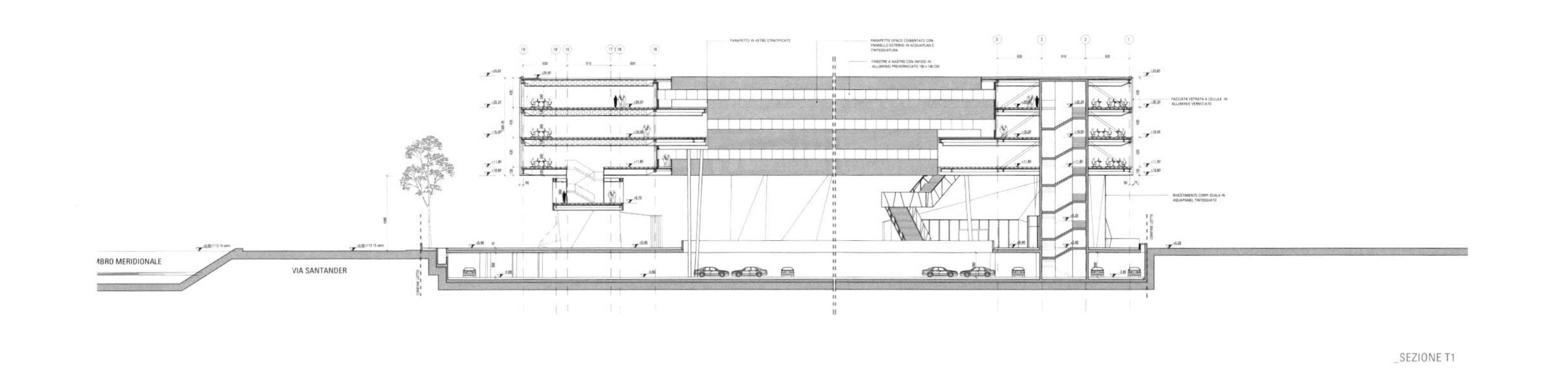

_SEZIONE T1

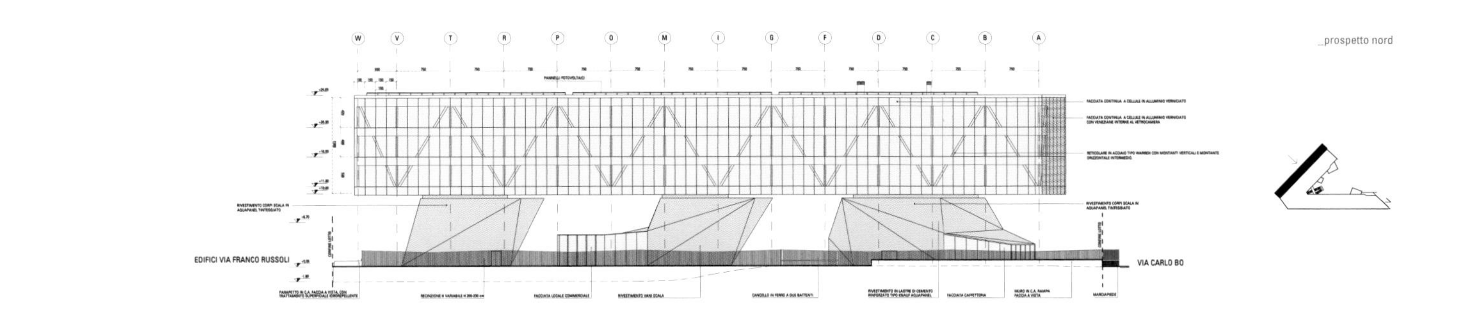

_prospetto nord

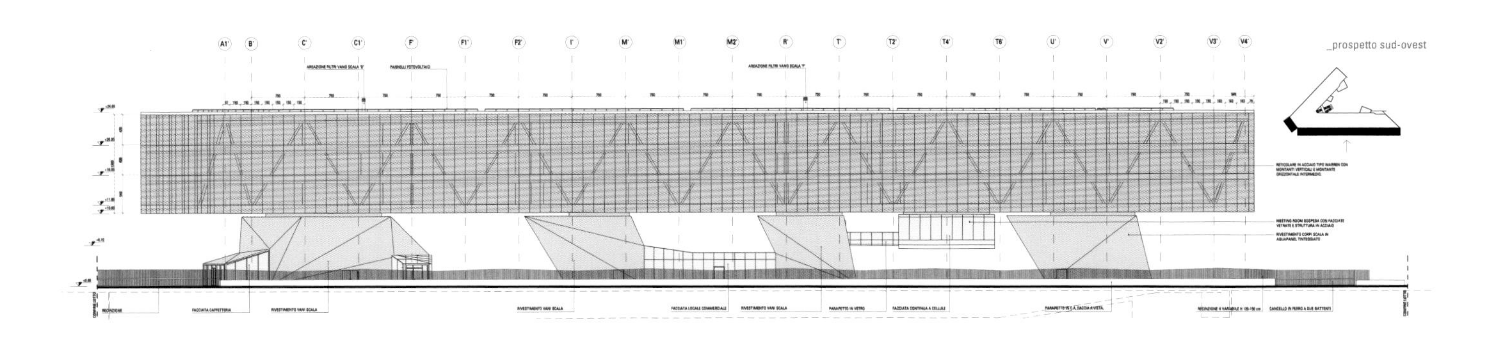

_prospetto sud-ovest

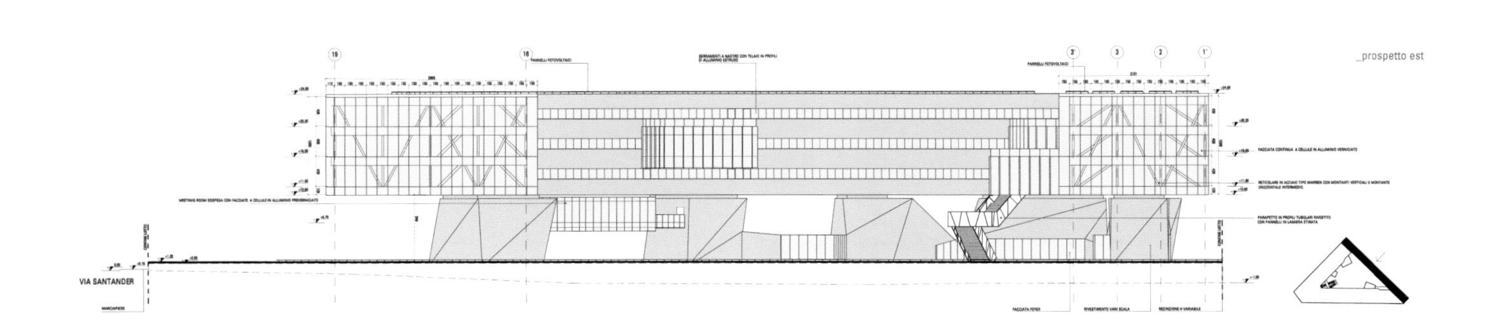

_prospetto est

卡尔顿世贸中心大厦

CARLTON WTC

ARCHITECT
Cees Dam, Diederik Dam

LOCATION
Almere, the Netherlands

PHOTOGRAPHER
Mathieuvan Ek, Luuk Kramer

FIRM
Dam & Partners Architecten

AREA
39,000 m^2

With a height of 120m, the Carlton tower is the tallest component of the Almere l'Hermitage office complex. This new business center includes three large slab-like buildings and is situated to the north of Almere Central Station, in the direct vicinity of the Mandela Park and nine other office complexes. Carlton WTC consists of a tall central volume flanked by two lower volumes and is supported by slender columns that create a passage from the station to the park. The entrances to the two large atria are also situated here. Vertical aluminium ribs placed in three different patterns dominate the facades, which become increasingly transparent towards to the top.

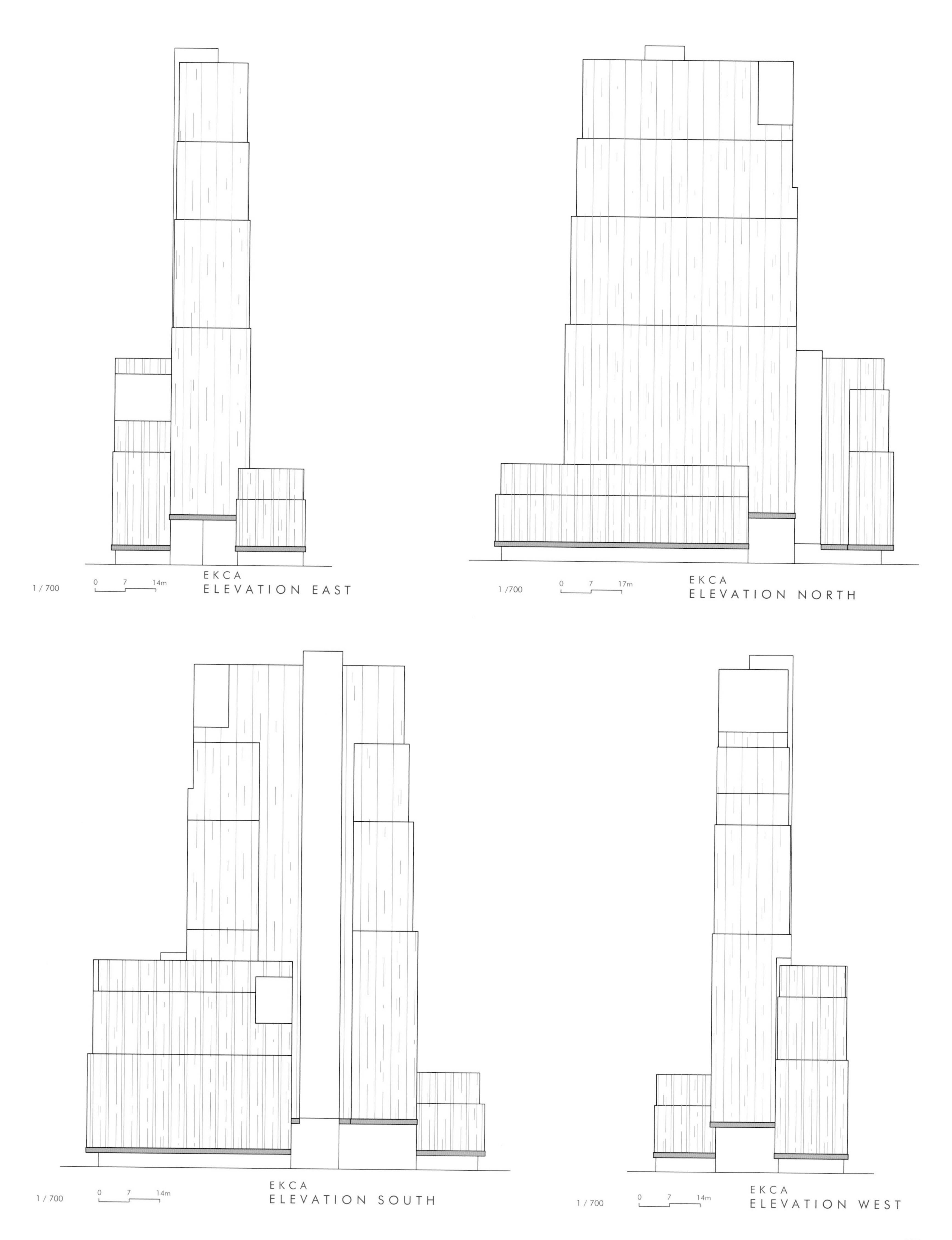
1 / 700
0
7
14m
EKCA
ELEVATION EAST
1 /700
0
7
17m
EKCA
ELEVATION NORTH
1 / 700
0
7
14m
EKCA
ELEVATION SOUTH
1 / 700
0
7
14m
EKCA
ELEVATION WEST

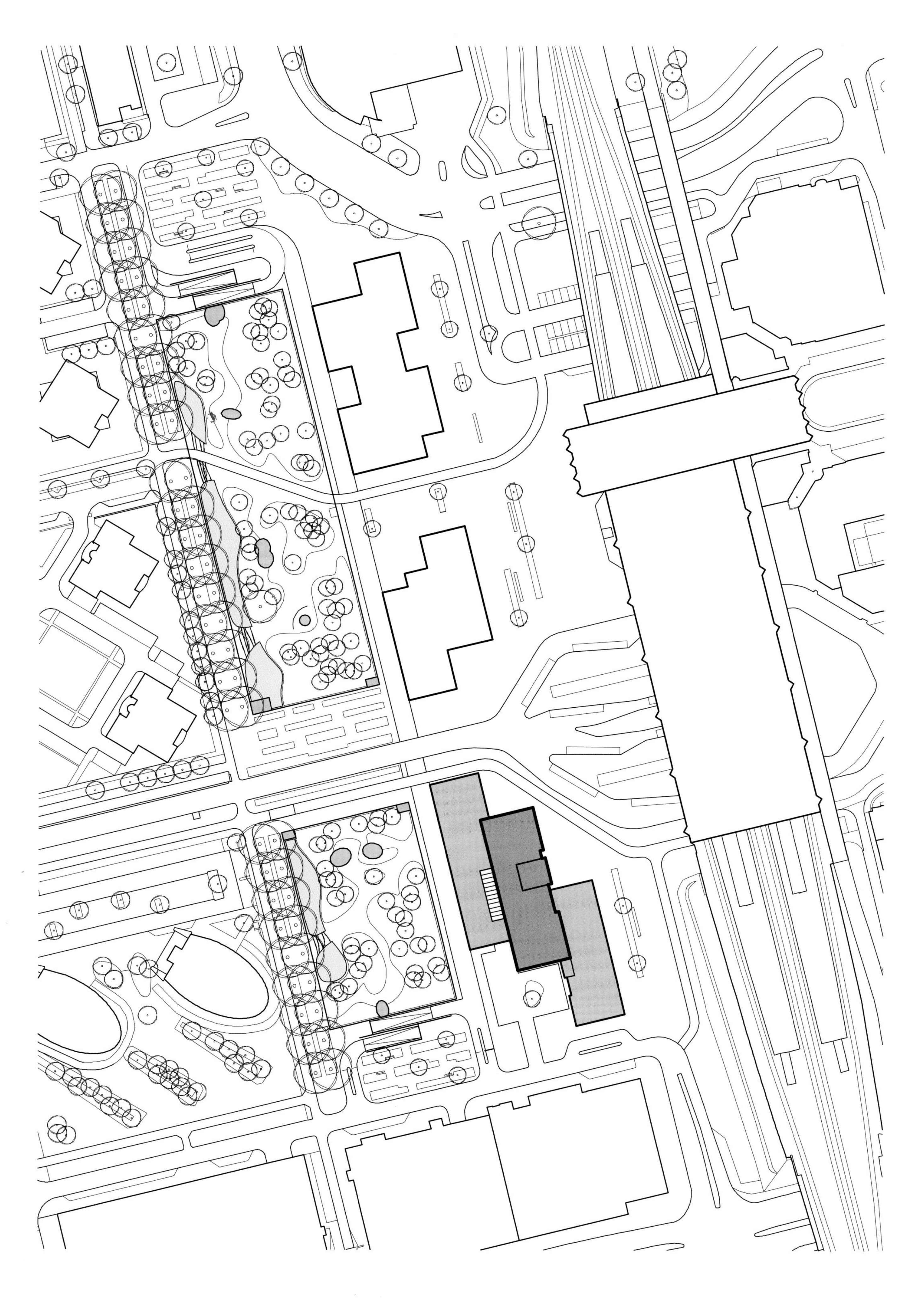

卡尔顿世贸中心大厦高 120 米，是阿尔梅勒市区办公建筑群中最高的建筑。这个新兴的商务中心包括三个巨大的板状建筑，坐落于阿尔梅勒中央火车站北部，毗邻曼德拉公园和其他九个办公建筑群。卡尔顿世贸中心大厦由一个高大的主楼和两侧的两个低矮的副楼组成，大楼旁有一条细长的通道可以从火车站通往公园。两个大的入口门廊也坐落在这里。外立面主要由三种不同图案的垂直铝条进行装饰，整个立面越往顶部越显得透明。

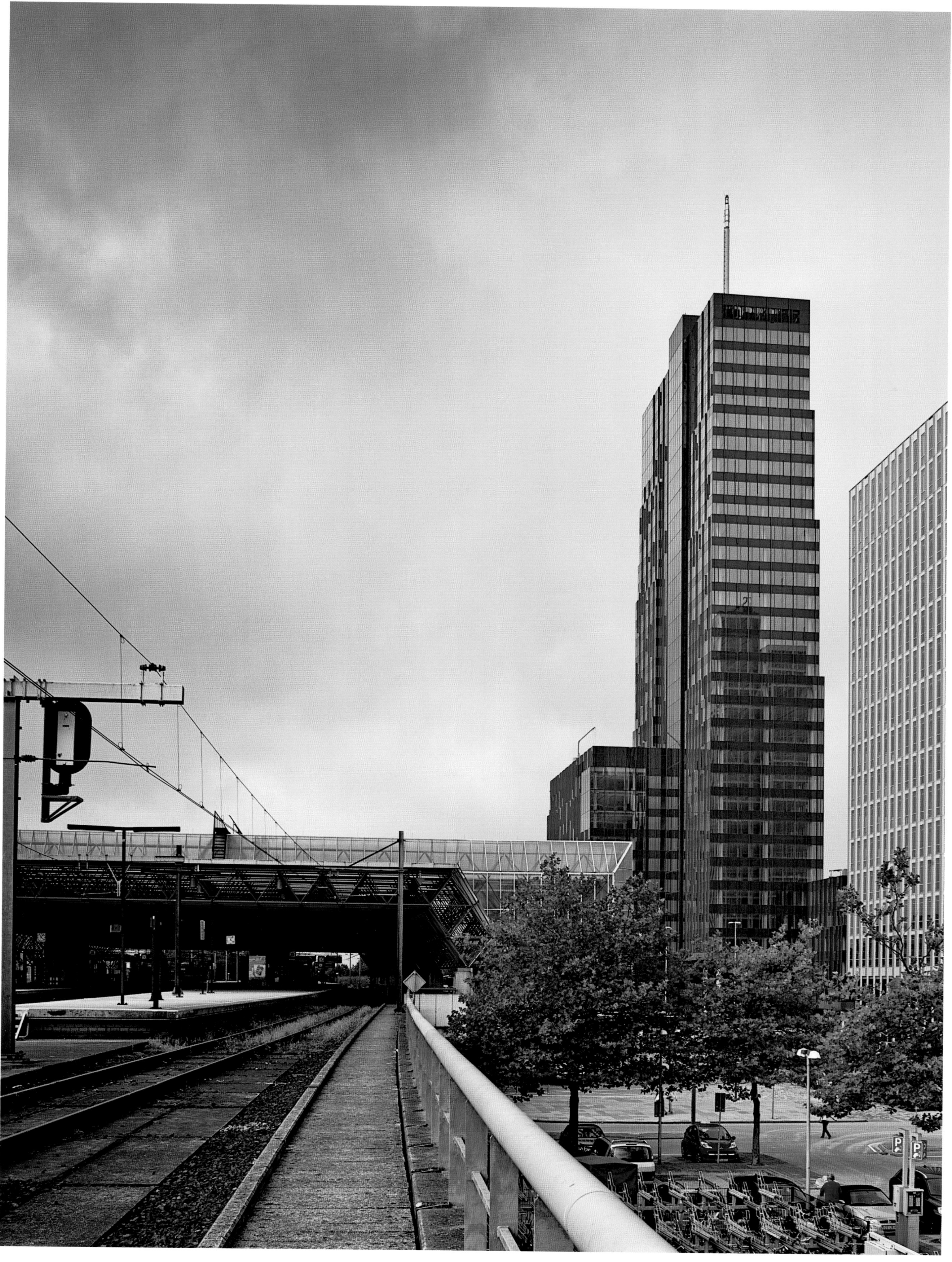

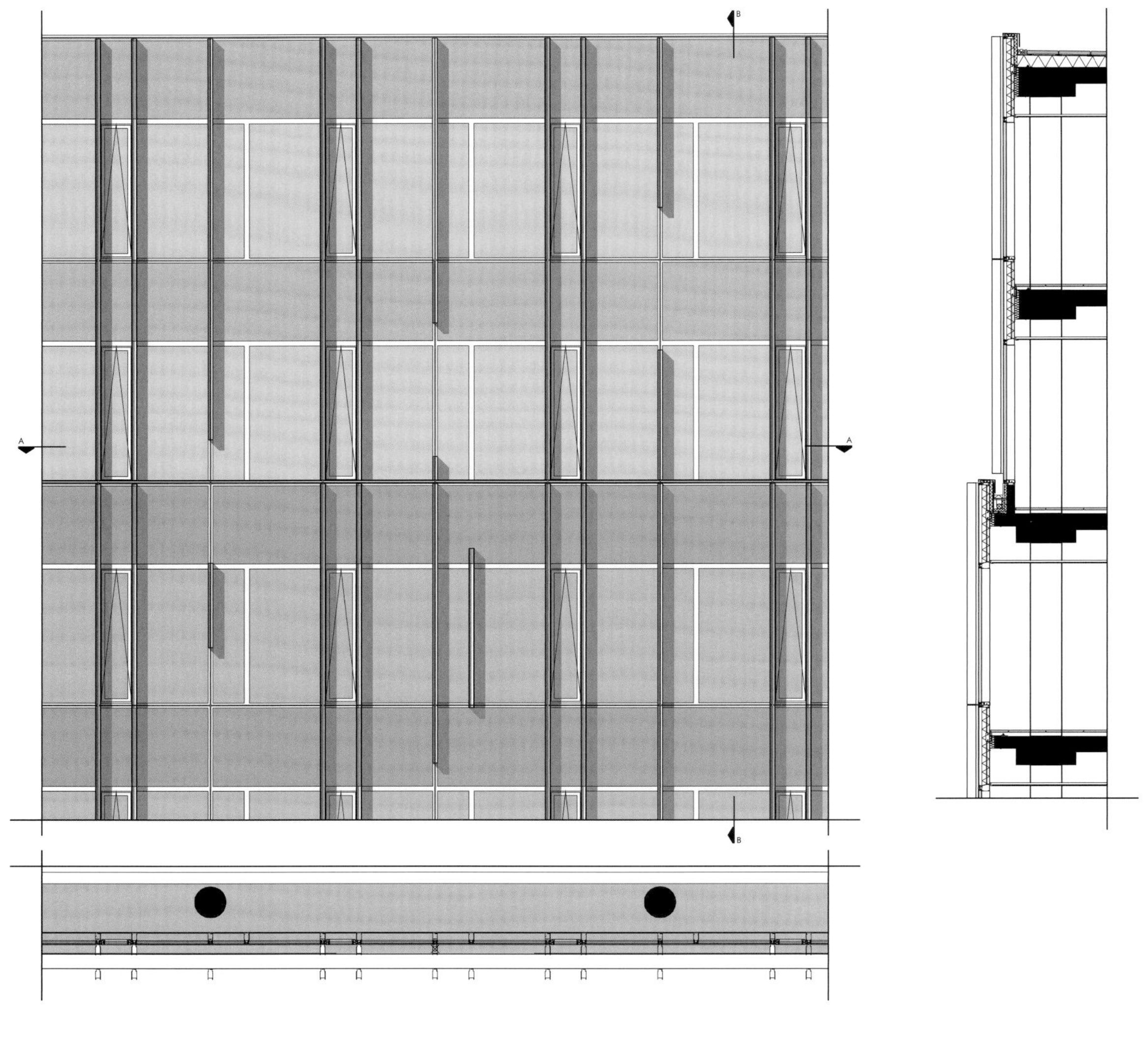
B
A
A
B

Instituut voor

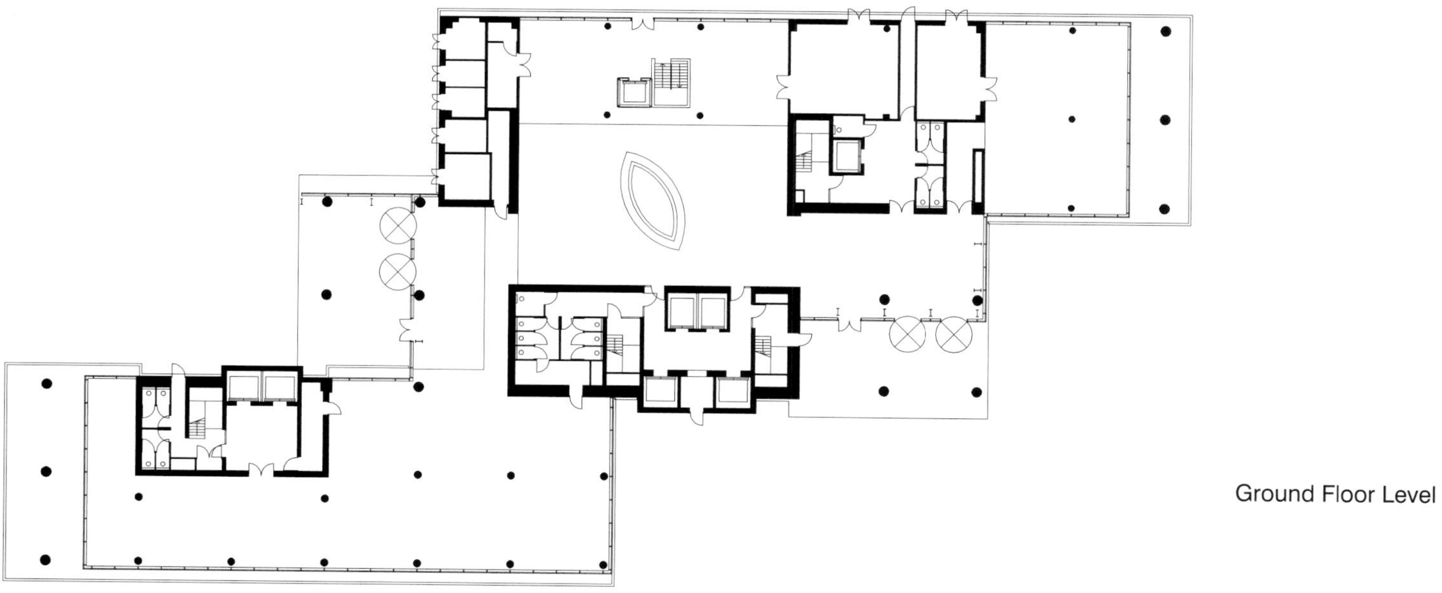

Ground Floor Level

千禧银行波兰总部

HEADQUARTERS OF MILLENIUM BANK, POLAND

ARCHITECT
Roman Dziedziejko, Mikolaj Kadlubowski, Michal Leszczynski, Krzysztof Mycielski

LOCATION
Warsaw, Poland

FIRM
Grupa 5 Architects

AREA
32,150 m²

PHOTOGRAPHER
Marcin Czechowicz

The building consists of three independent segments standing on a one storey plinth housing office lobbies, retail and restaurants. The exterior facades are stone and glass, whereas the interior facades are glazed.

Commercial offices and retail building A (phase 1 of the development) constitutes the east face of the central courtyard at the crossing of two main internal streets of the "Eko-Park" development in Warsaw.

A ground floor covered galleria connects the west and east parts of the building as well as the courtytard with the street entrance. The upper floors consist of three longitudinal office blocks with green courtyards in between them, accessible to office workers. Independent entrances to office lobbies on the ground floor are located directly underneath each office block. There is also an additional office space on the ground floor, between left and middle entrance hall. The floor plan can be easily divided into 1,2 or 3 office spaces with separate service areas and toilets.

Between office buildings there are colorful, glass boxes hanging on the facades, cantilevered over the green courtyards. Spaces in these boxes can be used as small conference rooms or offices. The facades of office buildings are designed as a combination of geometrically arranged panels of stone, glass and aluminium of different sizes. Every office room has a full height window with a glass balustrade in front of it. Vertical stripes of windows are interspersed with stone stripes. The inner facades shall be fully glazed to let maximum amount of light inside. Boxes shall be clad with color-printed glass. Three office buildings are connected by two-storey, underground parking with about 300 parking spaces. There are also about 30 parking spaces on the street level situated close to the main entrances. Parking space factor equals about 50 p.s. for 1,000m² of whole office and retail rentable area.

Millennium
bank

Millenniumpark

east elevation

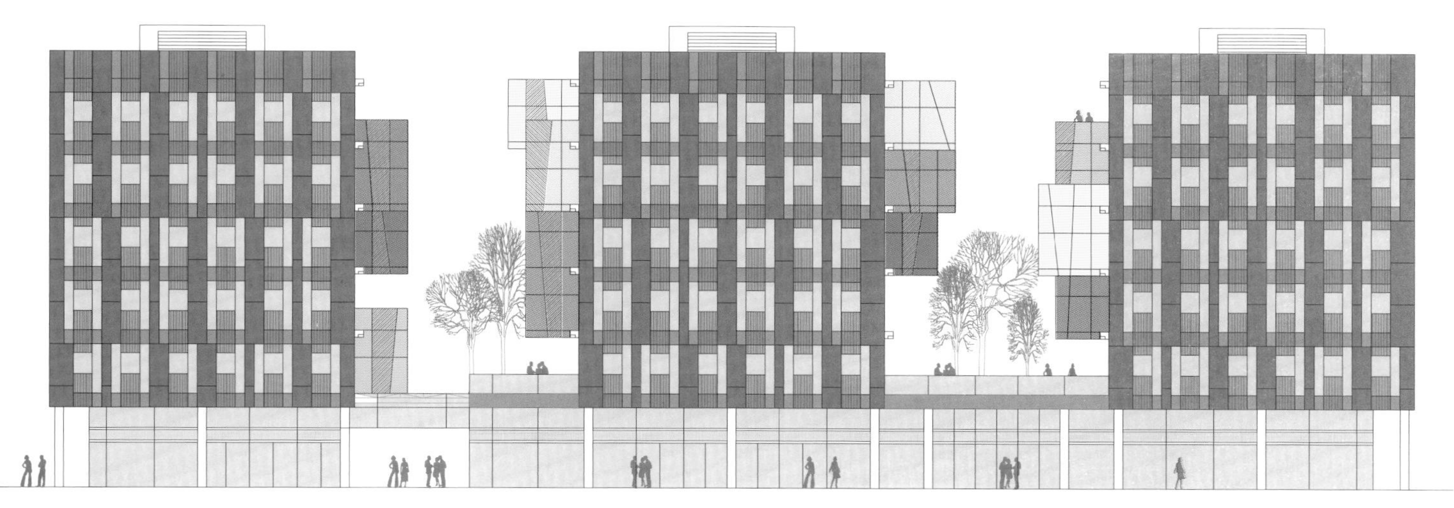

west elevation

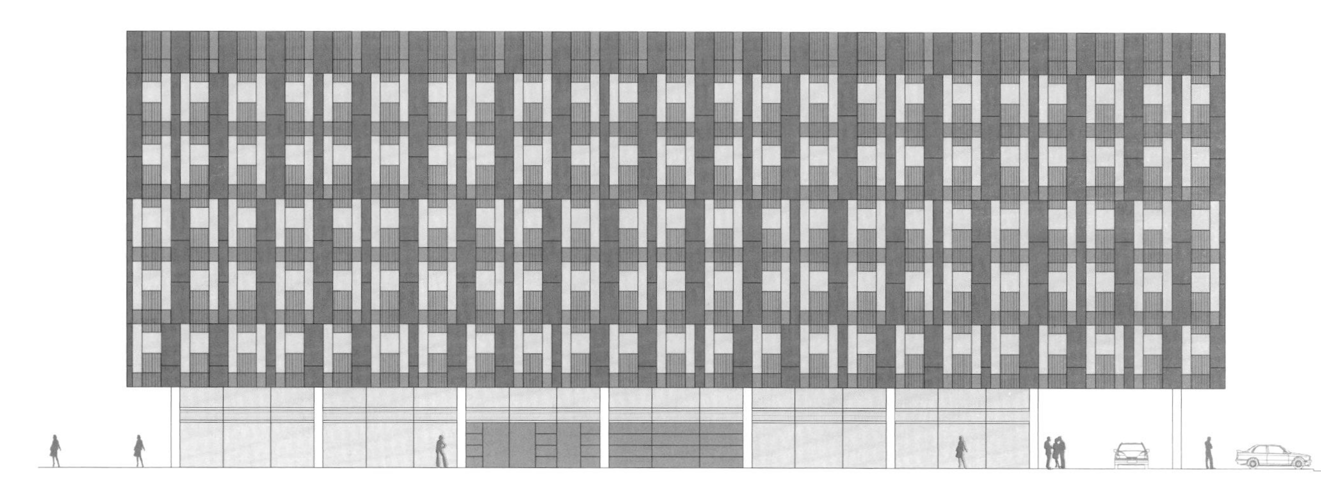

south elevation

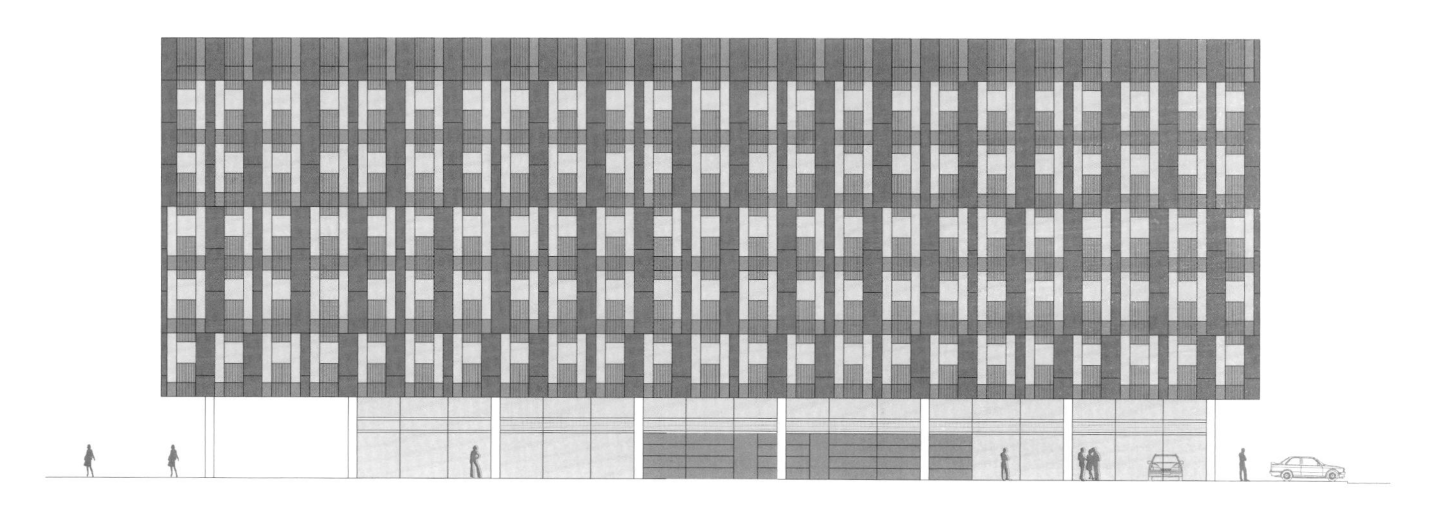

north elevation

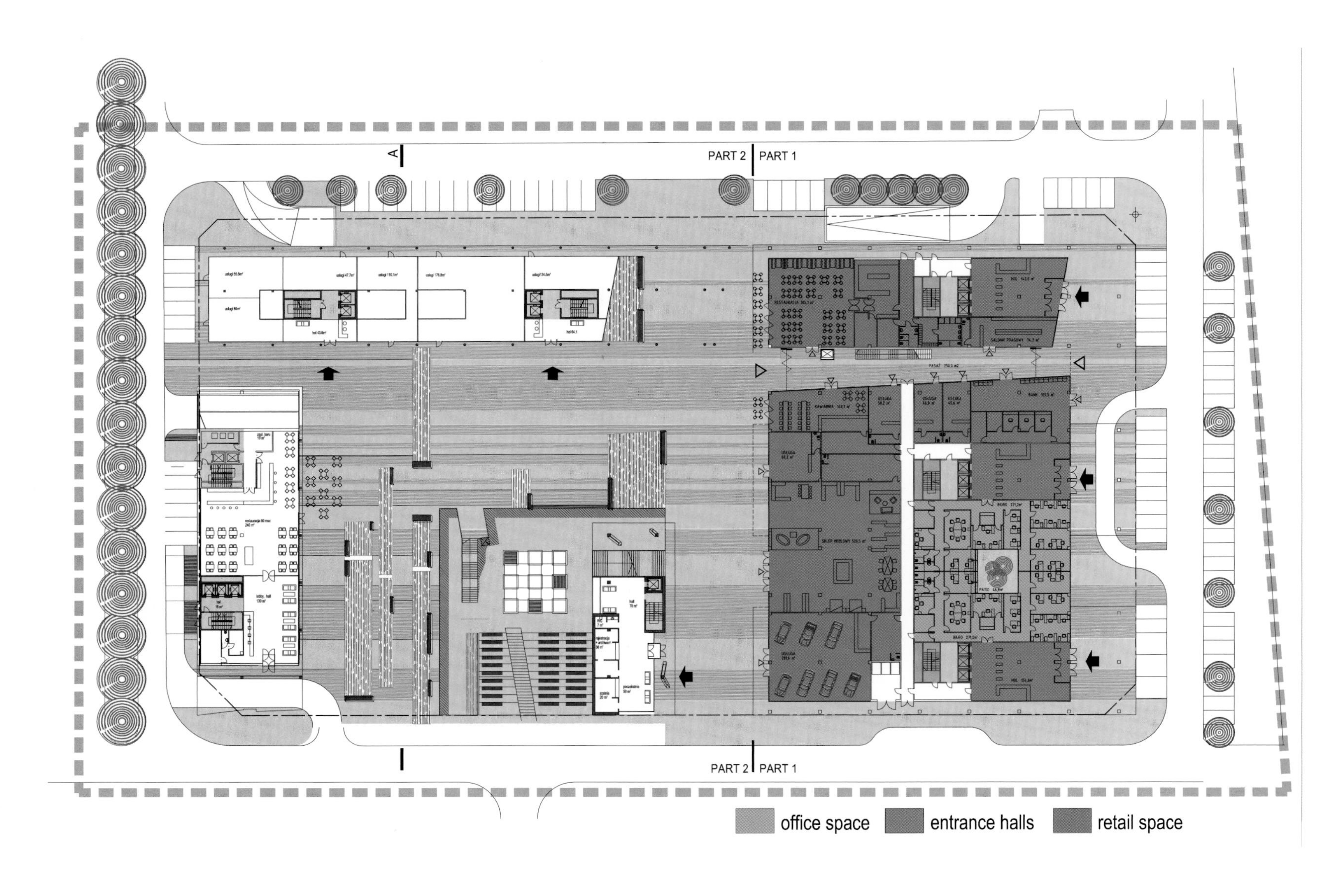
PART 2
PART 1
A
PART 2
PART 1
office space
entrance halls
retail space

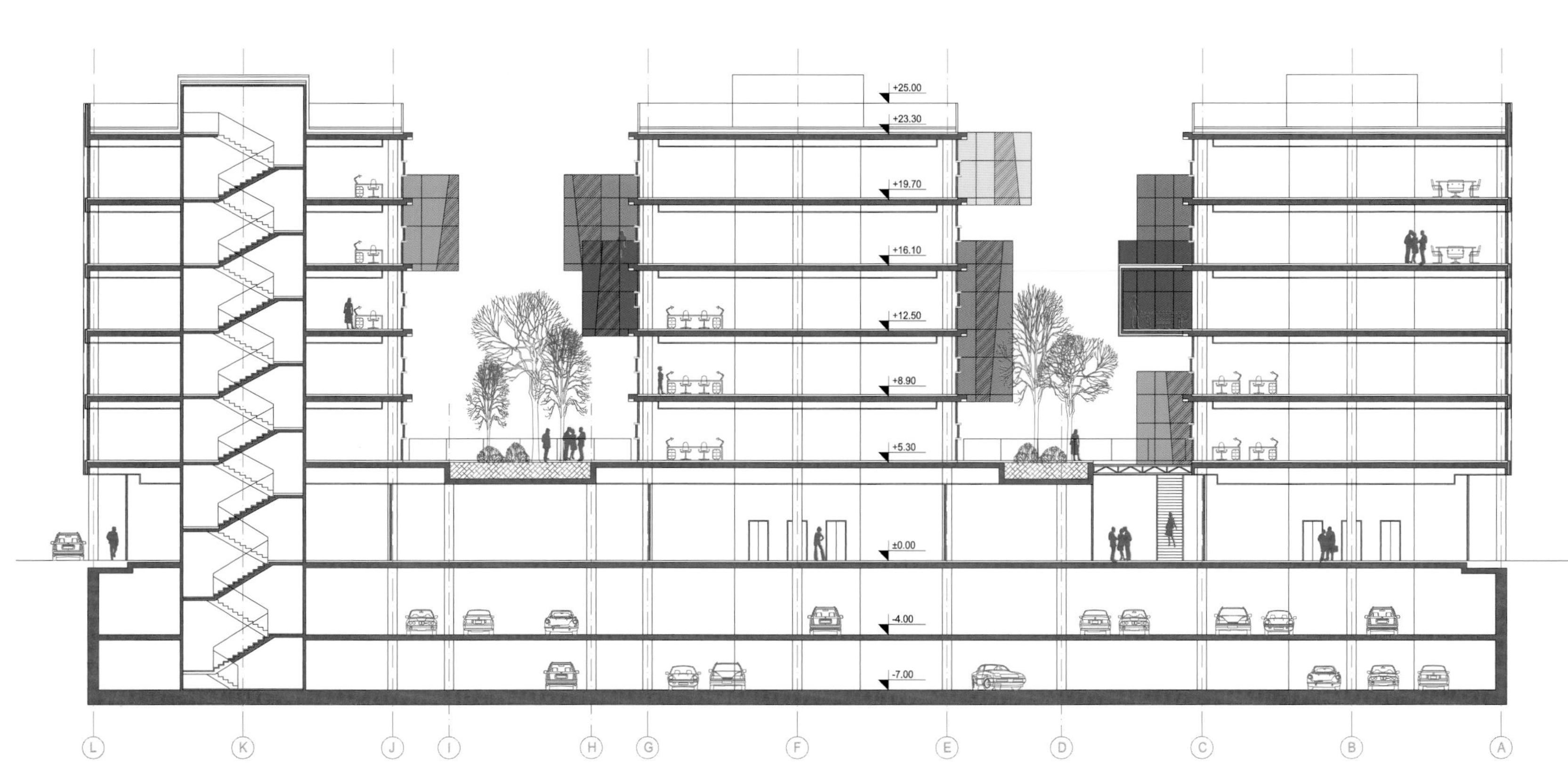
+25.00
+23.30
+19.70
+16.10
+12.50
+8.90
+5.30
±0.00
-4.00
-7.00
L
K
J
I
H
G
F
E
D
C
B
A

section A-A

M
Millennium
Millennium
Millennium
Millennium
Millennium

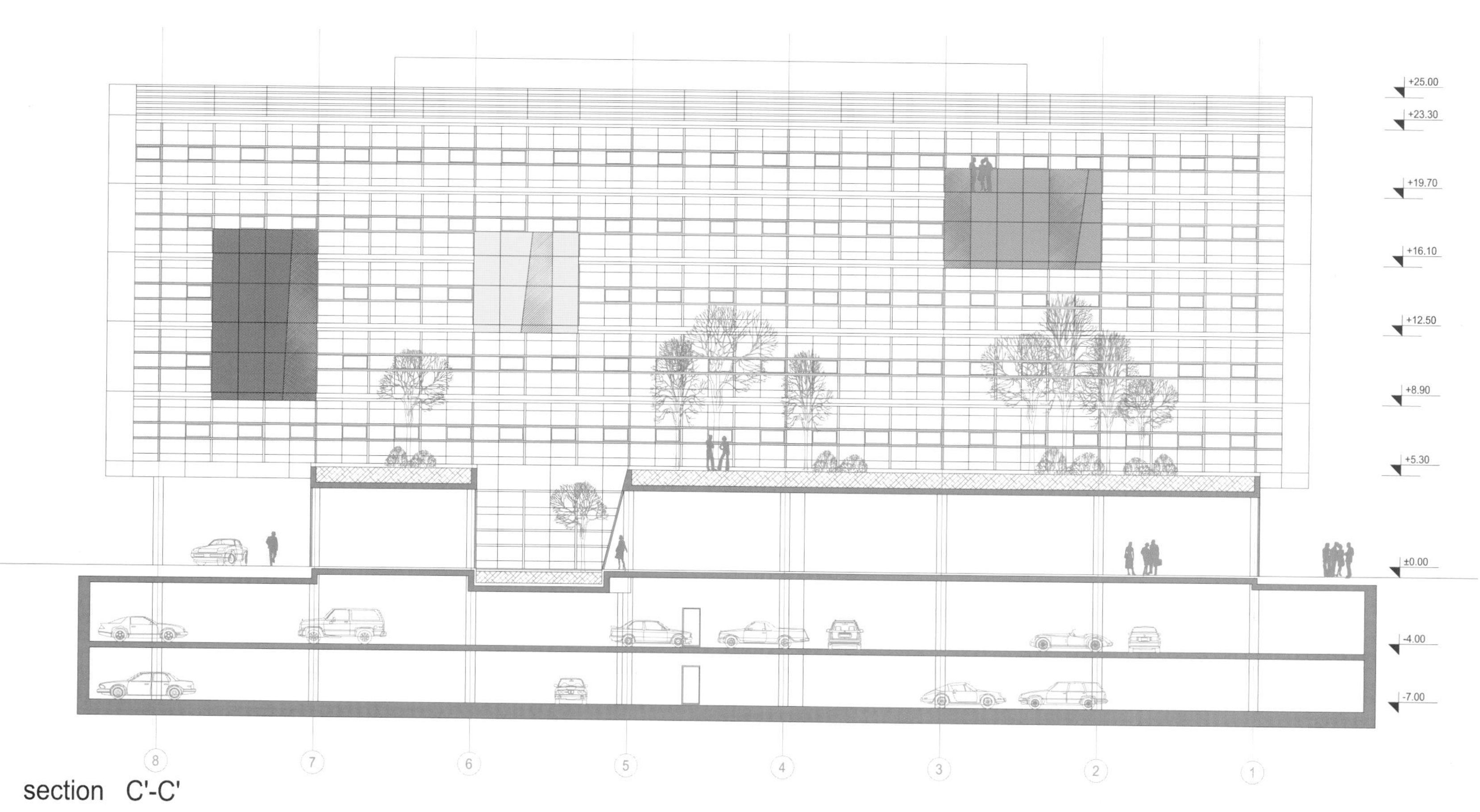

+25.00
+23.30
+19.70
+16.10
+12.50
+8.90
+5.30
±0.00
-4.00
-7.00
8
7
6
5
4
3
2
1

section C'-C'

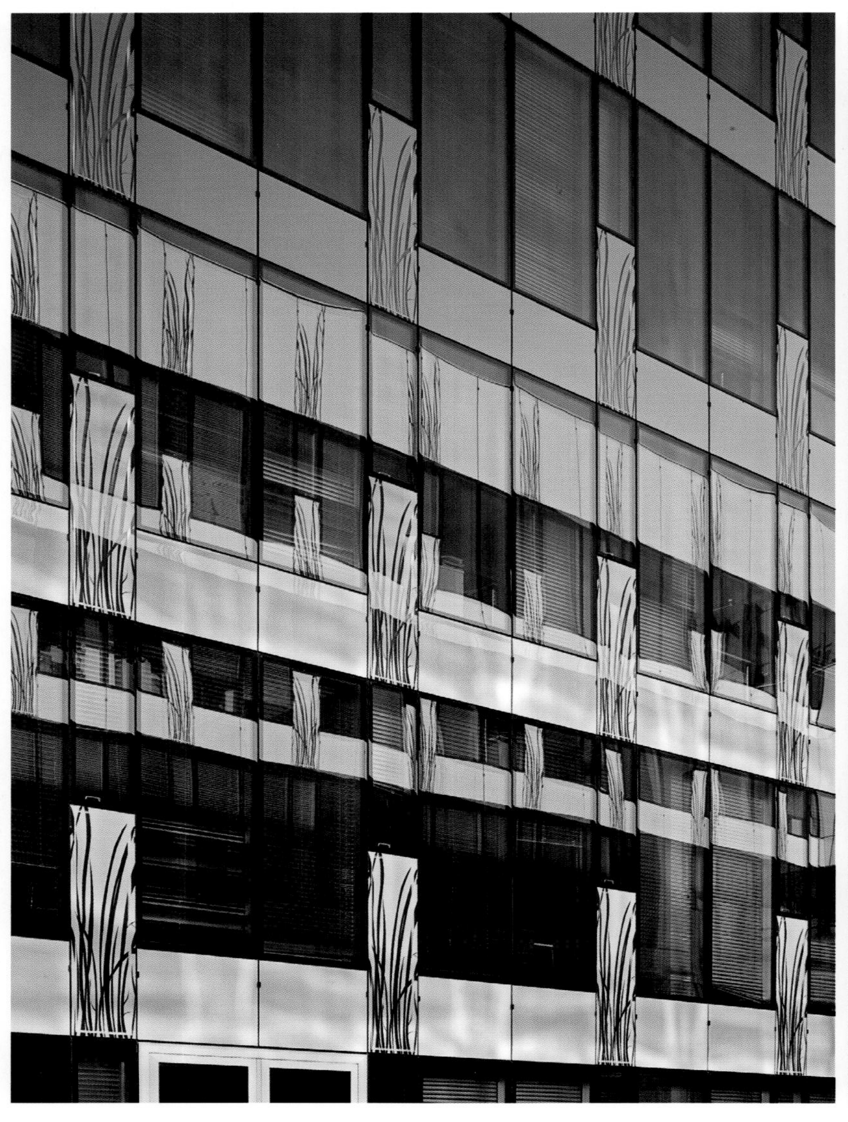

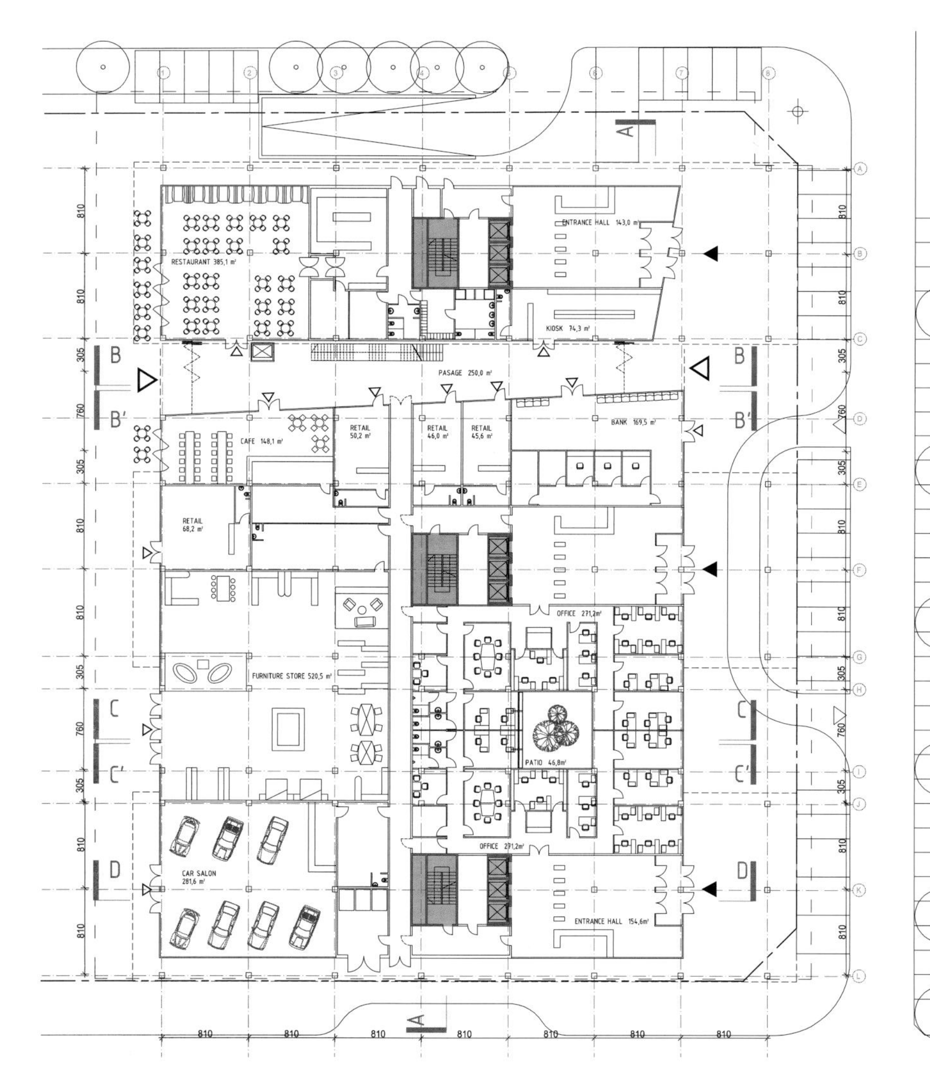

千禧银行波兰总部建筑包括三个独立的办公单元，矗立在同一个建筑基座上，基座内部设有办公大厅、零售店和餐馆等服务区域。大厦的外墙由石头和玻璃构成，使得内部空间十分明亮。

商业办公室和以零售店铺为主的A栋（设计图中标示项目1）构成了大厦的东侧部分，这里紧邻中心庭院广场，位于华沙著名的埃克公园开发区的两条主要街道的交叉路口处。

一条带有天棚的长廊贯穿大厦的东西两侧，另一面是临街的庭院广场。大厦上方是三个纵向分布的办公单元，中间利用绿色的庭院相互分隔，同时方便办公人员们通行。每个办公单元在楼下的基座中都设有独立的大门。在基座中，还设立了独立的办公区域，位于中间靠左侧的门厅内。基座内部的空间可临时分为一个、两个、或者三个办公空间，每个空间都有独立的服务区和洗手间。

办公单元之间的外墙上，设计了巨大的色彩斑斓的玻璃厢房作为装饰，悬挂在绿色庭院的上方。这里可以作为小型会议室和办公室。办公单元的外墙上，则利用不同规格的石头、玻璃和铝板进行组合，构成了一幅别具特色的几何图案。每个办公室里都设有通高的玻璃窗，上面设有玻璃扶手。窗户呈垂直的条状分布，中间有条形石块进行点缀，这样大厦内部就能够获得足够的自然光照。玻璃厢房外面覆盖着一层彩色玻璃。三个办公单元在一层和地下层相互连通，地下建有一个可容纳300辆汽车的停车场。地上还有30个停车位，位于大厦主入口附近。对于整个办公大楼和零售区域来说，占地1000平方米的停车空间系数约为50p.s.。

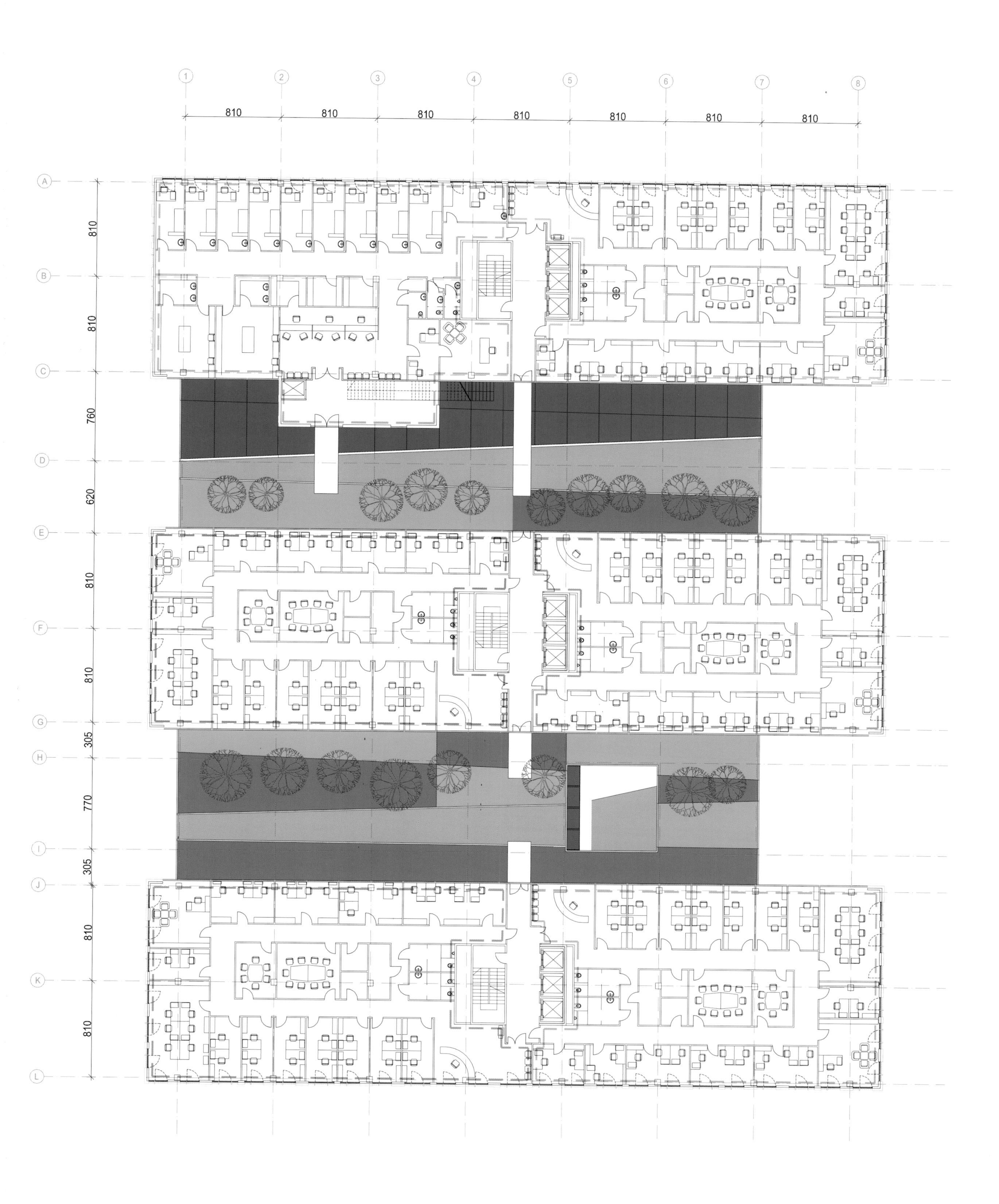

1
2
3
4
5
6
7
8
810
810
810
810
810
810
810
A
B
C
D
E
F
G
H
I
J
K
L
810
810
760
620
810
810
305
770
305
810
810

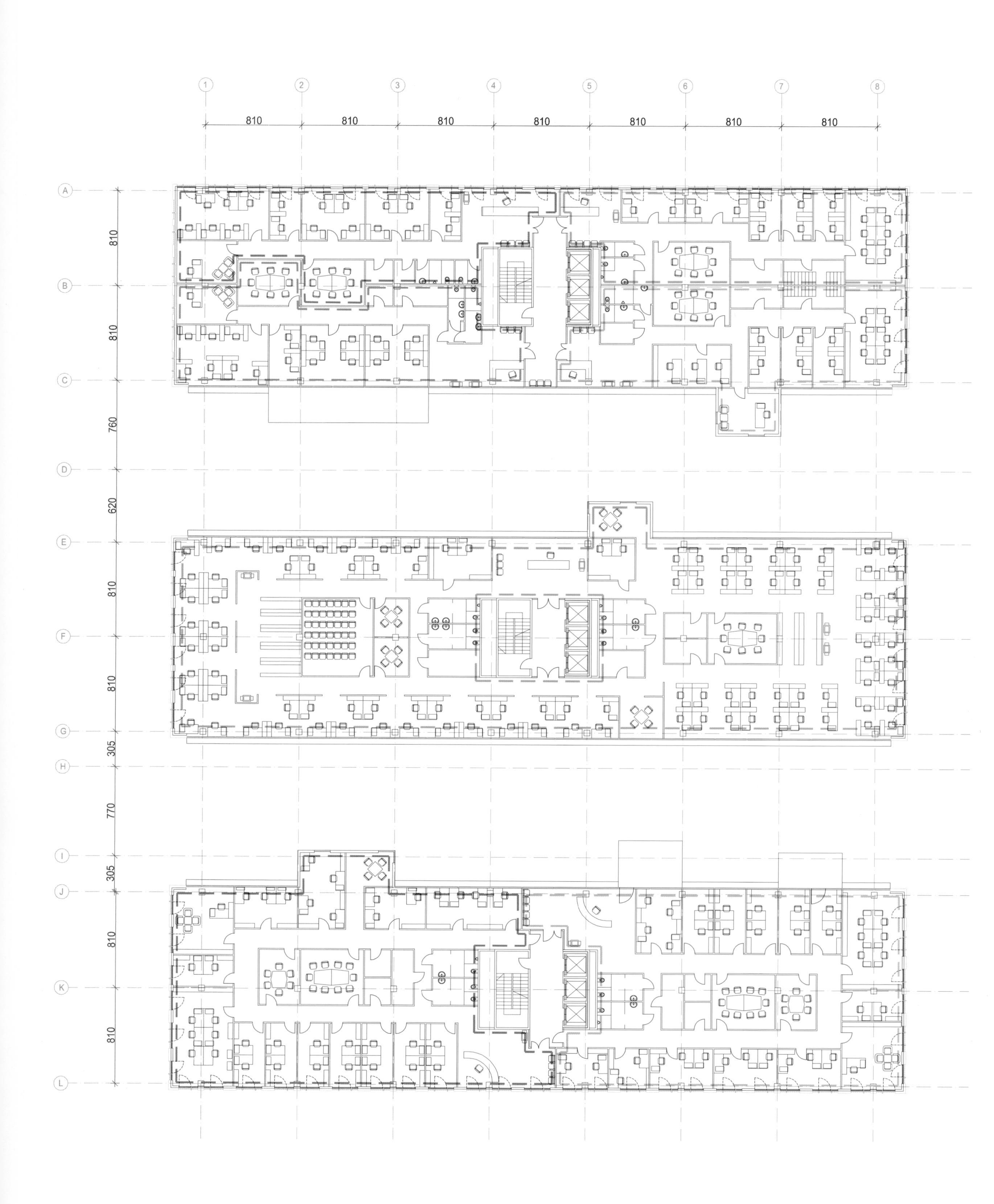

1
2
3
4
5
6
7
8
810
810
810
810
810
810
810
A
B
C
D
E
F
G
H
I
J
K
L
810
810
760
620
810
810
305
770
305
810
810

A2
Millennium
bank
M

乔治大街 400 号办公大楼

400 GEORGE STREET

ARCHITECT
Cox Rayner Architects

LOCATION
Brisbane, Queensland, Australia

PHOTOGRAPHER
Christopher Frederick Jones

400 George Street is both an office tower and a public space response to an important Brisbane CBD corner. The key design idea was to "erode" the building base to create a dynamic urban space enlivened by public life at both ground and elevated levels. Above these levels, the tower is honestly expressed for its purpose, such that the public base forms the dominant visual and spatial characteristic. To enrich the experience of the base, three artists were engaged to collaborate on the spatial development – Donna Marcus, Kenji Uranishi and Gemma Smith. Marcus's concept of the work "Trickle" resulted in a three-storey high volume or "urban room". This initiative led to the idea of an elevated food court cantilevering above George Street, it contrasting with other towers in the area by its people interaction with the public realm. In this way, 400 George Street responds to the architects' earlier initiative at Brisbane Magistrates Court to give back public use of the ground plane to the city, demonstrating how this can be achieved on the tightest of commercial office sites.

HOTEL
YMCA

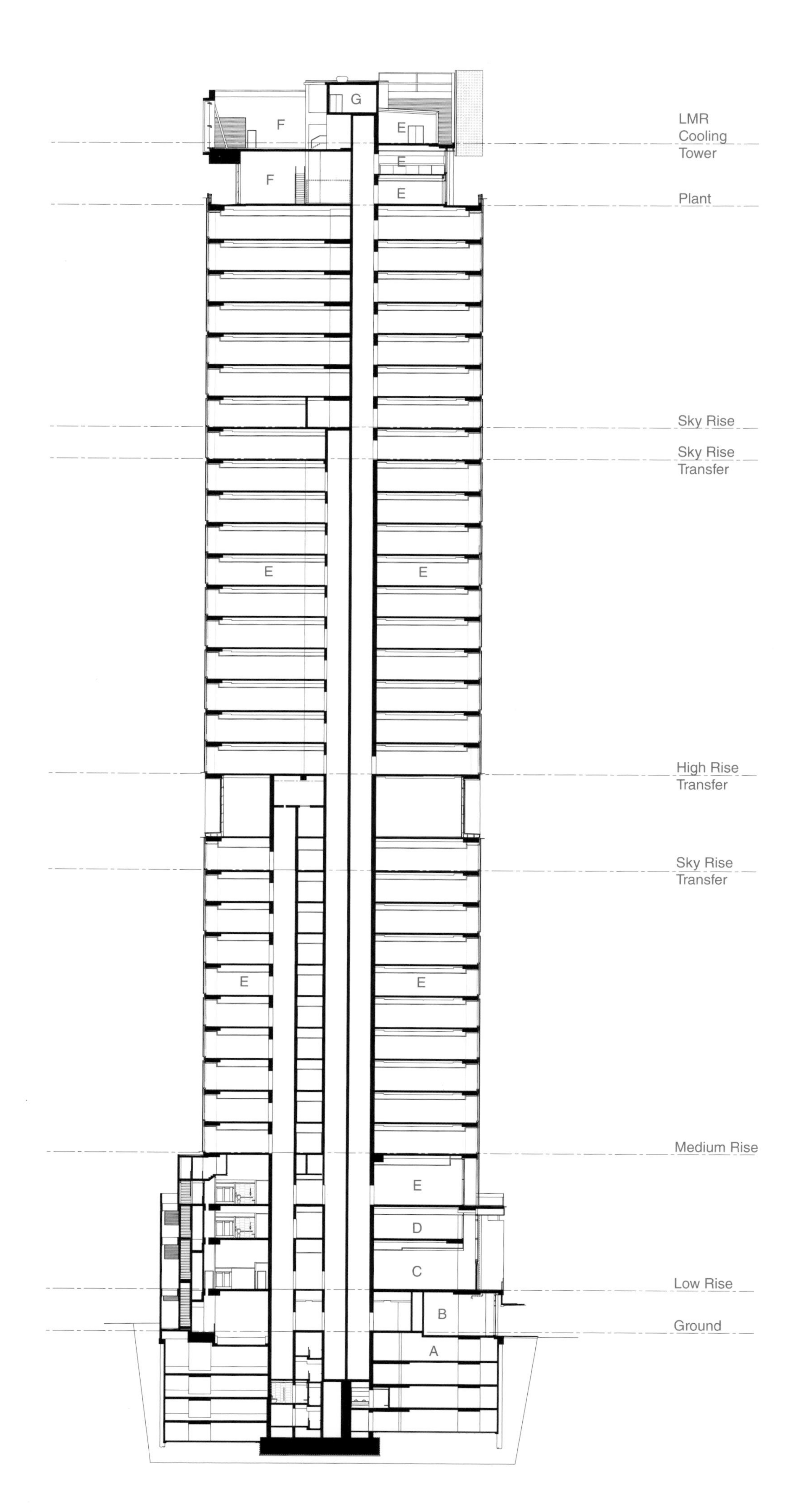

A Car Park
B Retail
C Food Hall
D Childcare
E Office
F Plant
G LMR

ACCOMMODATION

hotel
400 FOOD
STORE
WHITE

GEORGE

24-34

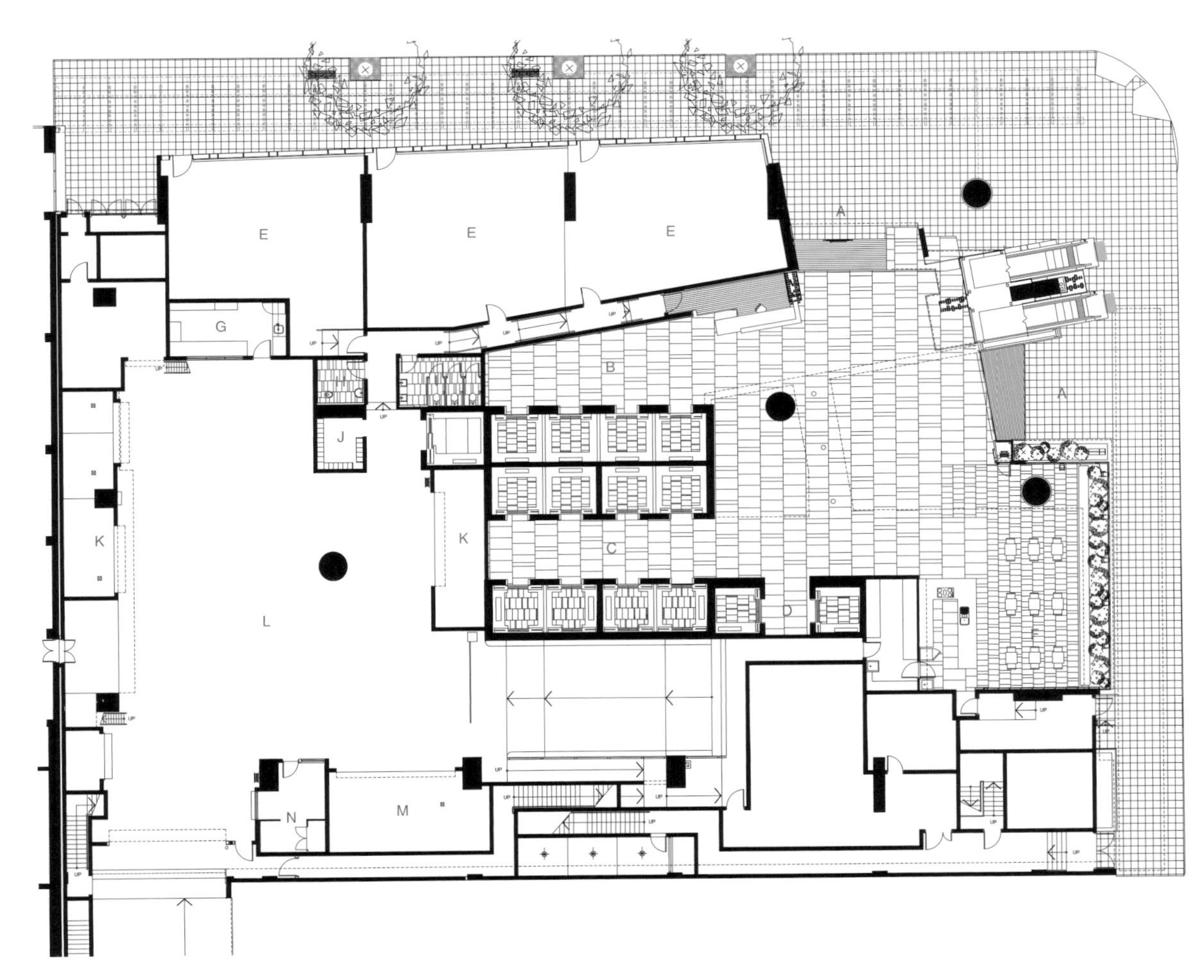

A Building Entry
B Sky Rise Lift Foyer
C Mid Rise / High Rise Lift Foyer
D Low Rise Lift Foyer
E Retail
F Cafe
G Building Management Room
H PWD W/C
I Uni-sex W/C
J Mail Room
K Recycled Waste Storage Area
L Loading Dock
M Refuse Storage Area
N Dock Management Security Room

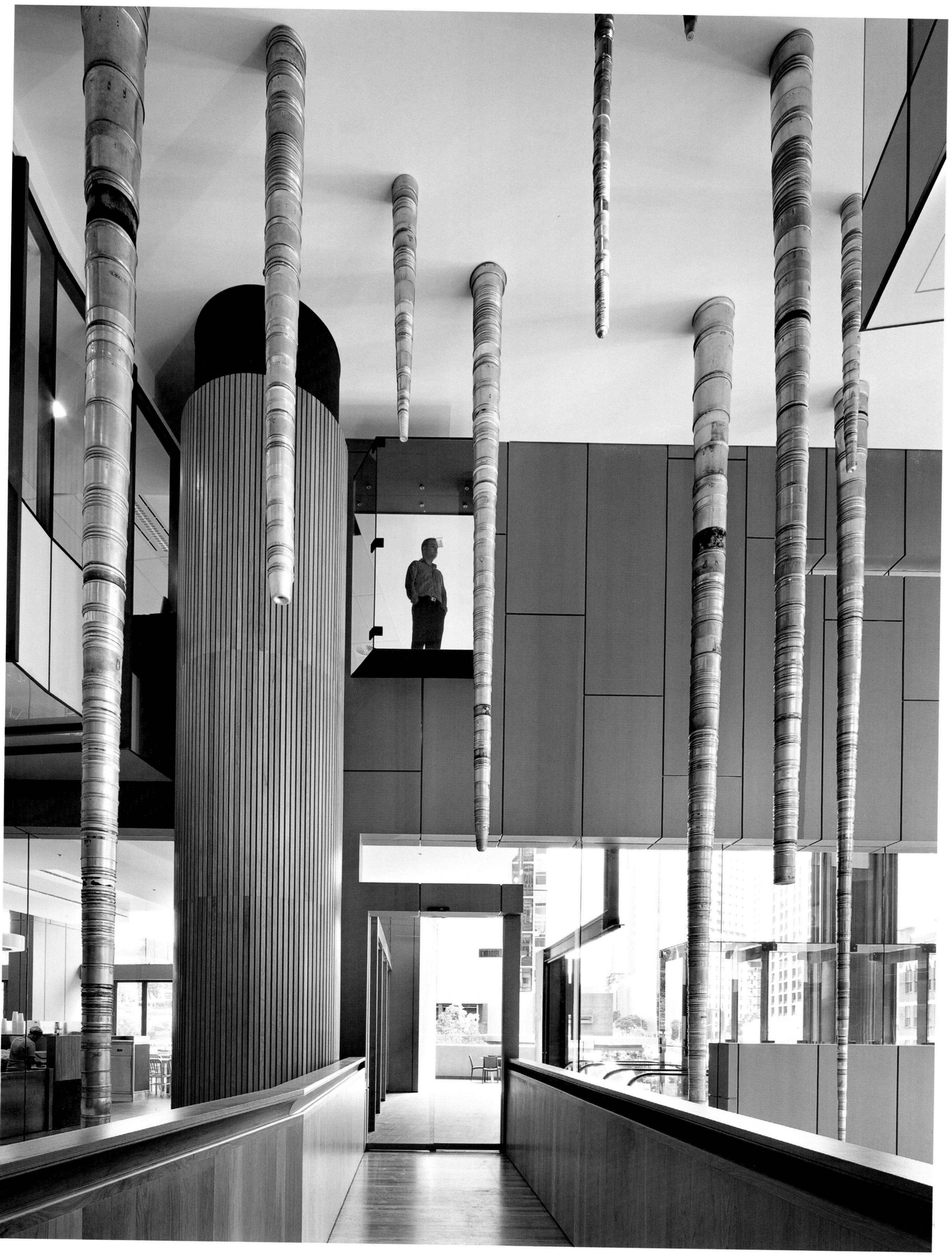

400 GEORGE

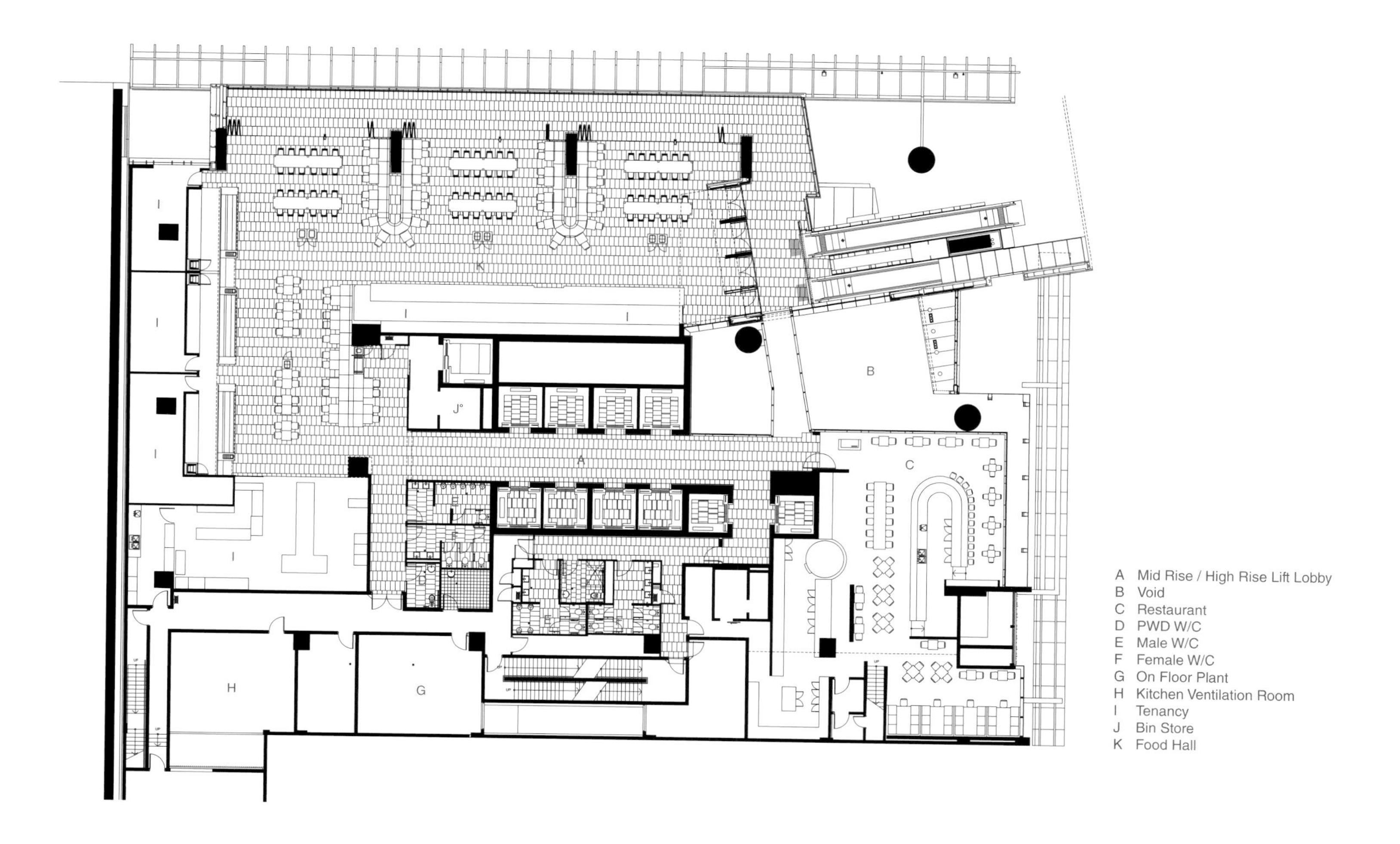

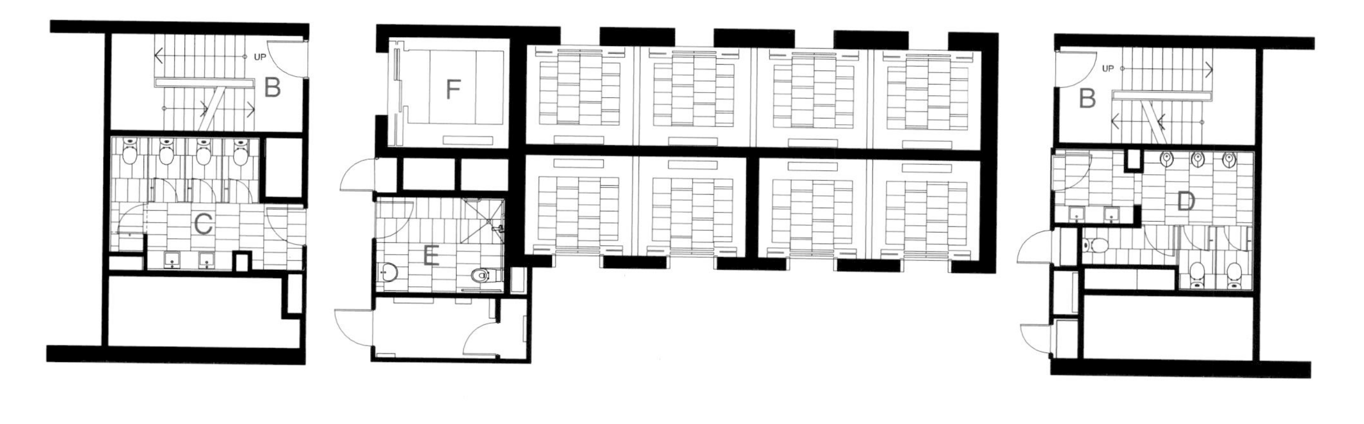

乔治大街 400 号既是一座写字楼，又是一个公共场所，与重要的布里斯班 CBD 一角遥相呼应。其关键的设计理念是“蚕食”建筑底座来创造一个动态的城市空间，这个空间因地面和高楼上人们的活动而显得生机勃勃。从这些层面上看，写字楼的设计淋漓尽致地体现了其用意，比如公共底座形成了主要的视觉和空间特点。为了丰富并完善底座的创作，三位艺术家在空间发展上协同合作——唐娜・马库斯、浦西贤治和杰玛・史密斯。马库斯的艺术作品《涓流》的创作理念成就了一座三层高的底座或“城市空间”。这个空间变成了一个架高的美食街，悬挑在乔治大街上，街道上到处都是熙熙攘攘的人群，与其他大厦门前的冷清形成鲜明的对比。就这样，乔治大街 400 号办公大楼响应建筑师们在布里斯班地方法院提出的早期倡议，将公共使用的地平面归还给城市，大厦的设计也正体现了这一点。

RCS

RCS MEDIAGROUP HEADQUARTERS – BUILDING C

ARCHITECT
Stefano Boeri, Gianandrea Barreca, Giovanni La Varra

FIRM
Boeri Studio

LOCATION
Milan, Italy

AREA
91,000 m²

PHOTOGRAPHER
Paolo Rosselli

RCS Mediagroup, a major Italian publishing house, has moved its offices to an area northeast of Milan, where some offices and printing facilities were already located. A call for tenders was issued that included the arranging of the area (91,000 sq. m), the restoration of an existing building and the construction of three new office buildings.

Building C, the first to be made, is composed of a low structure of 5 floors above ground and by a tower with a height of 80 meters. The development plan of the building has an open courtyard, whose lower part is parallel to Via Rizzoli; at its head, the courtyard faces toward the river Lambro and rises to form a square tower visible from the nearby motorway. The entire building is coated evenly on all sides by a double glazing of glass panels backed by aluminium point supports. It is a large and statuesque building, simple in its geometric structure and changing colors depending on time of day, weather conditions and seasons.

About 1,100 people live every day in the interior of the building, which is organized in modular offices and open-space environments characterized by flexible systems and light-weight and innovative partitions which are highly expressive of the corporate image.

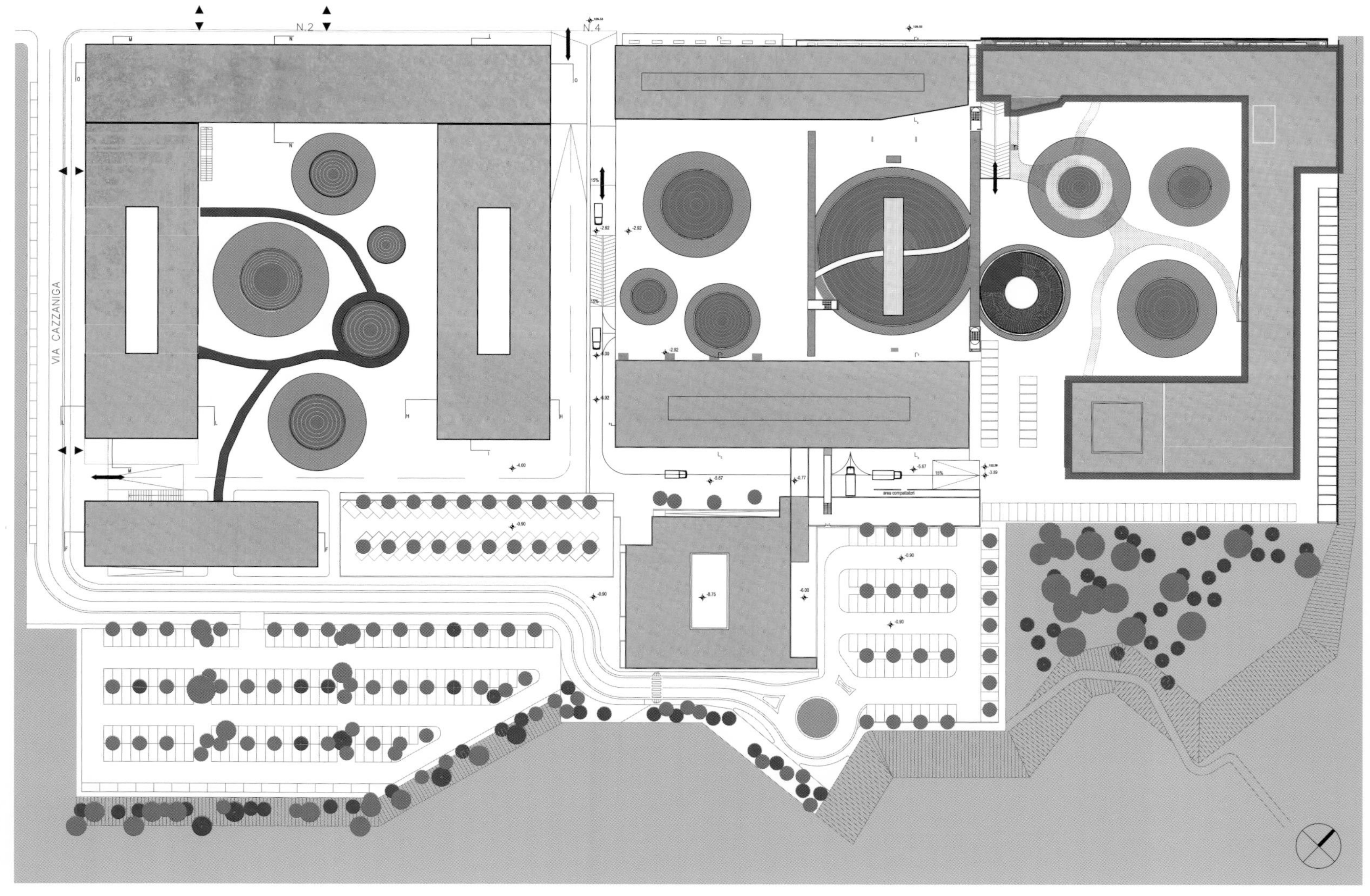
N.2
N.4
VIA CAZZANIGA

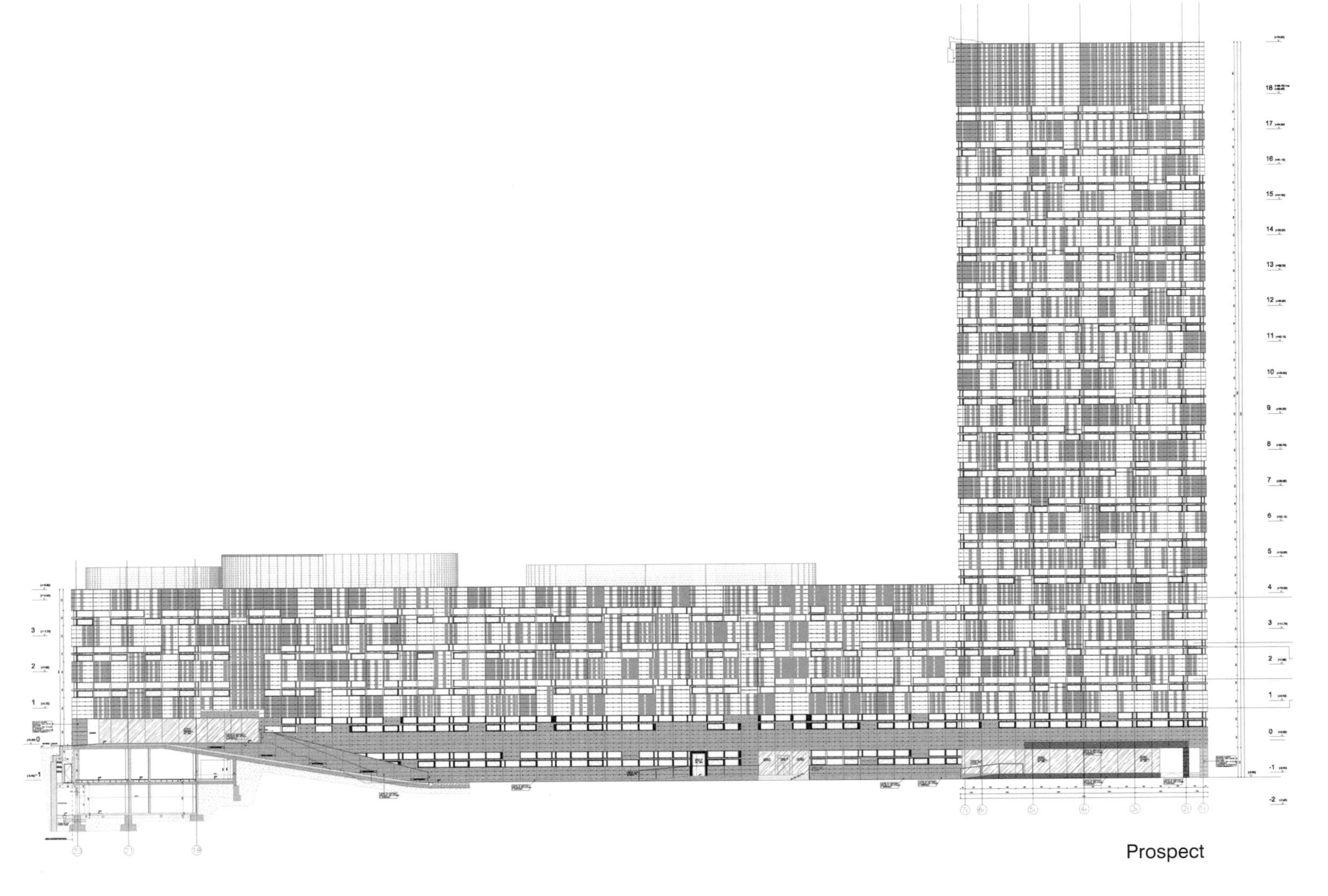

Prospect

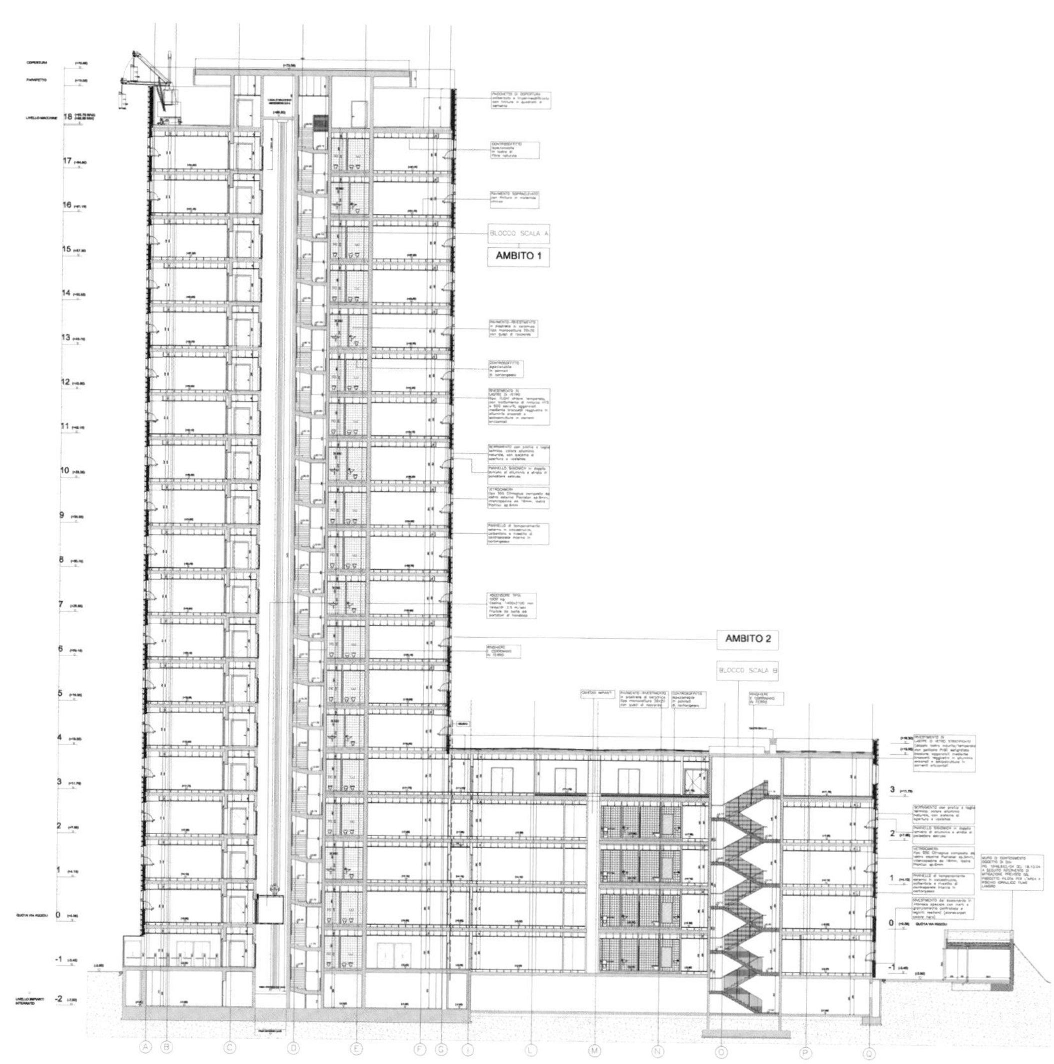

Section

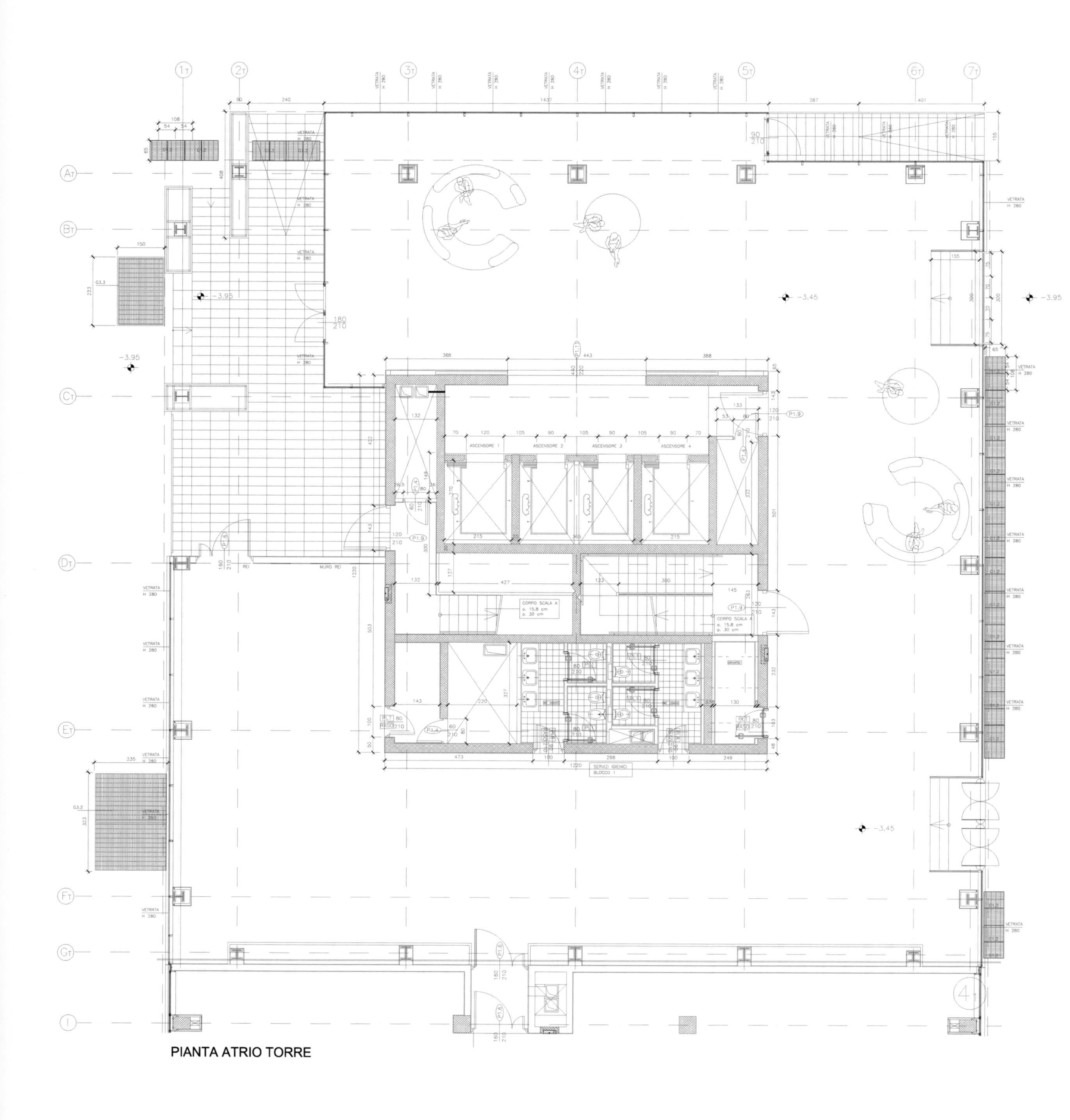

PIANTA ATRIO TORRE

RCS 传媒集团是意大利一家重要的出版社，现已将其办公楼搬到了米兰的东北角，那里已有一些先期搬去的办公室和印刷设备。RCS 传媒集团发布了一个招标布告，包括对该地 91 000 平方米的面积进行规划，改造现有大楼和建设三个新的办公大楼。

C 座楼是第一座要建的大楼，它由地面上一座五层楼高的矮层建筑和一座 80 米高的塔楼组成。大楼的开发规划中设计有一个开放的庭院，庭院下半部分与 Via Rizzoli 出版社平行；庭院前端正对着兰布罗河，广场上的塔楼就从这里拔地而起，从附近的高速公路上就能远远地看到它。整座大楼的外立面全部安装了铝型材外框的双层玻璃板。这是一座巨大而轮廓优美的建筑，它拥有简单的几何体结构，其颜色依据每天的不同时刻，以及天气和季节的变化而变化。

每天大约有 1100 人在这座大楼里办公，整座大楼设计有灵活多变的模块化办公室和开放空间，还有充分体现企业形象的休闲区与创新区。

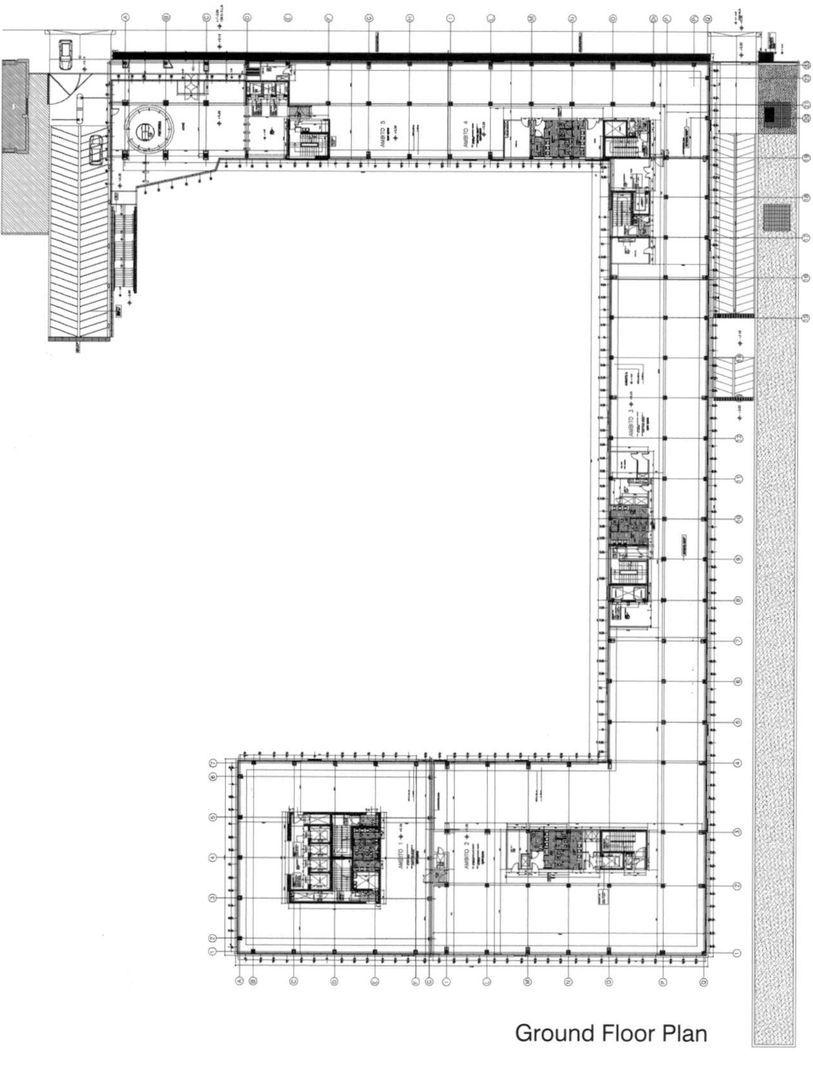

Ground Floor Plan

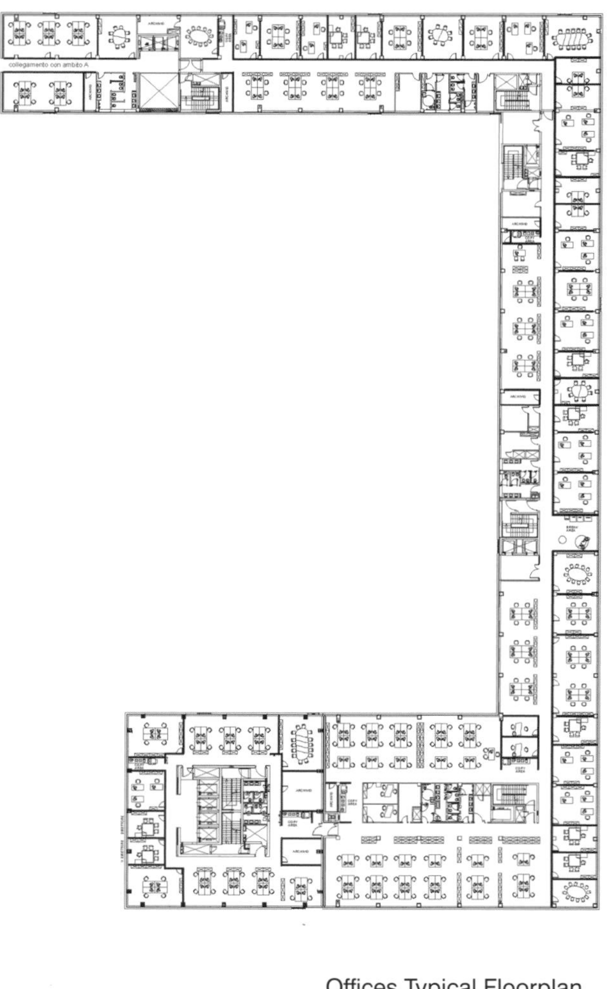

Offices Typical Floorplan

SKALIA

ARCHITECT
Arditti+RDT Arquitectos

FIRM
Ing. Alfredo Martinez

LOCATION
Zapopan, Mexico

AREA
30,000 m²

The building consists of a tower of 16 levels for offices with core services, wide stairs for service, main elevators, all rooms come with air conditioning, cleaning station and recordable pipeline installation. Also there is a mezzanine for private offices and shops for customer service, and the main lobby has two levels of commercial space with access from Mario Pani Street. There are also 6 levels of underground parking and a heliport on the roof.

The geometric form starts from a simple square, which is offset towards the facade to visually form two geometrical shapes. One of them is distorted in the finishing angle on top of the building generating in these levels, terraces for users.

The main access is emphasized with a bold glass triangle of large dimension that gives the user a very interesting spatial sense of space, passing by wooden decks with mirrors of water.

Its main lobby is located in the center of the ground floor, with high ceiling. The lobby represents a link with the mezzanine floor by means of a glass bridge, which parts symmetrically this space.

Skalia considers ecological characteristics that provide sustainability to the building and to the city. The building is designed with modules of 1.22m x 1.22m generating a better use of the materials. Its terraces allow a natural, solar block for plants while also generating social spaces and smoking areas for the users of the building.

Its design is based on their guidelines to generate savings in the thermal load in the interior of the building. Skalia meets the highest standards of systematization, being one of the best, intelligent buildings in Mexico.

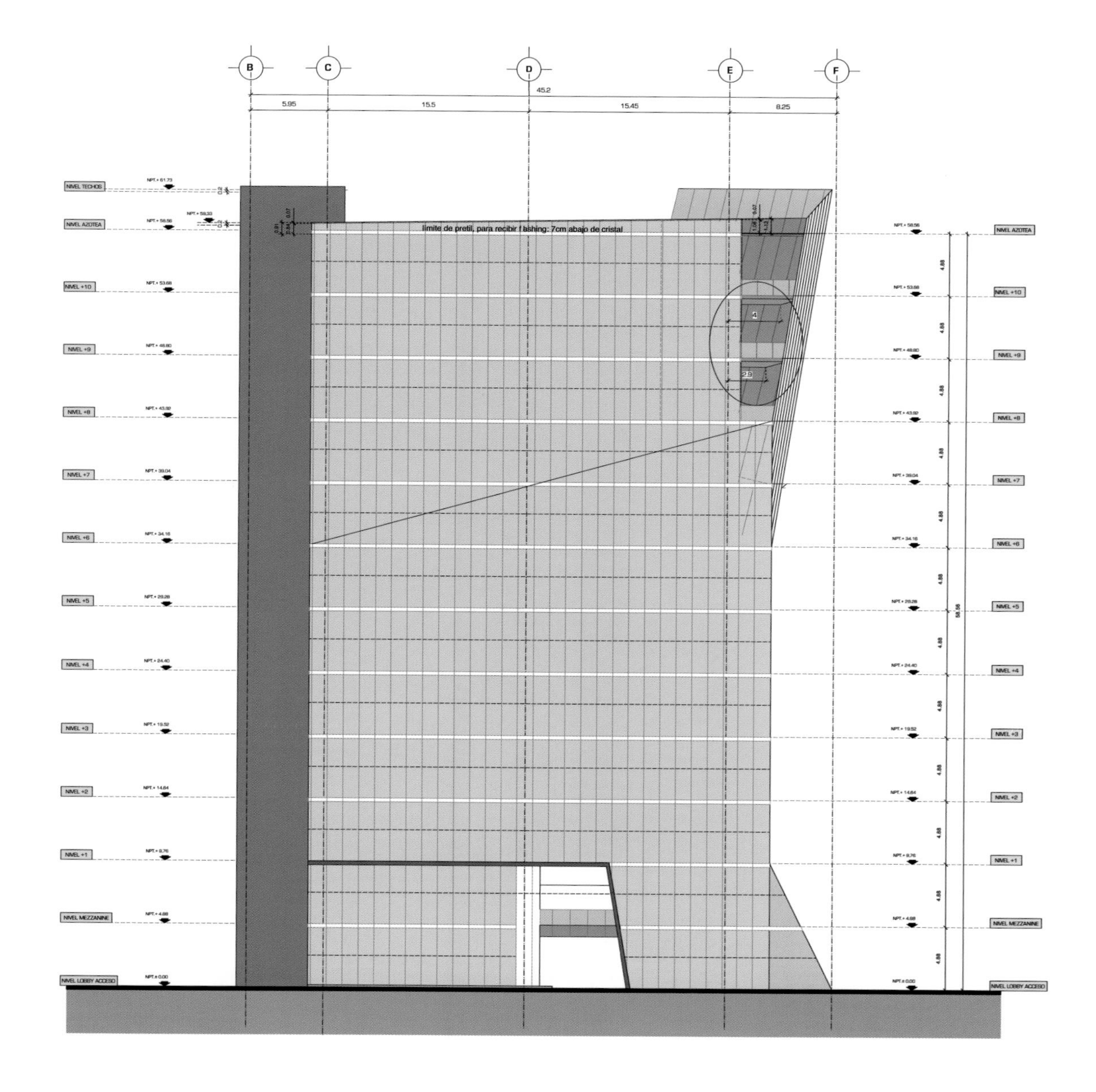

Oficinas Disponibles
(33) 3640 4143
044 33 3170 3132
VALET
SALIDA

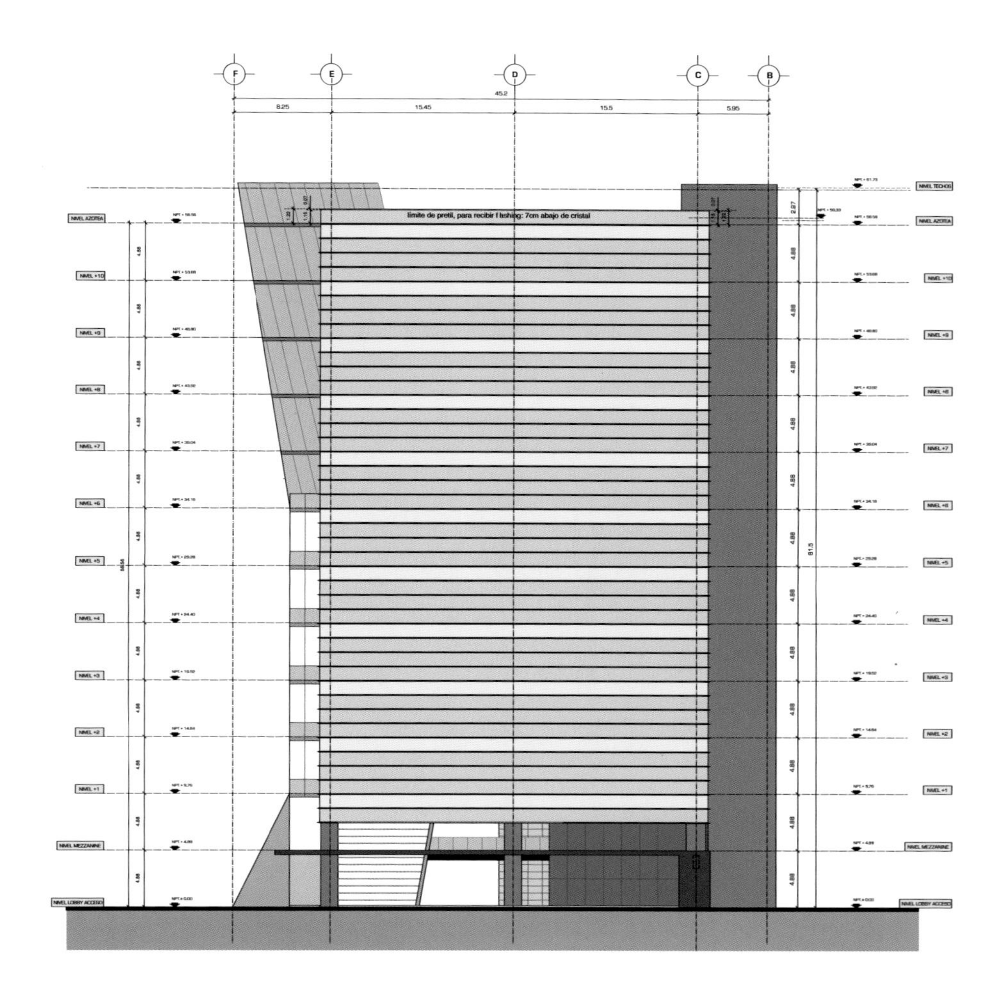
F
E
D
C
B
45.2
8.25
15.45
15.5
5.95
límite de pretil, para recibir f lashing: 7cm abajo de cristal

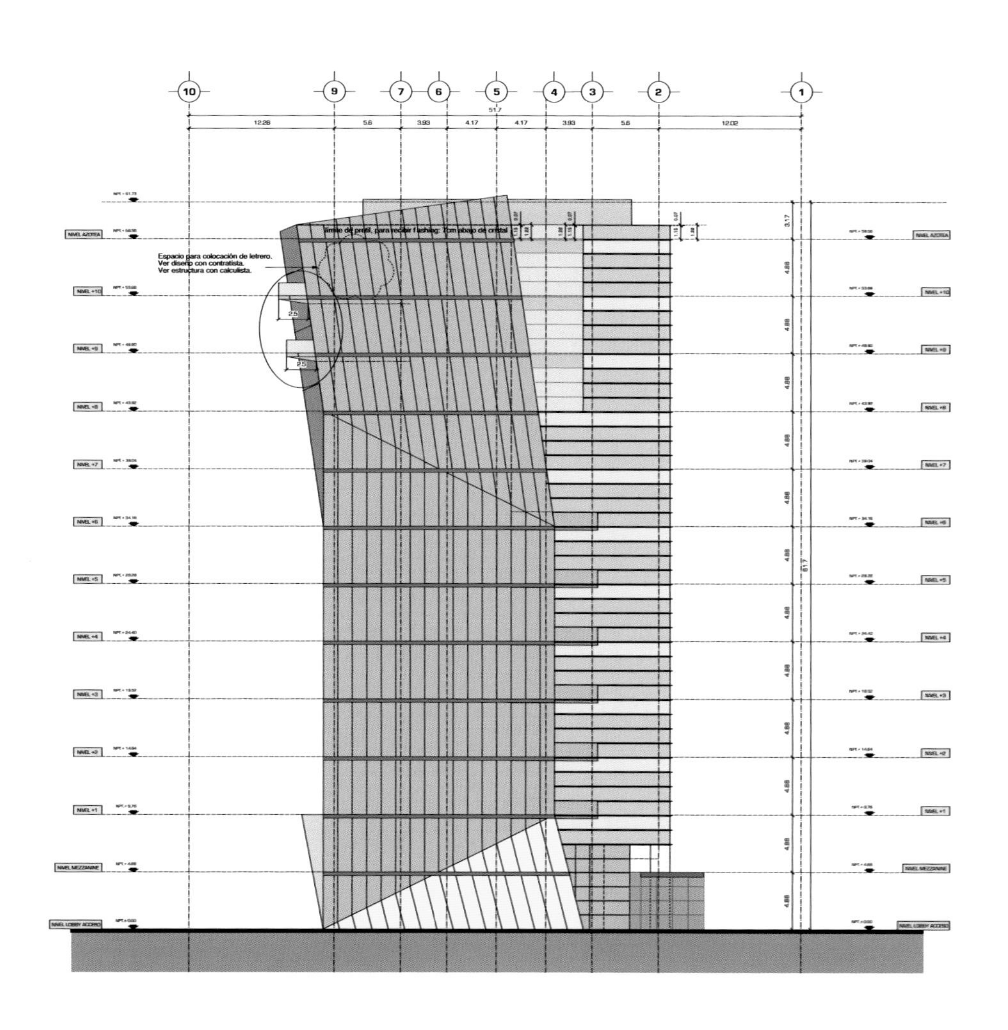
10
9
7
6
5
4
3
2
1
51.7
12.26
5.6
3.93
4.17
4.17
3.93
5.6
12.02
límite de pretil, para recibir f lashing: 7cm abajo de cristal
Espacio para colocación de letrero.
Ver diseño con contratista.
Ver estructura con calculista.
2.5
2.5

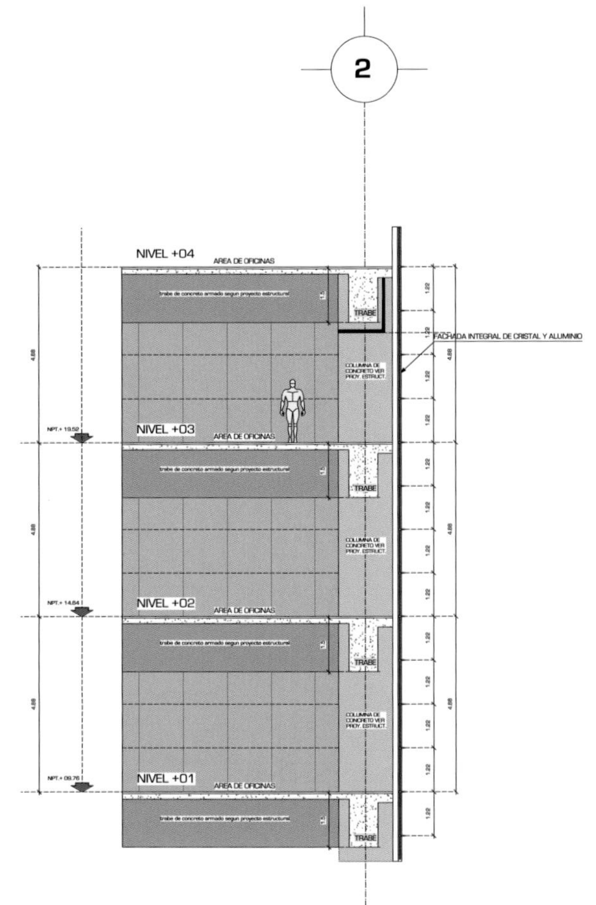

CORTE POR FACHADA 2

CORTE POR FACHADA 3

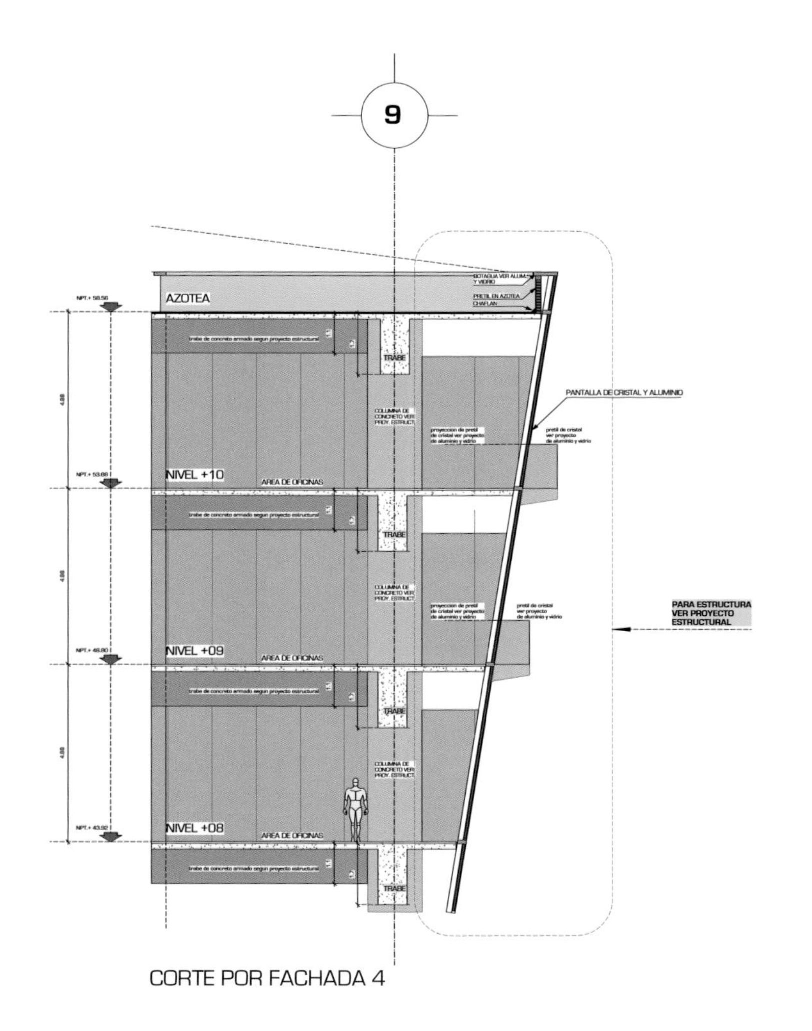

CORTE POR FACHADA 4

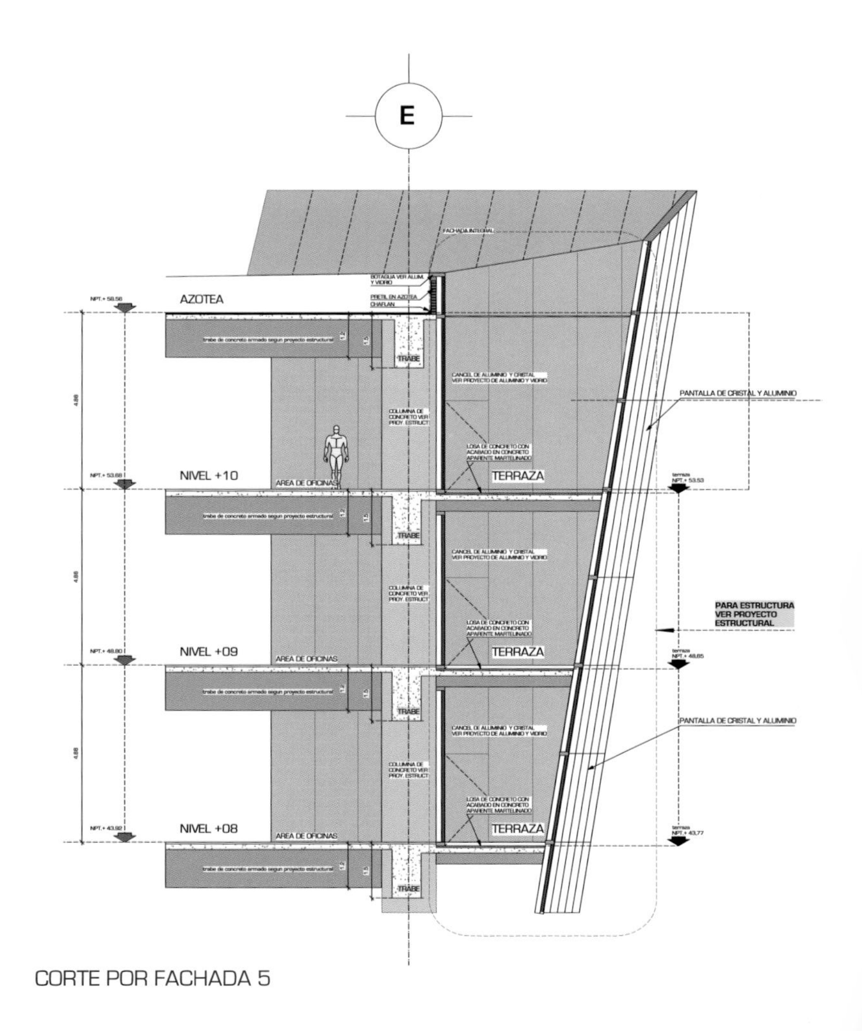

CORTE POR FACHADA 5

Ofi cin
04

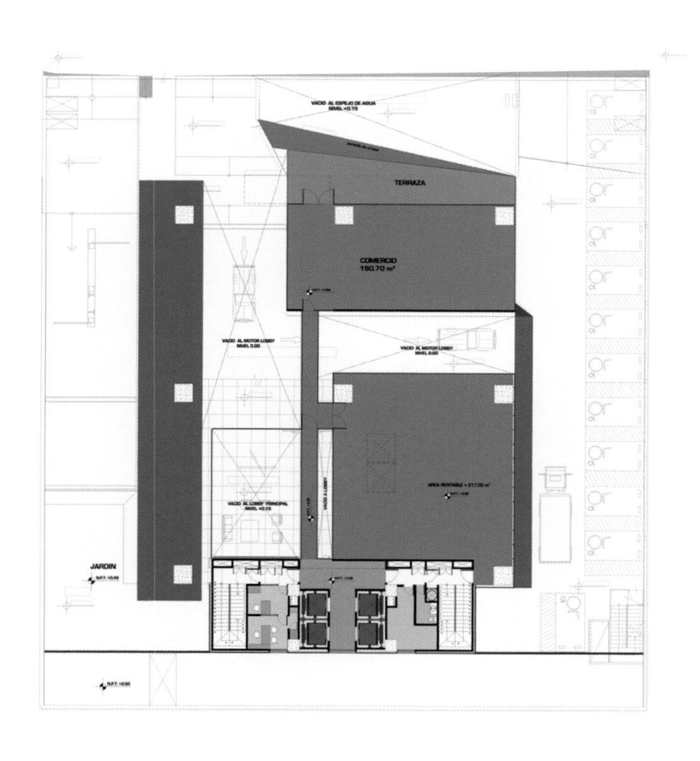
VACIO AL ESPEJO DE AGUA
NIVEL +0.15
TERRAZA
COMERCIO
190.70 m²
VACIO AL MOTOR LOBBY
NIVEL 0.00
VACIO AL MOTOR LOBBY
NIVEL 0.00
VACIO AL LOBBY PRINCIPAL
NIVEL +0.15
VACIO A LOBBY
JARDIN

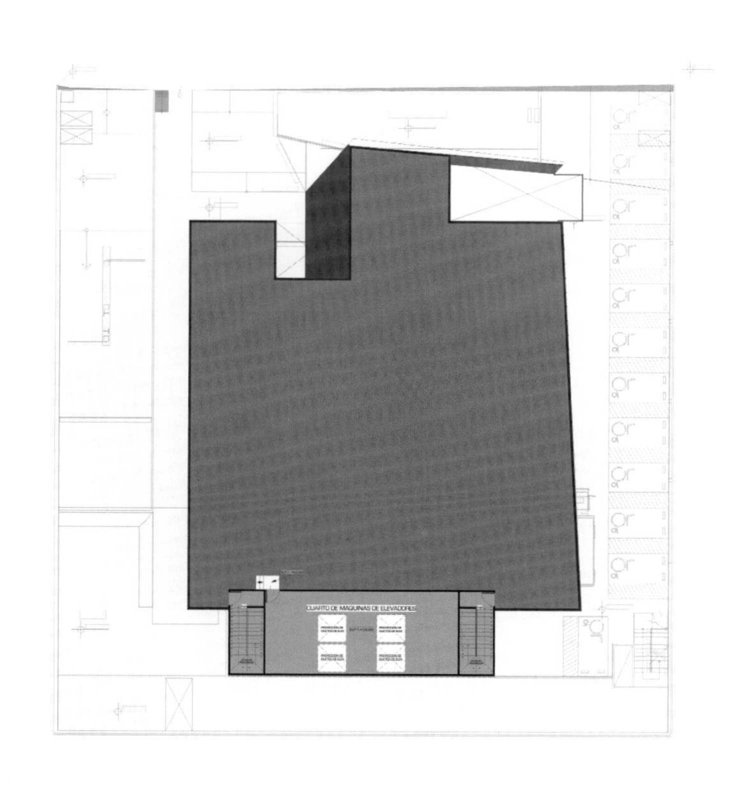
CUARTO DE MAQUINAS DE ELEVADORES

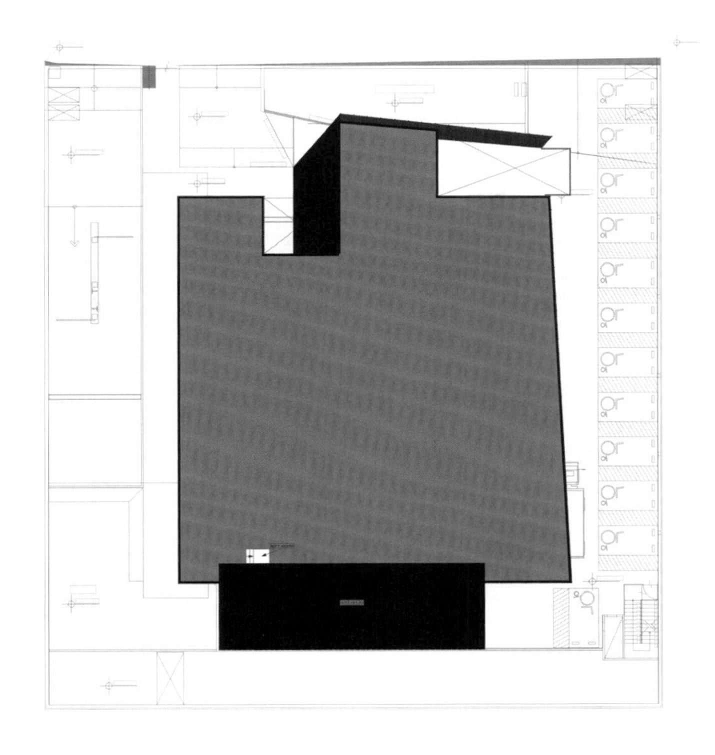

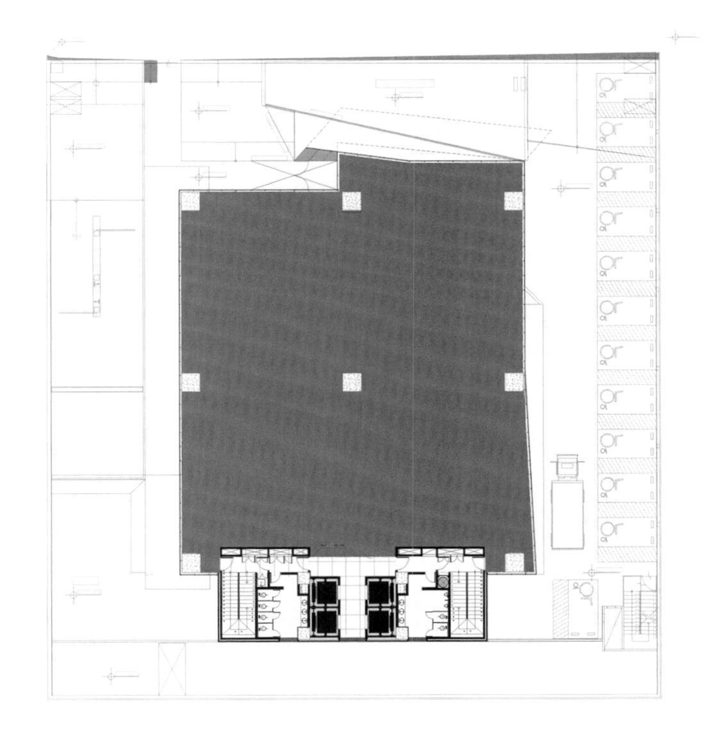

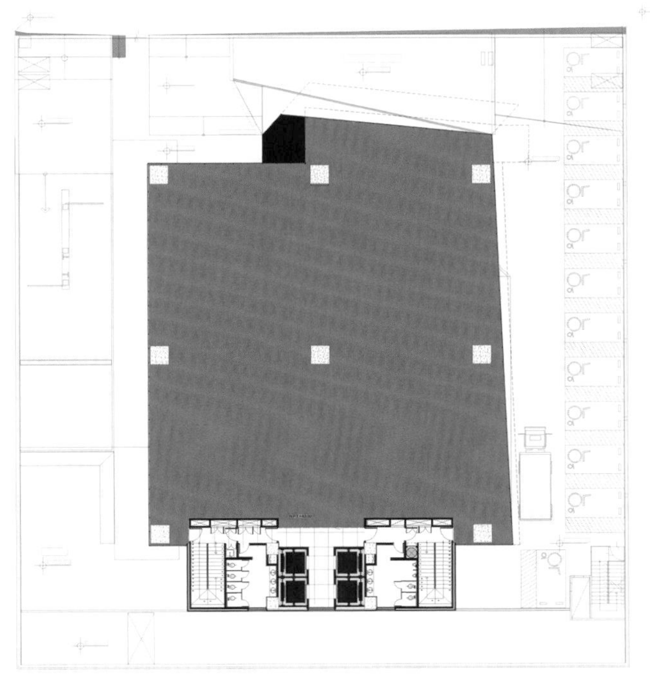

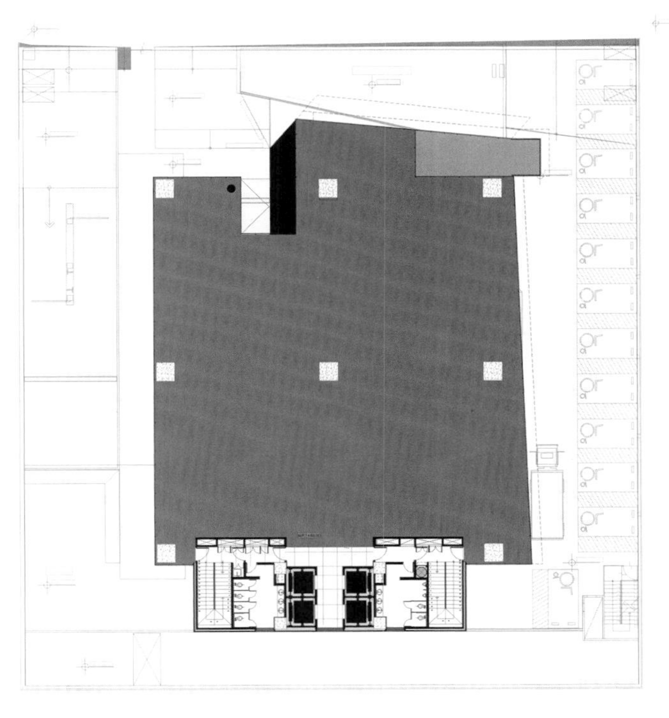

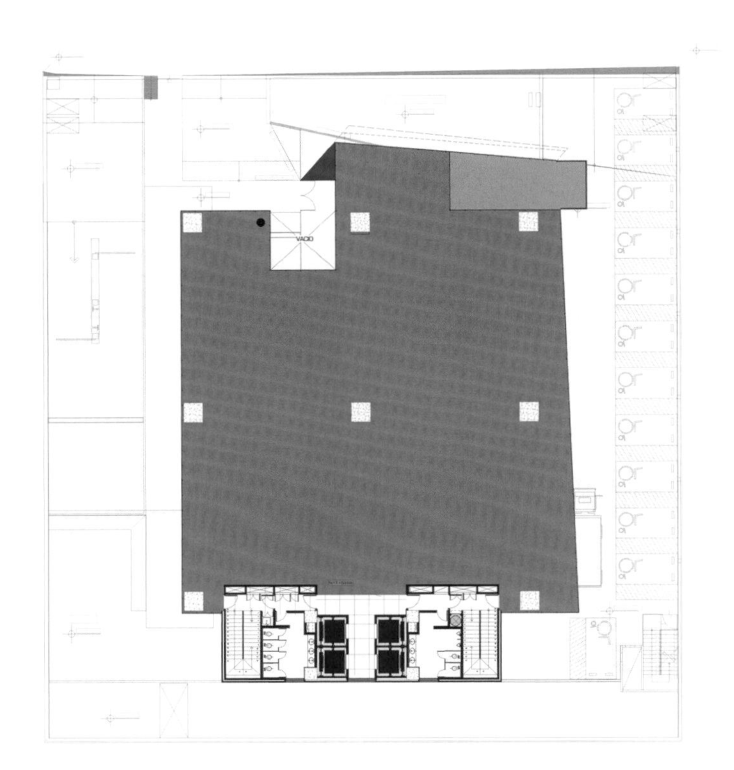
VACIO

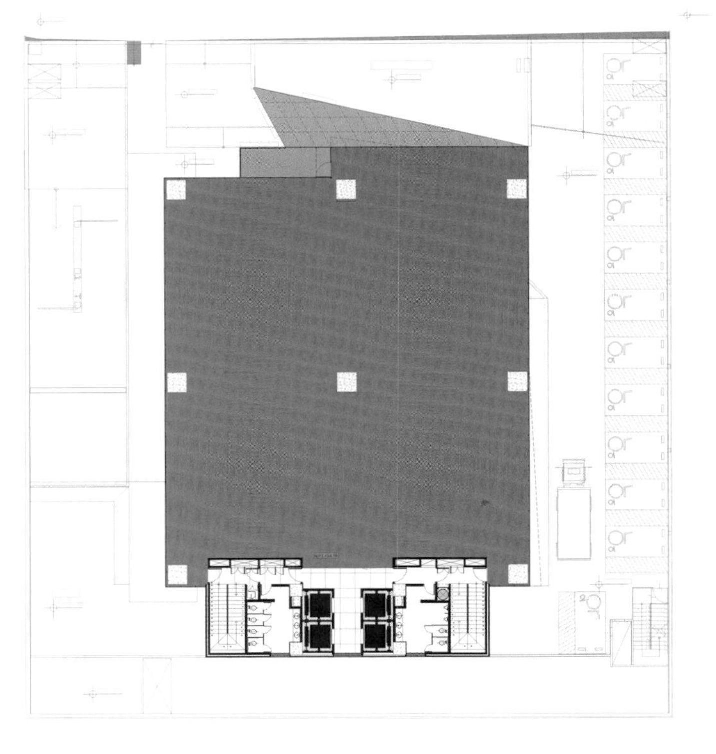

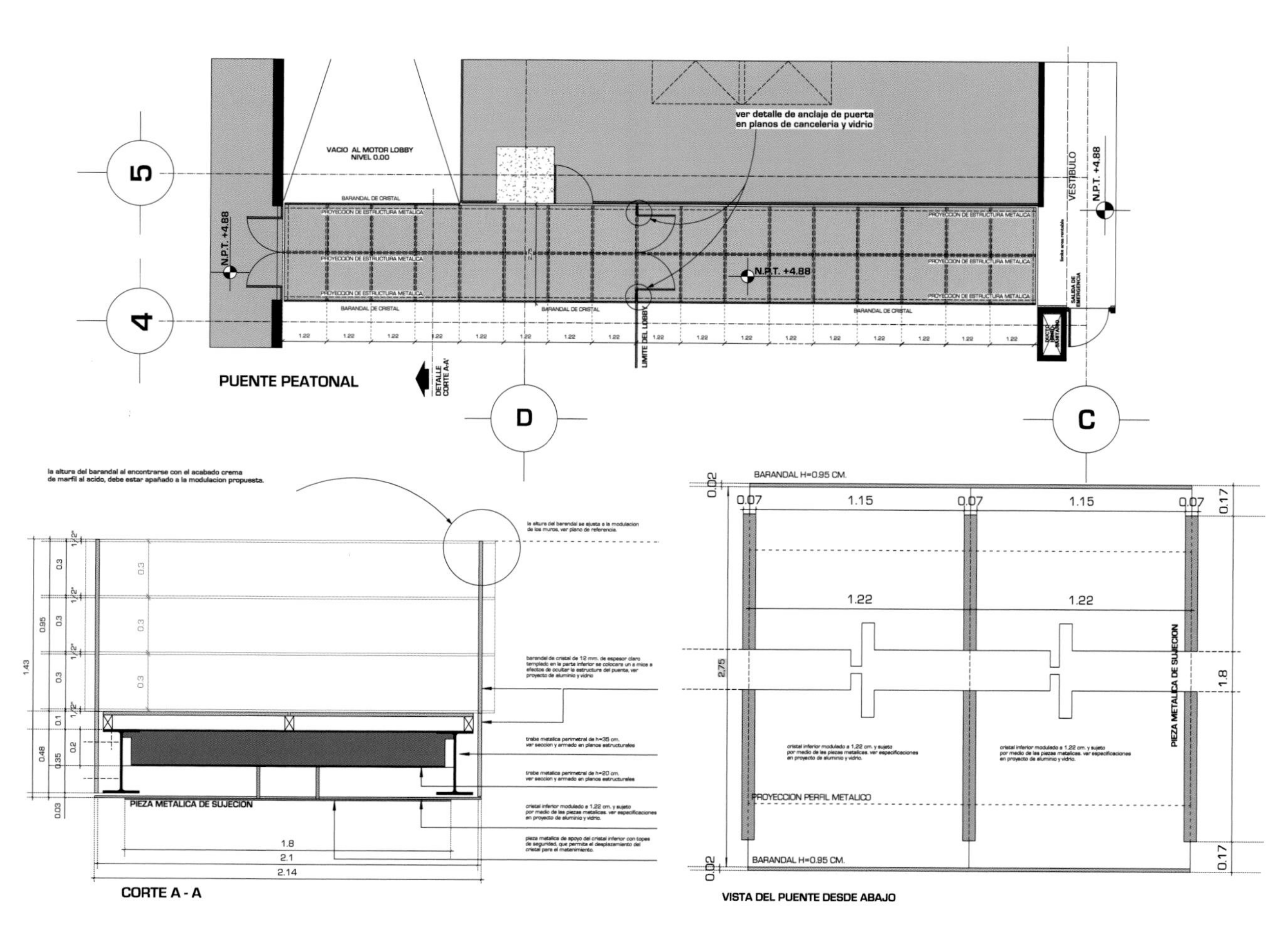

ver detalle de anclaje de puerta
en planos de cancelería y vidrio
VACIO AL MOTOR LOBBY
NIVEL 0.00
BARANDAL DE CRISTAL
PROYECCION DE ESTRUCTURA METALICA
N.P.T. +4.88
VESTIBULO
LIMITE DEL LOBBY
5
4
D
C
PUENTE PEATONAL
DETALLE CORTE A-A'
la altura del barandal al encontrarse con el acabado crema
de marfil al acido, debe estar apeñado a la modulacion propuesta.
PIEZA METALICA DE SUJECION
1.8
2.1
2.14
CORTE A - A
BARANDAL H=0.95 CM.
0.07
1.15
1.22
0.17
2.75
PIEZA METALICA DE SUJECION
PROYECCION PERFIL METALICO
VISTA DEL PUENTE DESDE ABAJO

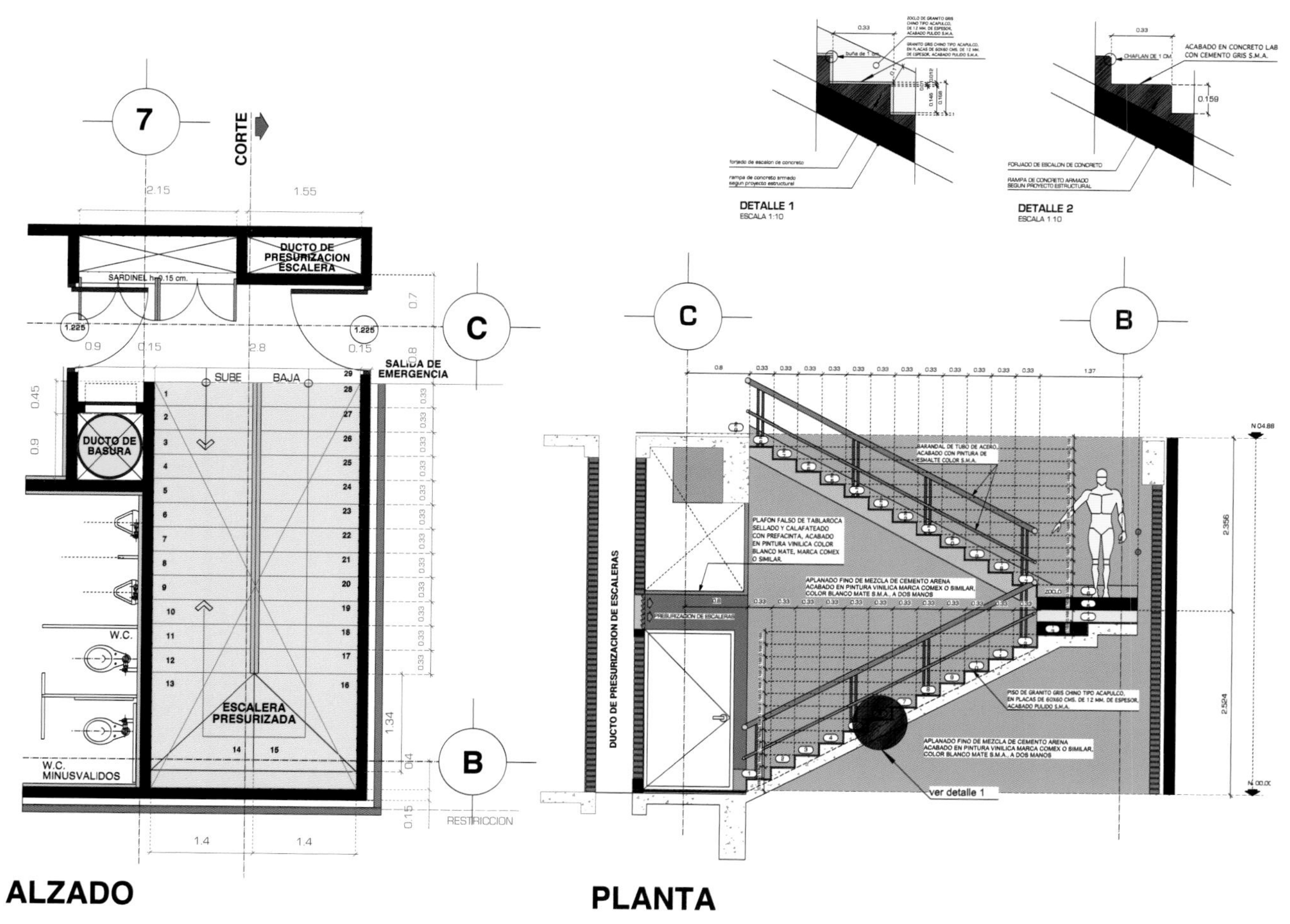

CORTE
DUCTO DE PRESURIZACION ESCALERA
SARDINEL h=0.15 cm.
SALIDA DE EMERGENCIA
SUBE
BAJA
DUCTO DE BASURA
W.C.
W.C. MINUSVALIDOS
ESCALERA PRESURIZADA
RESTRICCION
DETALLE 1
ESCALA 1:10
DETALLE 2
ESCALA 1:10
DUCTO DE PRESURIZACION DE ESCALERAS
ver detalle 1
ALZADO
PLANTA

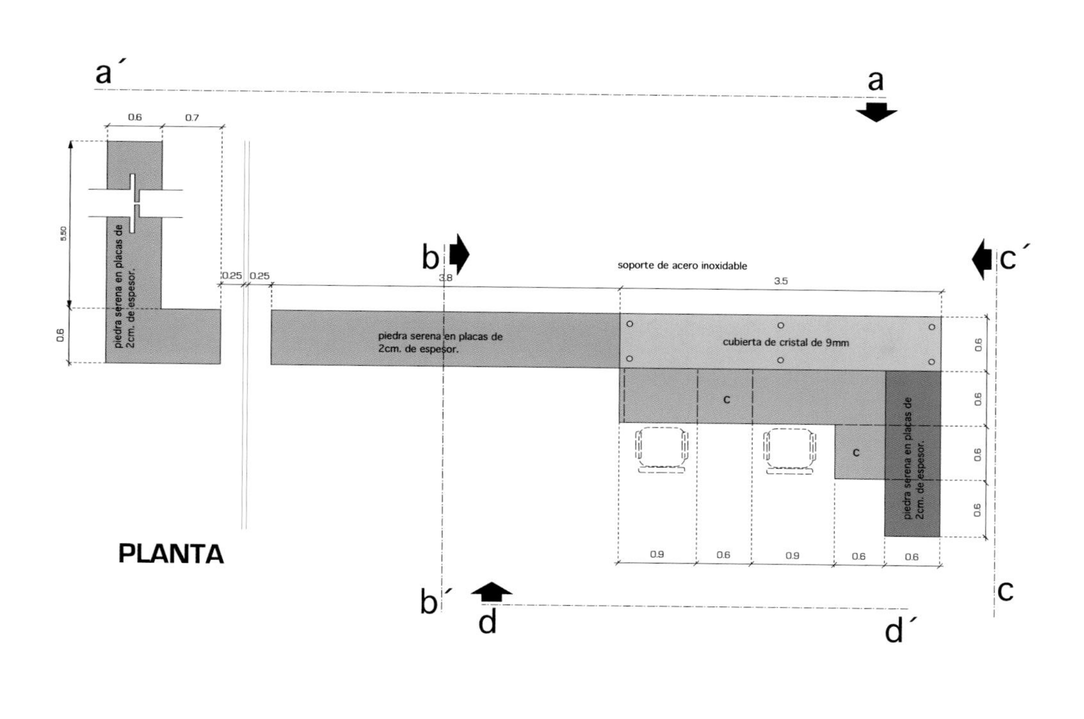

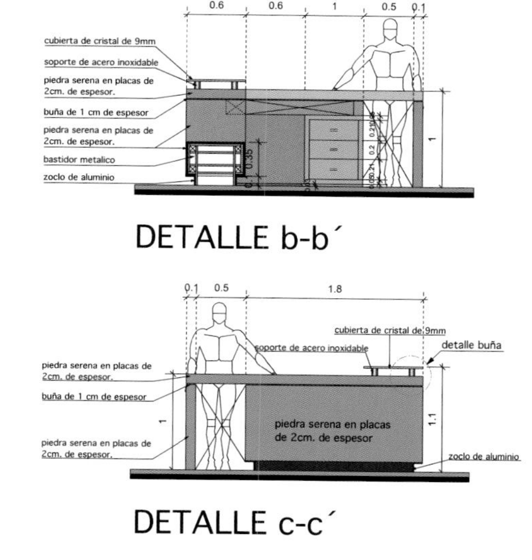

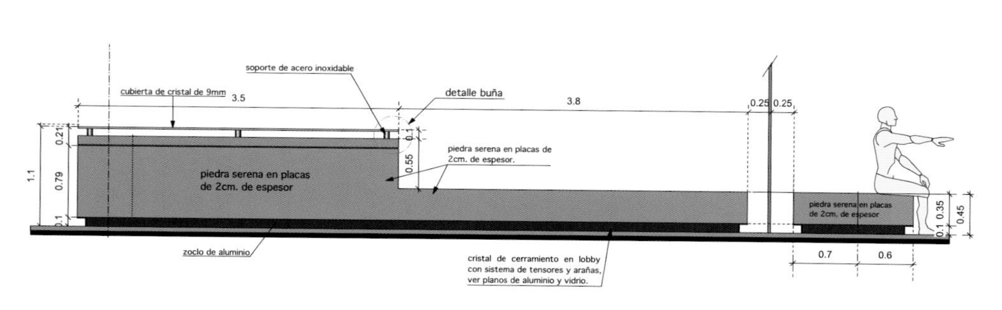

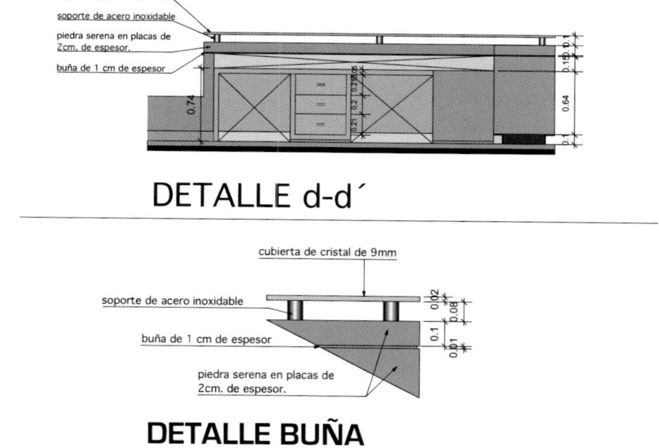

该建筑为一座 16 层高的办公大厦，内部有核心服务层、宽阔的服务通道与主要的楼梯、带有空调的房间、清洁站和可记录的管道装置。夹层有私人办公室和客户服务商店，大厅有可从 Mario Pani 大街直接进入的两个零售楼层，还有六层地下停车场以及屋顶的直升机停机坪。

办公大厦的几何造型以简单的正方体为基础，通过其在主立面上的位移，在视觉上构成两个几何体。其中一个几何体在上层以一定的角度向前倾斜，从而在顶层形成露台。

主入口因大胆使用了巨大的三角形玻璃结构而引人注目，当使用者走过木质平台和平静如镜的水池看到这个入口时，会生出想进入这个有趣的空间一探究竟的念头。

大堂位于一层的中央，有两层楼高，通过一座玻璃廊桥与夹层相连，玻璃廊桥将空间对称地一分为二。

Skalia 公司认为生态学特征能赋予建筑物以及整座城市可持续性。为有效利用建筑材料，在大楼的立面采用了特别设计的 1.22 米见方的模块。其台阶式的构造能有效地阻挡太阳的直射，同时区分公共区域与吸烟区域。

大楼的设计以环保节能为宗旨，以便最大化减少大楼内部的热负荷。Skalia 大厦的设计遵循了系统化设计的最高标准，使其当之无愧地成为墨西哥最出色的智能型办公大楼之一。

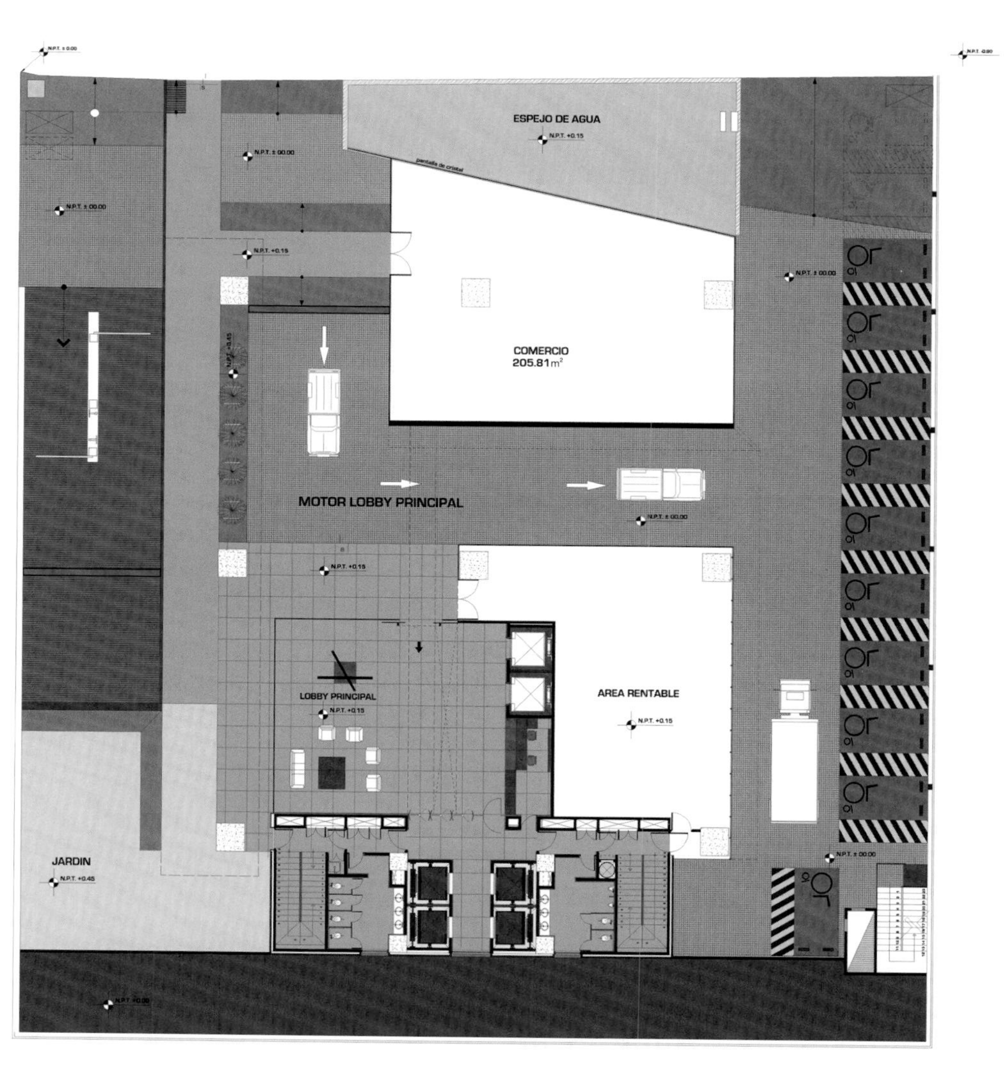

萨拉戈萨 2008 年博览会大厦的改建工程

REFURBISHMENT OF THE 2008 EXPO ZARAGOZA BUILDINGS

ARCHITECT
Estudio Lamela

LOCATION
Zaragoza, Spain

PHOTOGRAPHER
Daniel Schäfer

The project will make the 2008 Expo Zaragoza buildings become into offices and shopping areas.

The project consists of the refurbishment of the whole complex of "multi-pavilion buildings" in a large business park which will become the largest one in Spain. The designers can distinguish between 3 sub-groups with pre-existing forms which have to be maintained: the Ronda buildings – more linear in shape, but with undulating facades, the Ebro buildings – shape like a water droplet – and the Actur building. In all cases there are plans for intermediate floors to allow passage from two levels at a height of eight meters to four at four meters. There is also a "surgical" type of operation to open spaces and relocated cores necessary for future office use.

The project is noted for its commitment to environment and sustainable development. The project has sought above all as much recycling and renovation as possible, respecting the proposal and philosophy of Expo 2008. The project has sought to transform the roofline into useful and pleasant spaces for the user, from which you can see the river and part of the city that surrounds it.

The starting point was a structural skeleton and the project reuses some of the previous installations. The different volumes, jutting out with sinuous forms, remain consistent with its function and harmony with its essence, projecting color and dynamic outwards.

E4

F4
F3

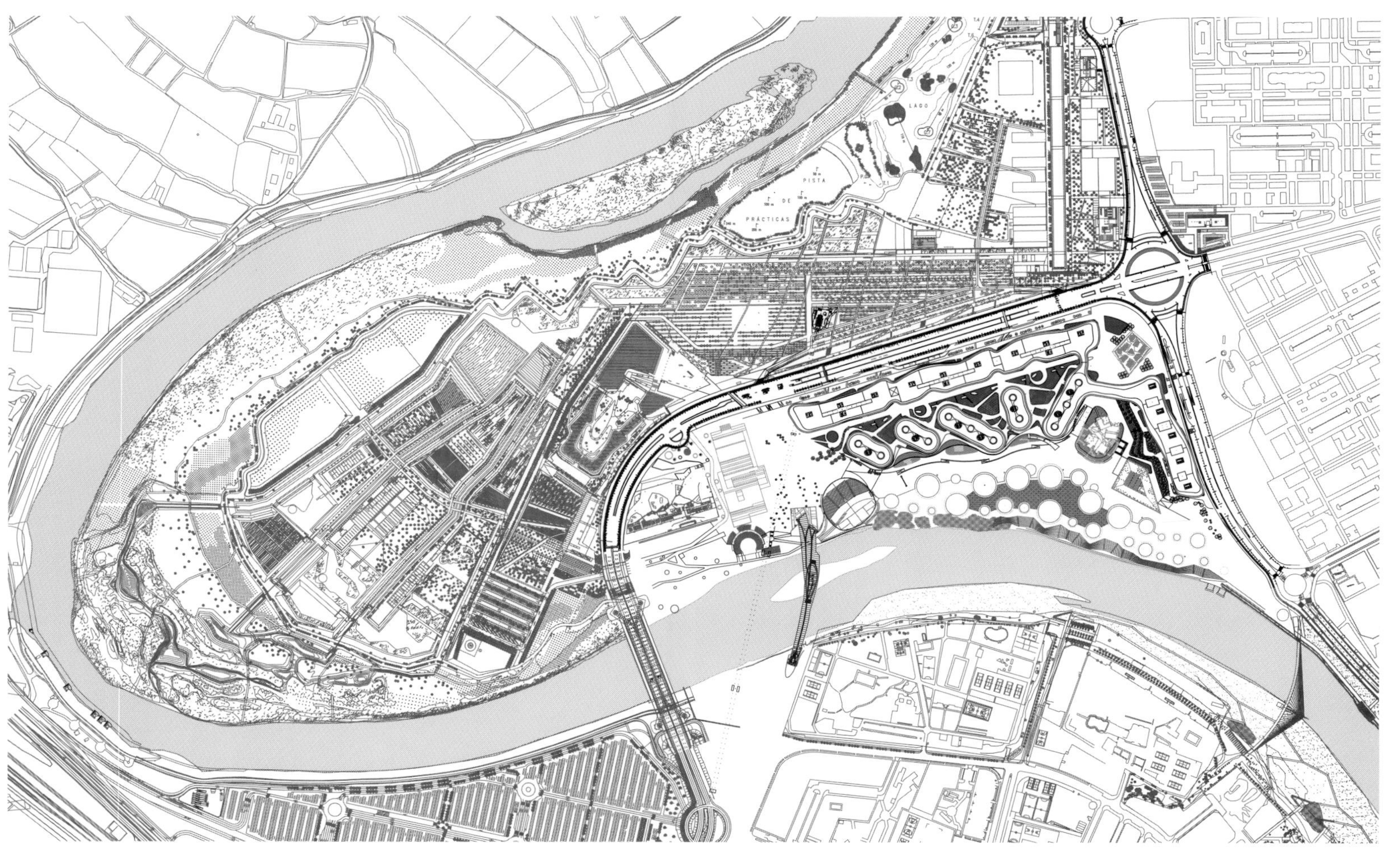

ALZADO GENERAL SUR EDIFICIO RONDA
ESCALA 1 / 750

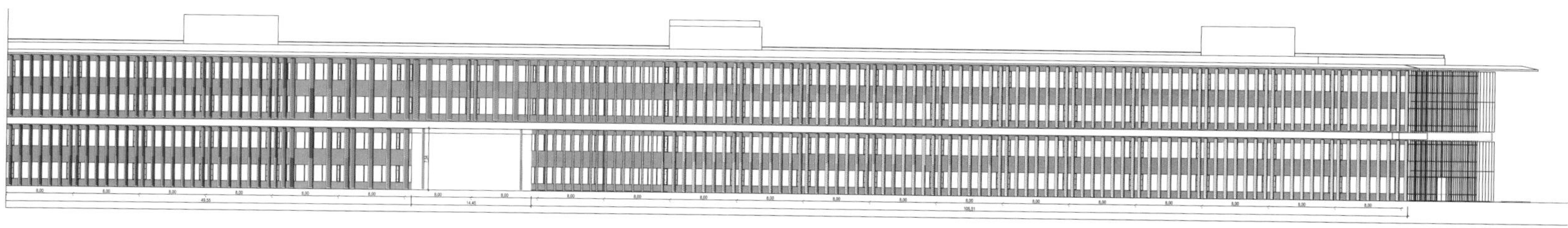

ALZADO GENERAL SUR EDIFICIO RONDA 1
ESCALA 1 / 300

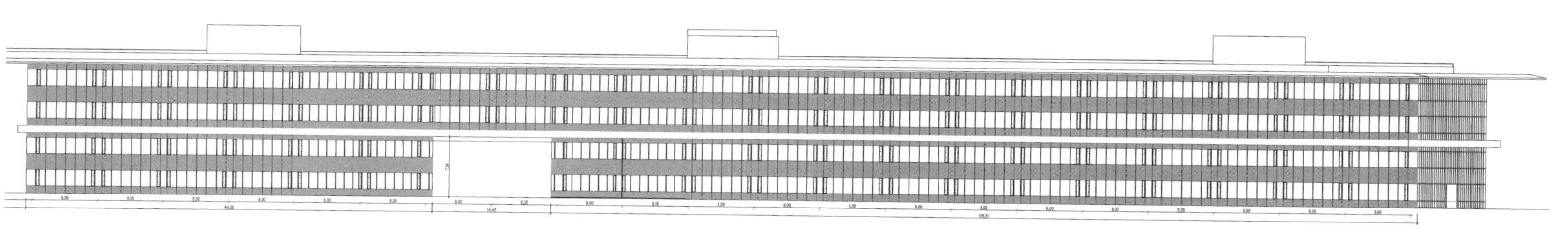

ALZADO FACHADA INTERIOR SUR EDIFICIO RONDA 1
ESCALA 1 / 300

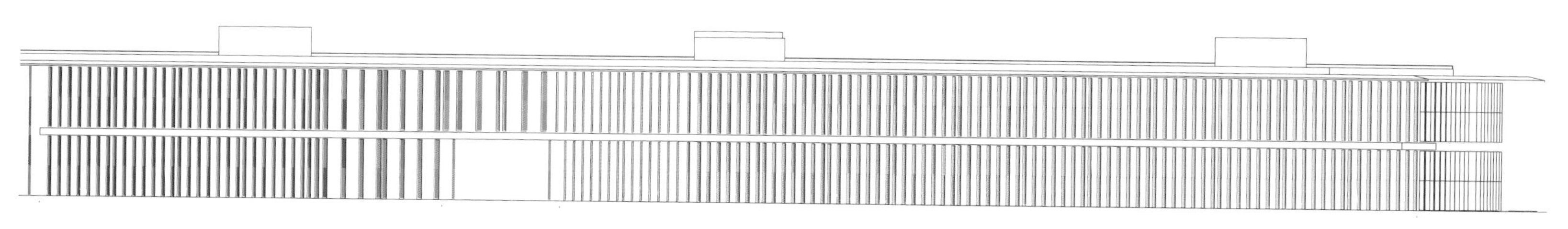

ALZADO LAMAS SUR EDIFICIO RONDA 1
ESCALA 1 / 300

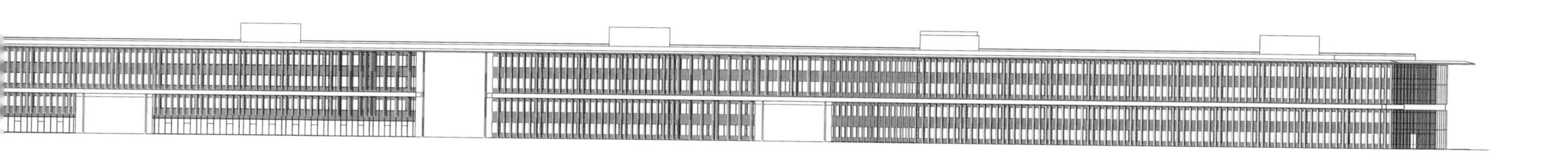

abr
2012
Edificio ExpoZaragoza

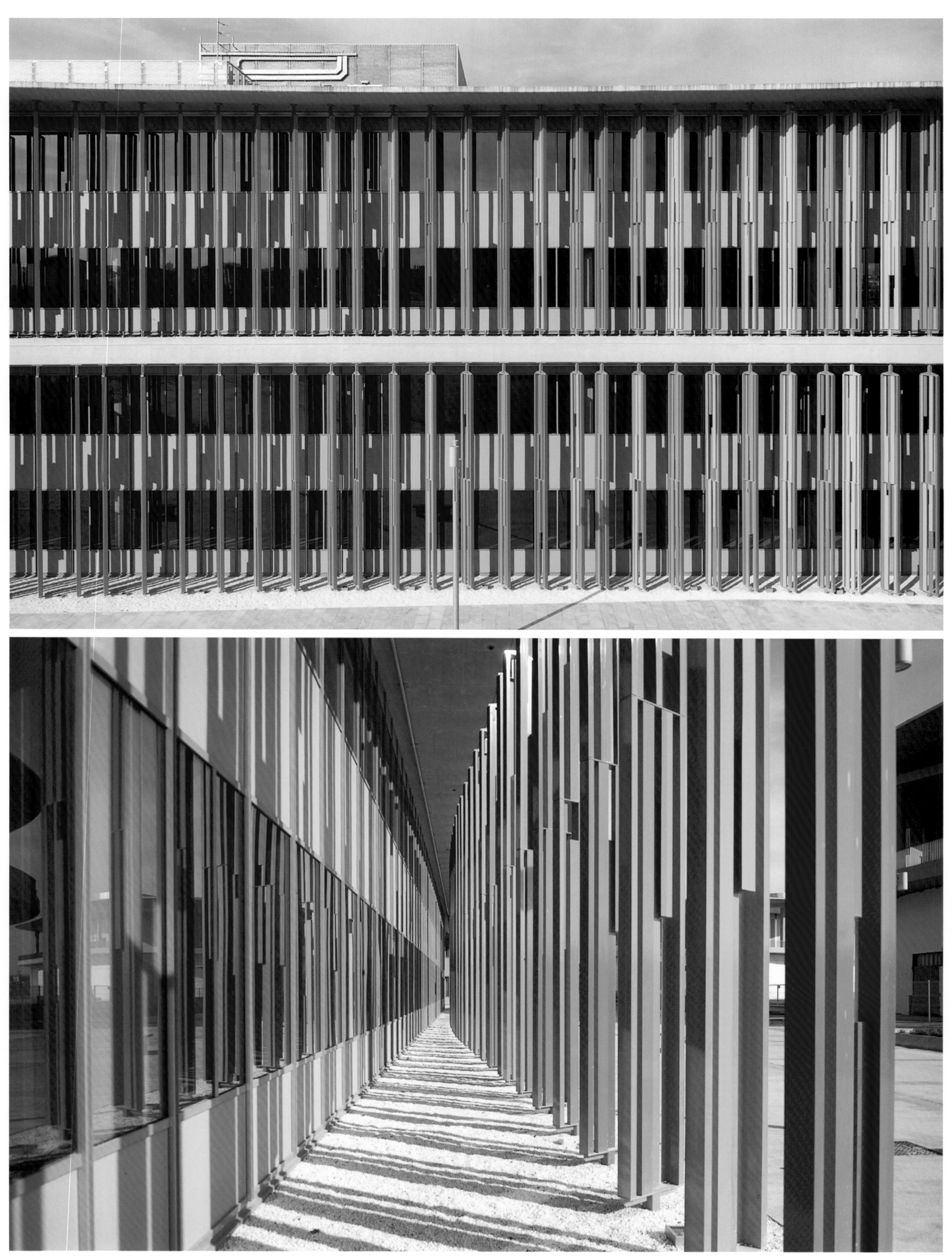

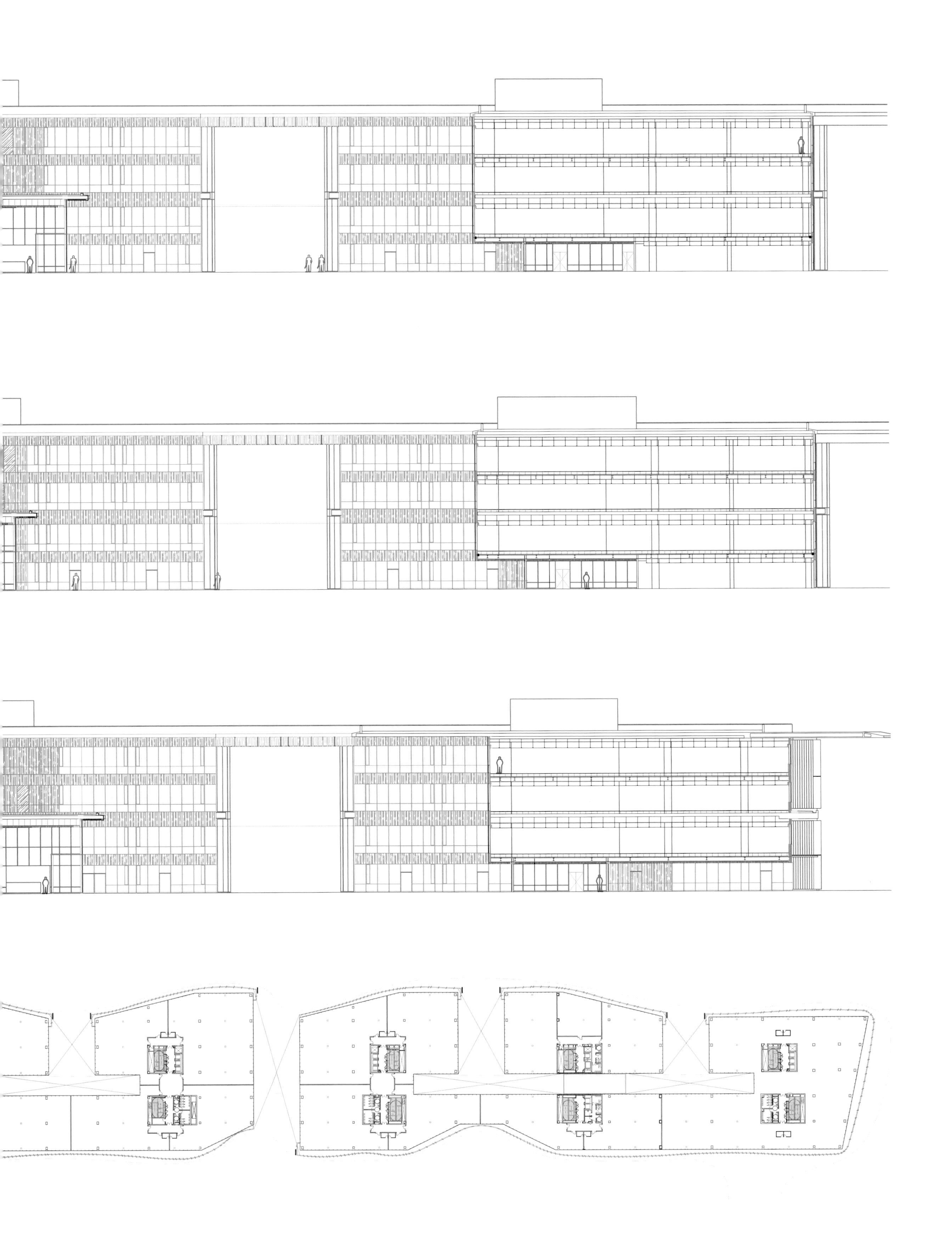

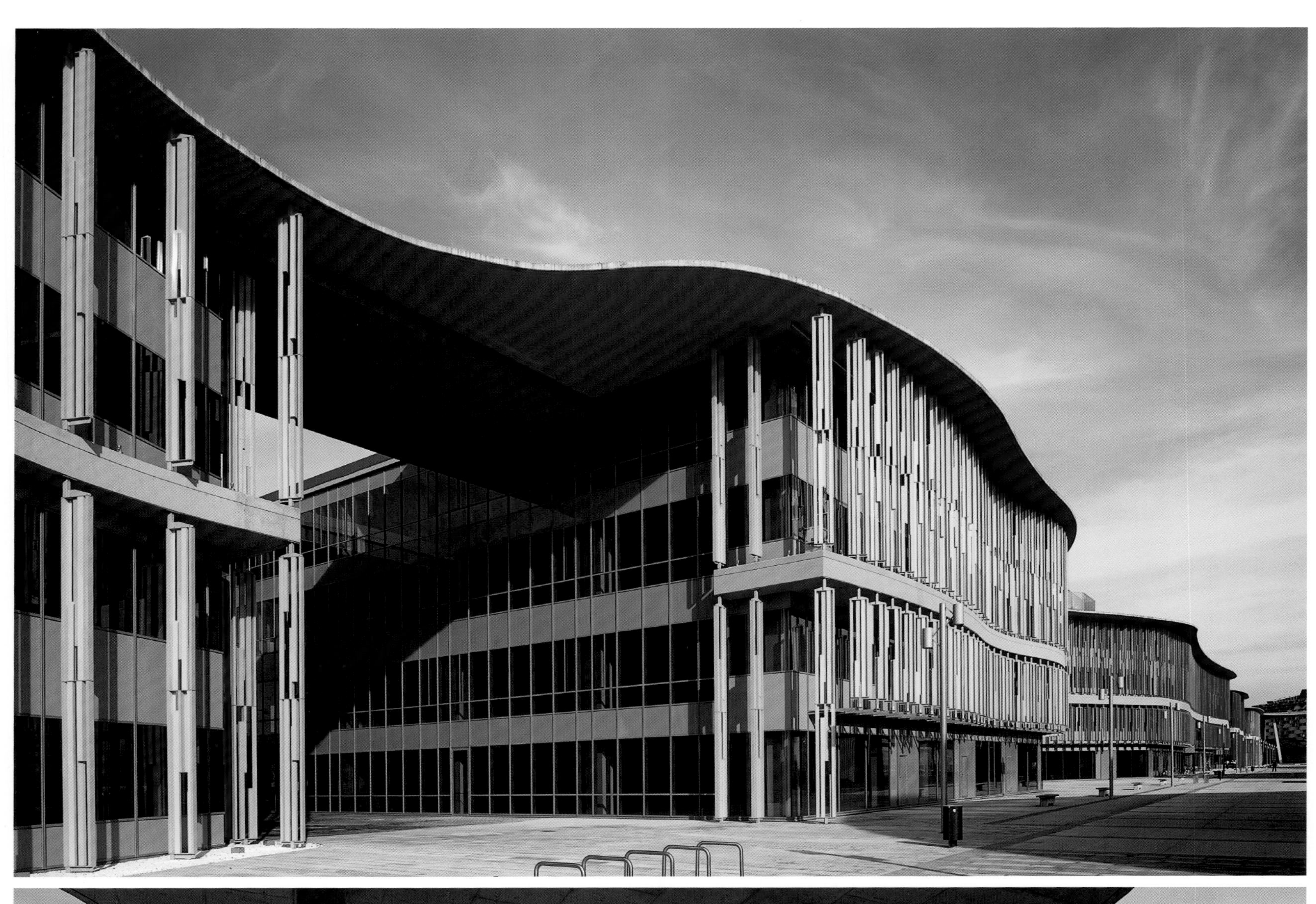

该改建工程要将萨拉戈萨 2008 年博览会大厦改造为一栋集办公和商业于一体的建筑。

整个工程要将这个包括多个展馆在内的超级综合体建筑，改造为西班牙最大的大型商业区。设计师们需要区分并保留三个原有的分区：郎达大厦，外形呈线形，外墙正面呈波浪状起伏；埃布罗大厦，水滴形外观；阿克图尔大厦。在所有的楼梯内部，设计师们需要规划出供人们通行的中间楼层，也就是将原有的两个八米高的楼层经过添加楼层，变为四个四米高的楼层。这是一种外科手术式的改建方案，需要开放并规划内部空间，为将来作办公用途使用时重新定位各个核心功能区域。

改建工程以环境保护和可持续发展而著称，力图探寻一种能够尽可能多地循环利用的改建方式，在一定程度上保留 2008 年博览会的理念和思想。设计师们还改造了屋顶的轮廓，从而获得了实用和美观的空间，参观者在这里能够看到远处的河流和四周城市的美景。

大厦设有一个钢骨结构的大门入口，这里利用了很多原有的建筑装置。各种不同的空间结构，以弯曲和蜿蜒的形态向四周延伸，融功能性和美观于和谐的一体，表现出华丽的色彩，动感十足。

63
63
B C
A
IMPASS
63B

65

271 17TH STREET

ARCHITECT
tvsdesign

LOCATION
Atlanta, Georgia, USA

AREA
46,452 m²

271 17th Street is the latest and largest office building addition to Atlantic Station, the heralded brownfield re-development master planned by tvsdesign and developed by AIG Global Real Estate. The project is a blend of the efficient Atlanta-style leasing floor-plate, an urban design attitude taken from the great business districts of New York and Chicago, and materials and form that recall Streamline Moderne design.

The building totals three storeys below street level and 25 storeys above, including 7 levels of structured parking, a lobby and retail level, and 20 storeys of office space. A six meters glass parapet provides a distinctive termination to the offset geometry and accommodates signage for the lead tenant. Service entries and utility areas are located three levels below the street, maximizing the available retail and lobby frontage.

Inserted into Atlanta's most active live/work/play environment, the 46,452 square meters building provides first-class commercial office space, excellent urban retail street frontage and a signature look for the most desirable corporate tenants in the Atlanta market.

BB&T
LIFE HAPPENS HERE
CITY SPORTS

BB&T
BB&T
271
SPEED
LIMIT
25

BB&T
BB&T
TARGET
271

BB&T
271

201

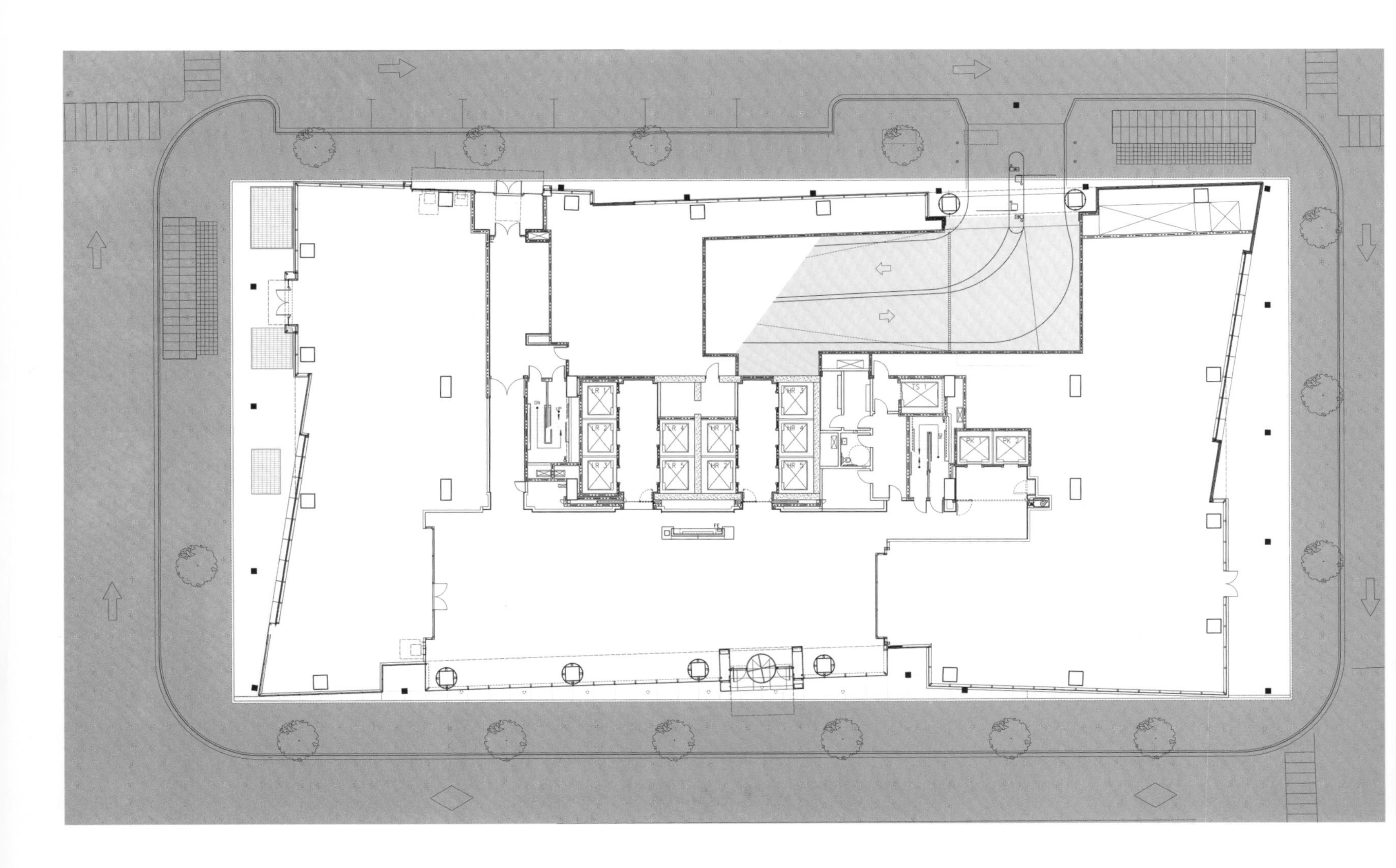

BB&T

BB&T
LIFE
HAPPENS
HERE

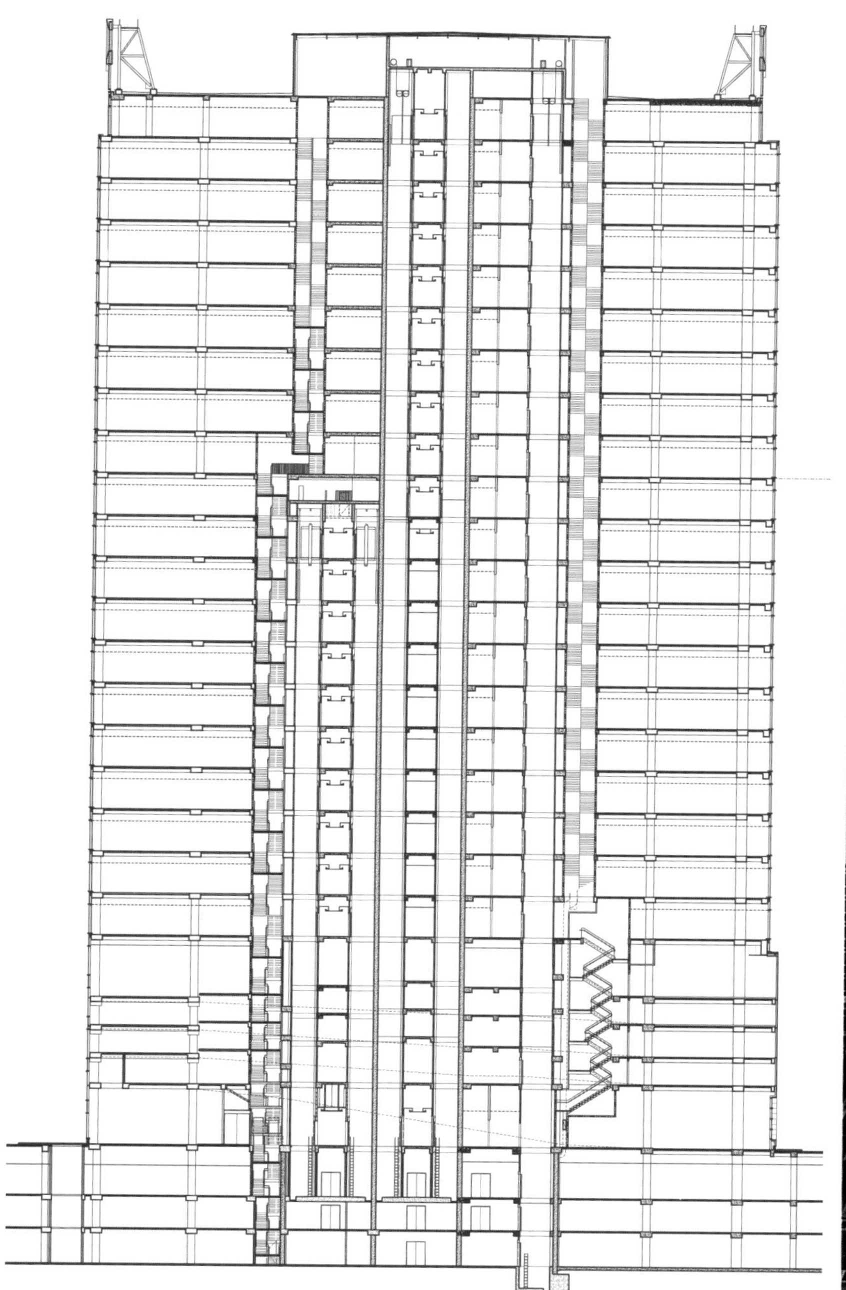

17 大街 271 号大楼是仅次于大西洋站的最新、最大的办公大楼，这片棕色地带的再开发项目由 tvs 设计公司进行总体规划，由 AIG 全球房地产公司进行开发。该项目借鉴了纽约和芝加哥大商业区的城市设计理念，采用了高效率的亚特兰大风格租赁楼面板，加上高科技的材料以及独特的外形，使整座大楼无不体现出流线型的现代建筑风格。

大楼由地下 3 层和地上 25 层组成，包括 7 层结构化停车场、一个大堂和零售层及 20 层办公空间。这个带壁阶的大楼一角有一堵 6 米高的玻璃护墙，护墙上的广告牌上有招租信息。服务区和生活区位于地下 3 层，这样的设计可以使大堂和零售区的面积最大化。

坐拥亚特兰大最活跃的融生活、工作、娱乐于一体的优越地理位置，46 452 平方米的大楼提供了一流的商业办公空间、最佳的城市零售临街店铺以及在亚特兰大市场寻找最理想的企业承租商的特色广告牌。

BB&T

271 17th Street

271
BB&T

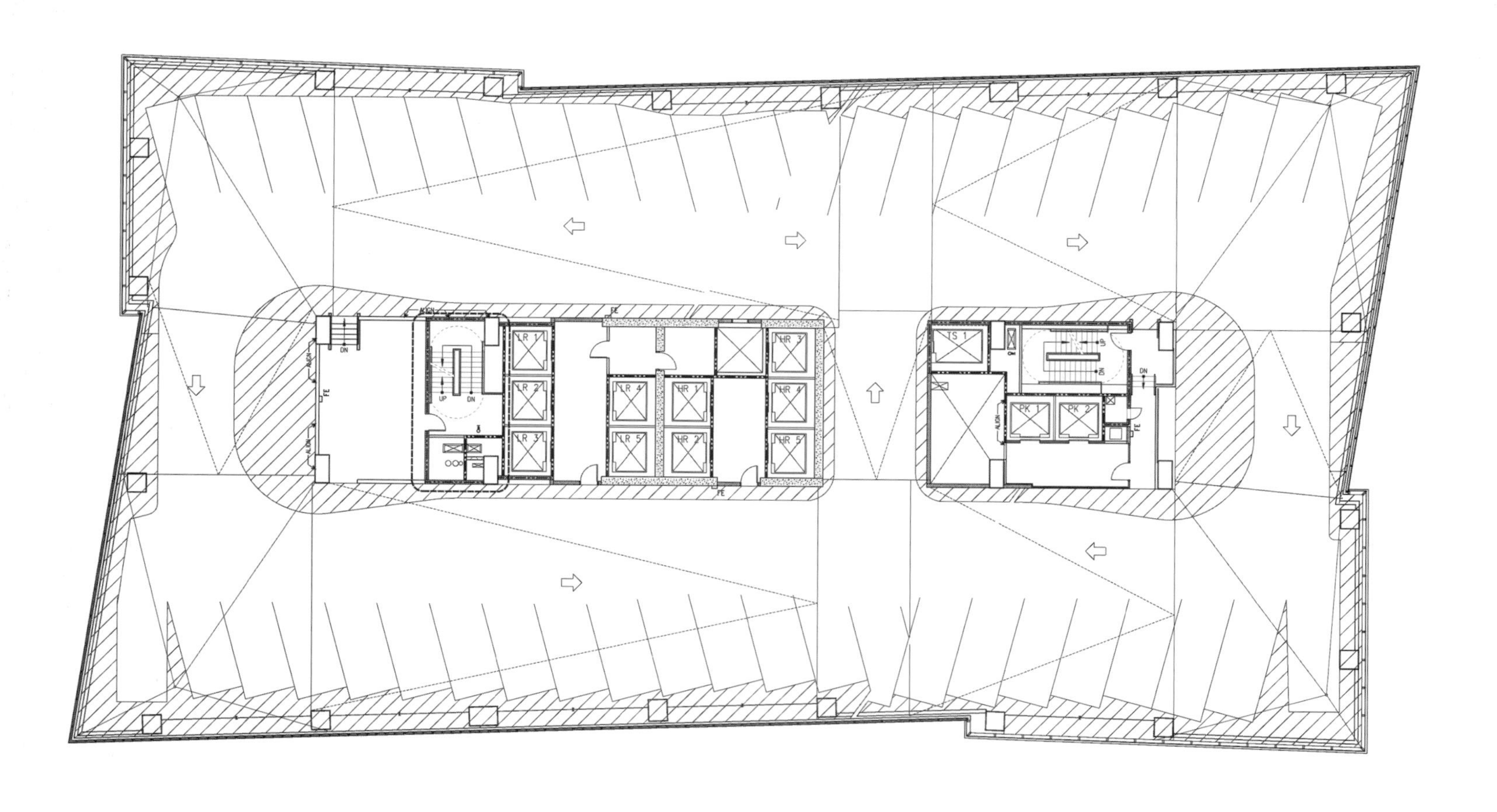
DN
ALIGN
FE
UP
DN
LR 1
LR 2
LR 3
LR 4
LR 5
HR 1
HR 2
HR 3
HR 4
HR 5
FE
TS 1
UP
DN
PK 1
PK 2
ALIGN
FE

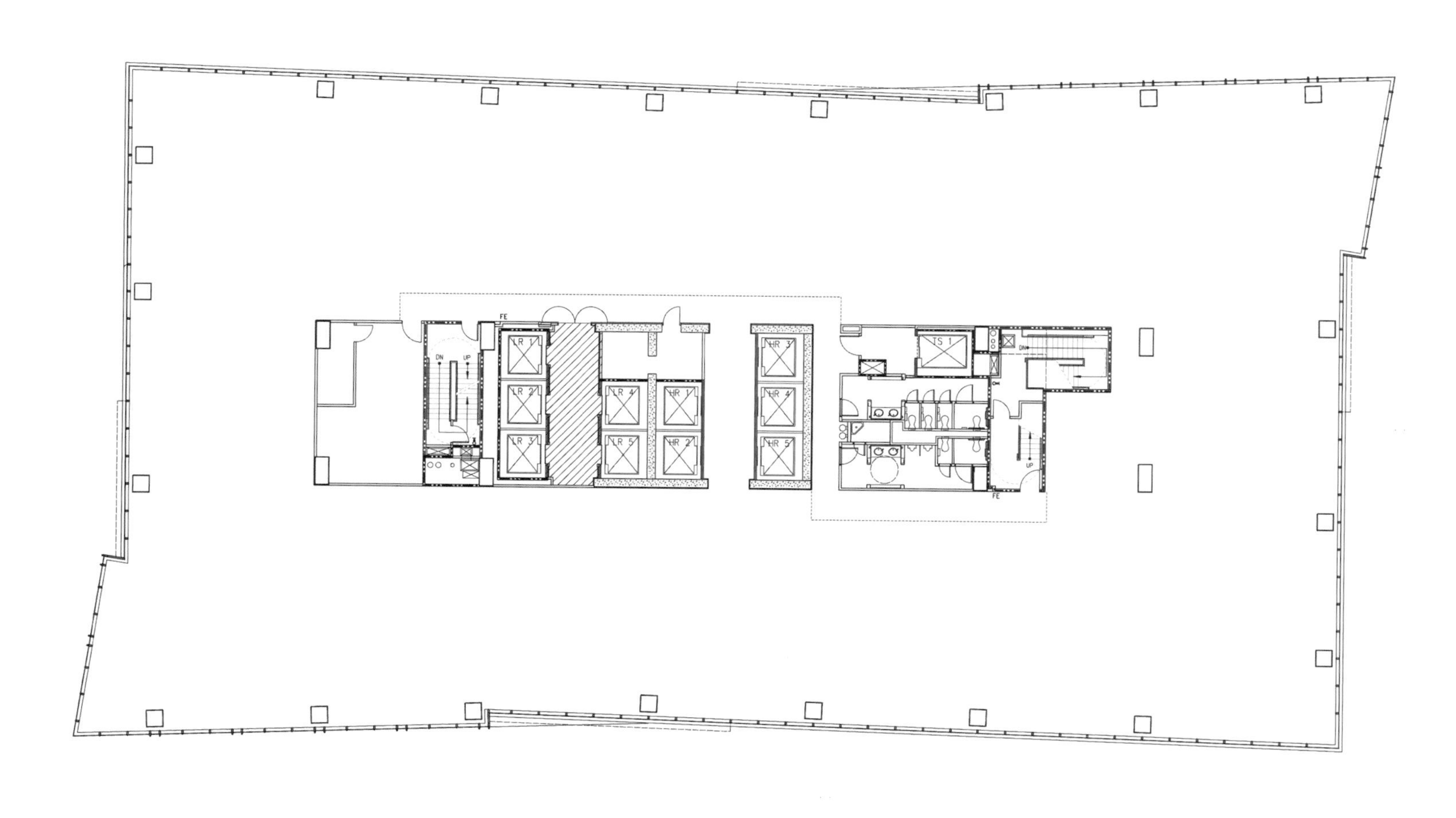
FE
DN
UP
LR 1
LR 2
LR 3
LR 4
LR 5
HR 1
HR 2
HR 3
HR 4
HR 5
TS 1

PARKING

PARKING
PARKING
16 – 25
16 – 25
BB&T

5 – 15
16 – 25

17½ STREET

5
15

加尔根沃德体育场

GALGENWAARD

ARCHITECT	FIRM	LOCATION	AREA	PHOTOGRAPHER
Cees Dam, Diederik Dam	Dam & Partners Architecten	Herculesplein, Utrecht, the Netherlands	30,000 m²	Luuk Kramer

In addition to the renovation and extension of the stadium, the redevelopment plan for Galgenwaard included, among other things, the new build of an office and parking space ensemble on its west side and a residential building, Apollo Residence, to the east of the stadium. To house the offices Dam & Partners Architecten designed a combination of one tower and three lower wings that spread out in the direction of the river, with the central entrance as a connecting element. The tower consists of two slabs: the white of marble and concrete on one side of the stadium, a titanium skin on the city side. These materials contrast with the rough slate and green glass of the lower wings facades.

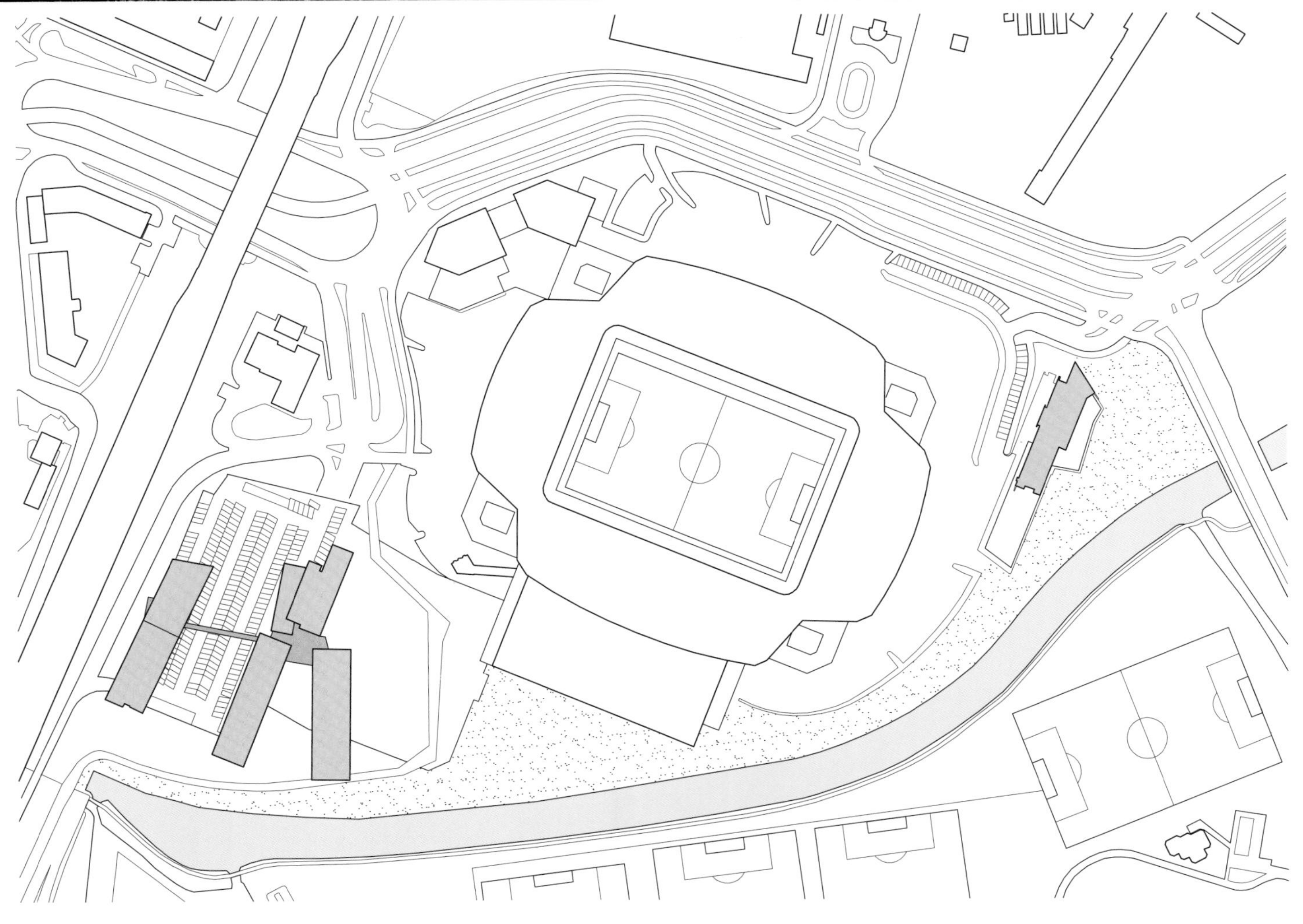
1 / 3000
0
30
60m

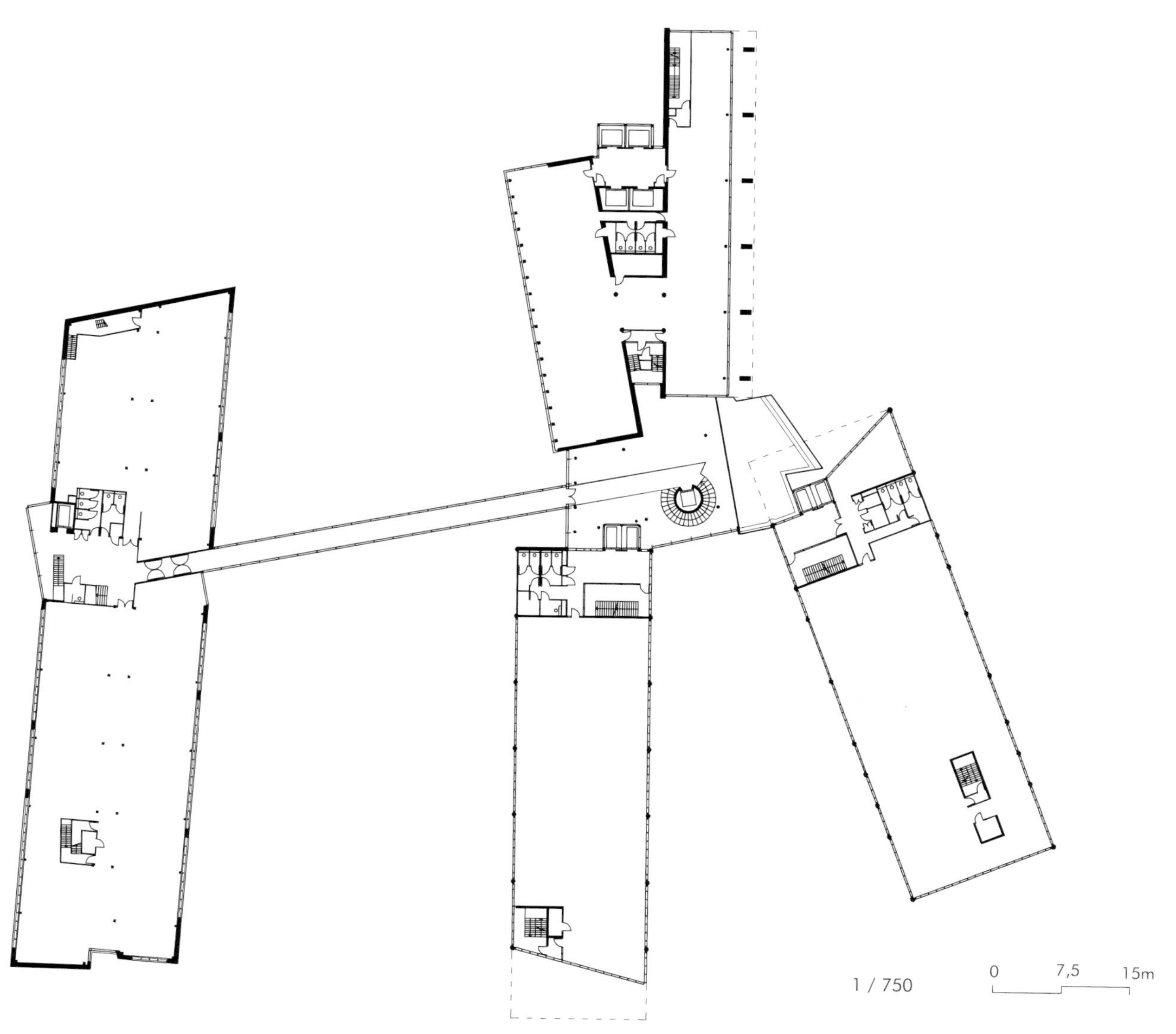
1 / 750
0
7,5
15m

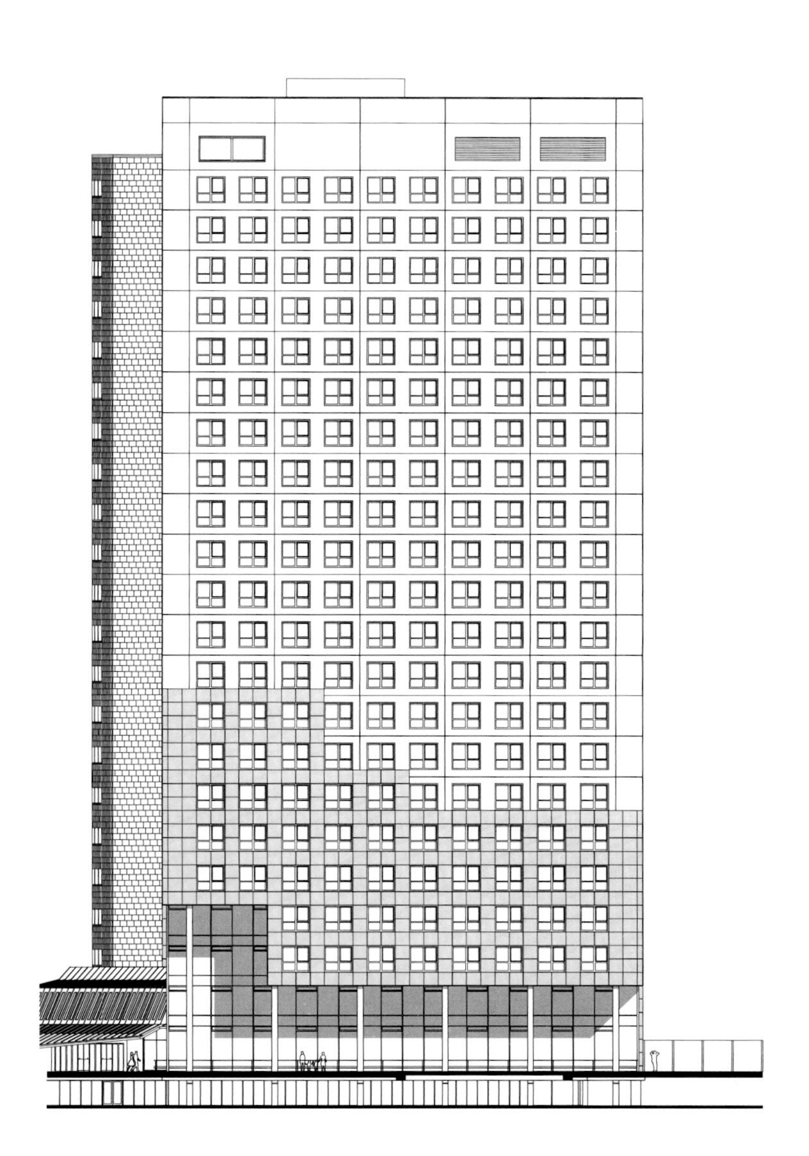

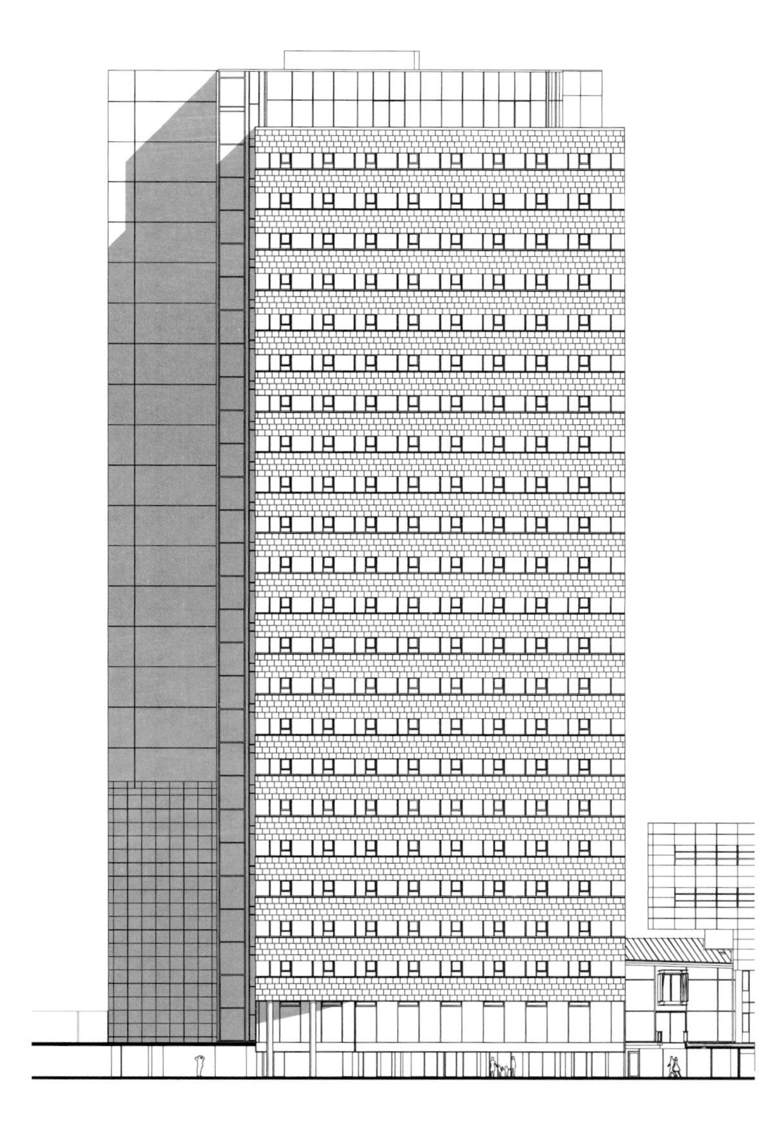

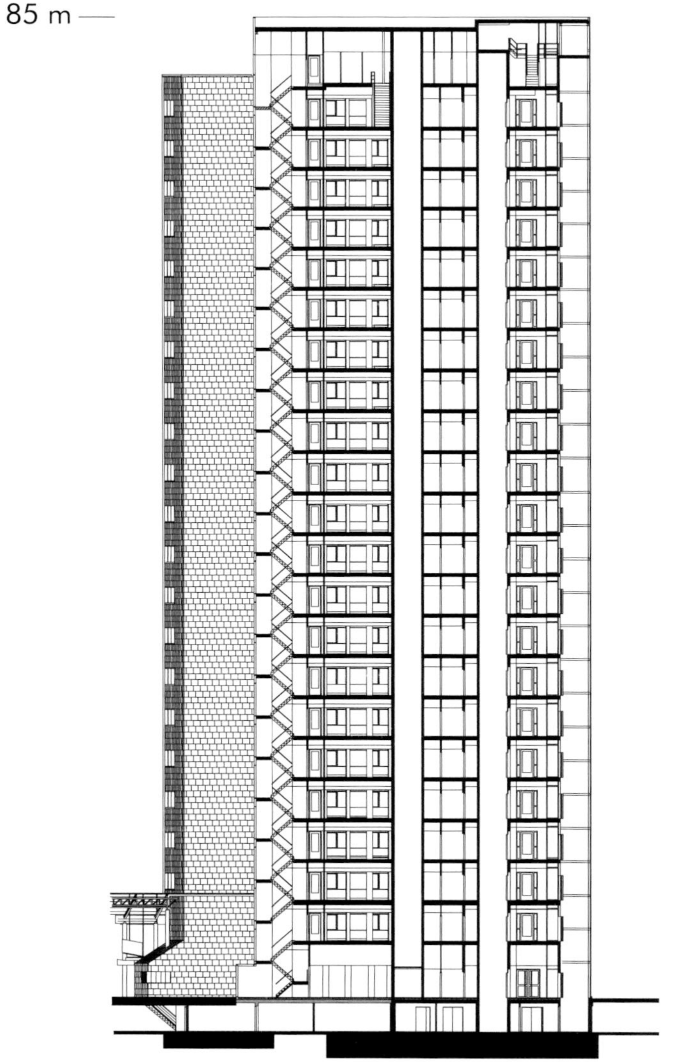

1 / 500　0　5　10m

加尔根沃德体育场的重建方案，除了场馆的改造和扩建外，还包括在它西侧新建一个办公区和一个停车场，以及在体育场的东侧建造一座居民楼——阿波罗住宅区。Dam & Partners Architecten 将办公区设计成一座塔楼和三个沿河方向展开的低翼造型建筑，中央入口将四者巧妙地连为一体。塔楼由两个板块构成：大理石和混凝土材质的白色建筑板块在体育场一侧，钛合金皮肤建筑板块在城市一侧。这些材料与粗糙的板岩和低翼造型建筑立面的绿色玻璃形成鲜明的对比。

D.O.S.
Herculesplein
92 t/m 98

KAFFEE PARTNER HEADQUARTERS

ARCHITECT
3deluxe

LOCATION
Osnabrück, Germany

AREA
28,600 m²

PHOTOGRAPHER
Emanuel Raab, Sascha Jahnke / 3deluxe

Following the international award-winning Leonardo Glass Cube 3deluxe has once again completed a piece of corporate architecture with an ambitious design and expressive formal language.

On all storeys, asymmetrically curved ribbons on the facade link the three edifices that make up the U-shaped complex with one another. They elegantly conceal the orthogonal grid of the concrete skeleton structure. Accurately assembled from 150 custom-made prefabricated concrete parts, each weighing six tons, the facade ribbons structure the corporate headquarters' 100-meter elevation facing the road while creating a sense of depth and dynamics. The latter culminates in the animated, vertically staggered composition of the administration wing: With its ceiling slabs, differently shaped and cantilevered on each floor, the corner building of the complex represents a prominent eye-catcher. In line with the building's surroundings, the dynamic facade was designed with a view to being observed at eye level and from a short distance. From the perspective of passers-by, the under surfaces of the overlapping storey slabs and facade ribbons present a varied interplay between shape, light, and shadow. At times, the light reflecting off the copper-colored window frames give the white surfaces a gentle metallic shimmer.

The strikingly designed corner building is not only the visual, but also the functional centerpiece of the corporate headquarters. In addition to individual offices and conference rooms it also includes communal lounge areas for informal conversation. The offices for members of the management board are located in a roof-top edifice that is clad with white metal panels and whose shape adopts the basic rectangular structure of the building complex. The executive suite features custom-made furniture and high-end fittings, whereas architectural elements recur in the interior design.

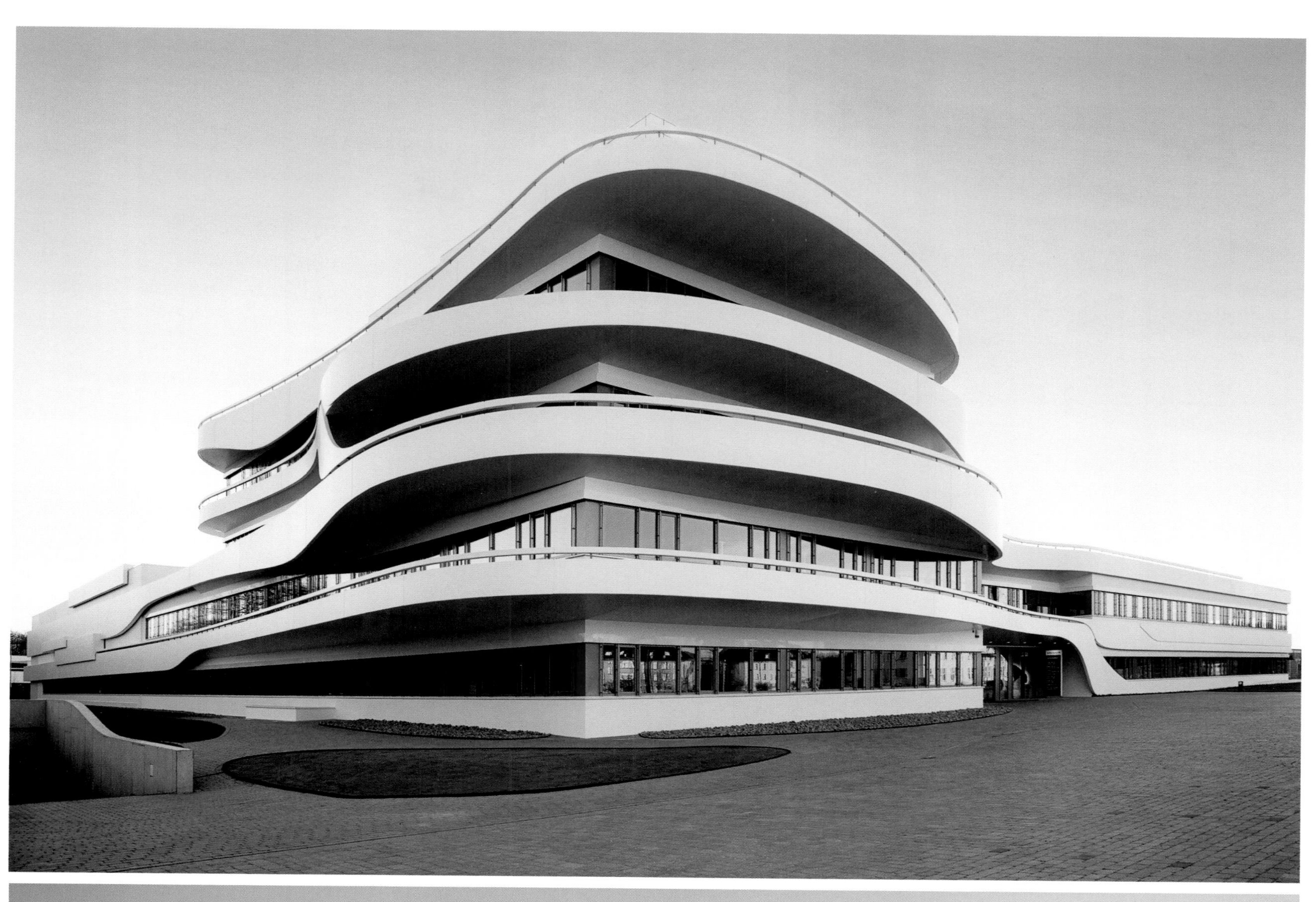

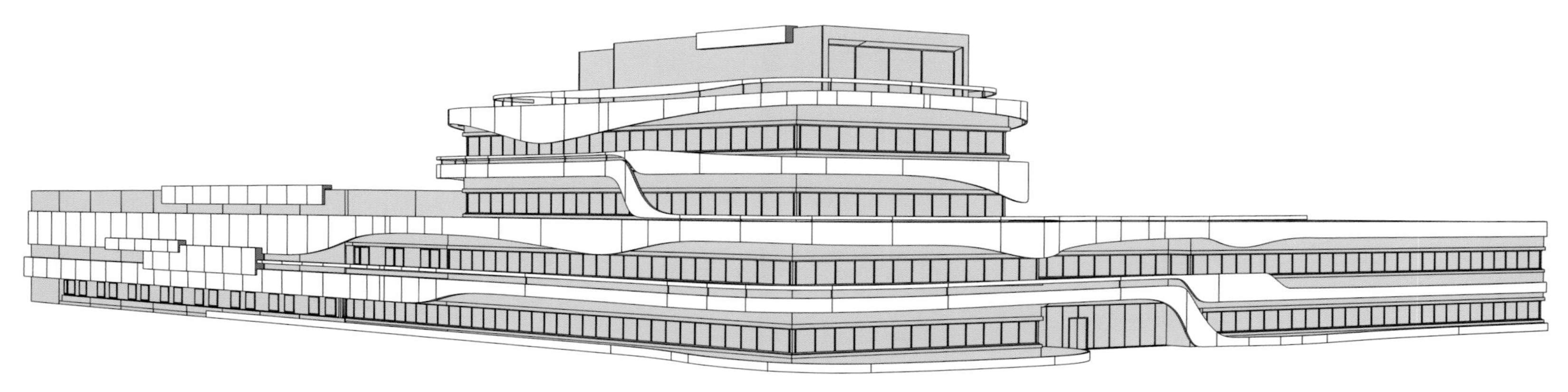

Perspective view: south east and north east façade

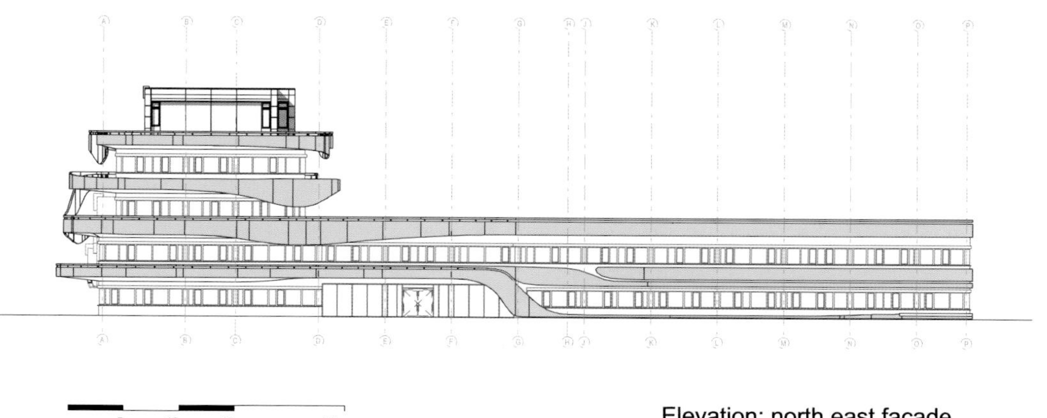

Elevation: north east façade

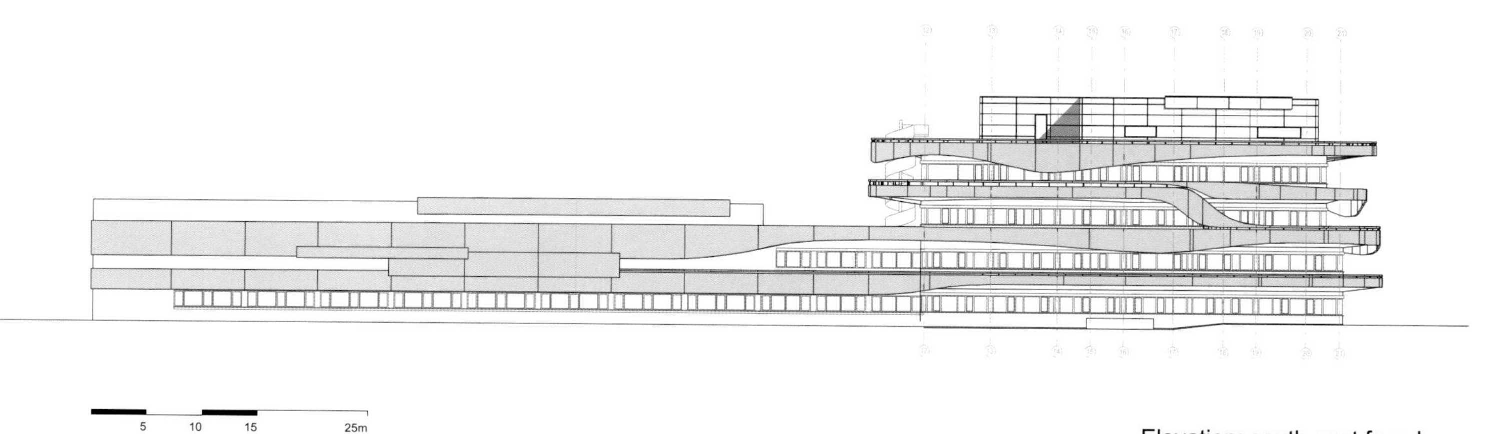

Elevation: south east façade

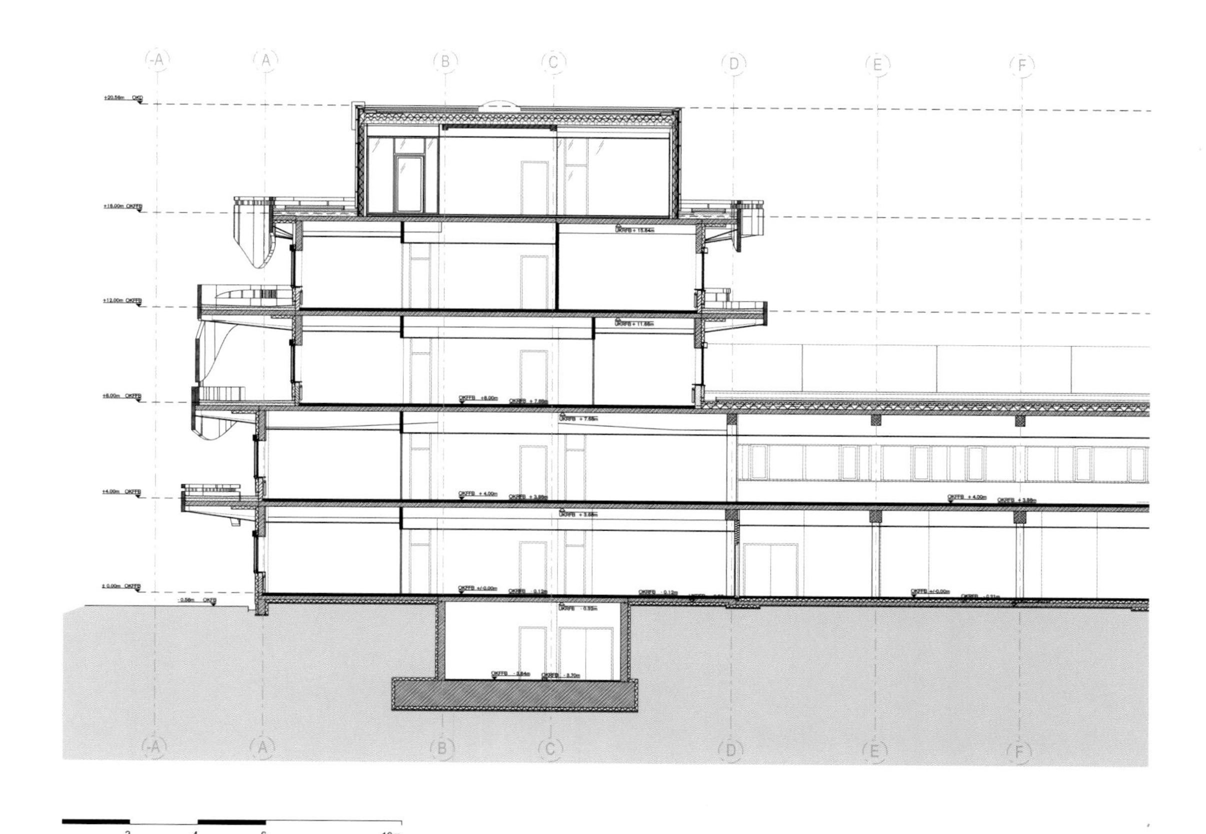

Section across the administration tract

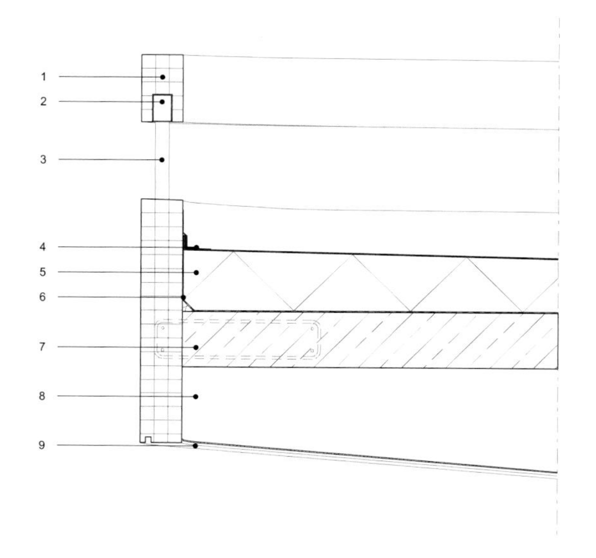

1. Pre-fabricated concrete part, cast with overhanging reinforced concrete floor
2. Hollow steel profile
3. Stainless steel pipe V4A
4. Membrane roof Vedafin S
5. Thermal insulation: 2% down-grade insulation WLG 040, self-compensating thermal insulation 120 mm, WLG 040
6. Bituminous roof sealing
7. Reinforced concrete floor, 200 mm
8. Reinforced concrete downstand beam
9. Barrisol stretch ceiling

Cross section detail: overhanging façade element

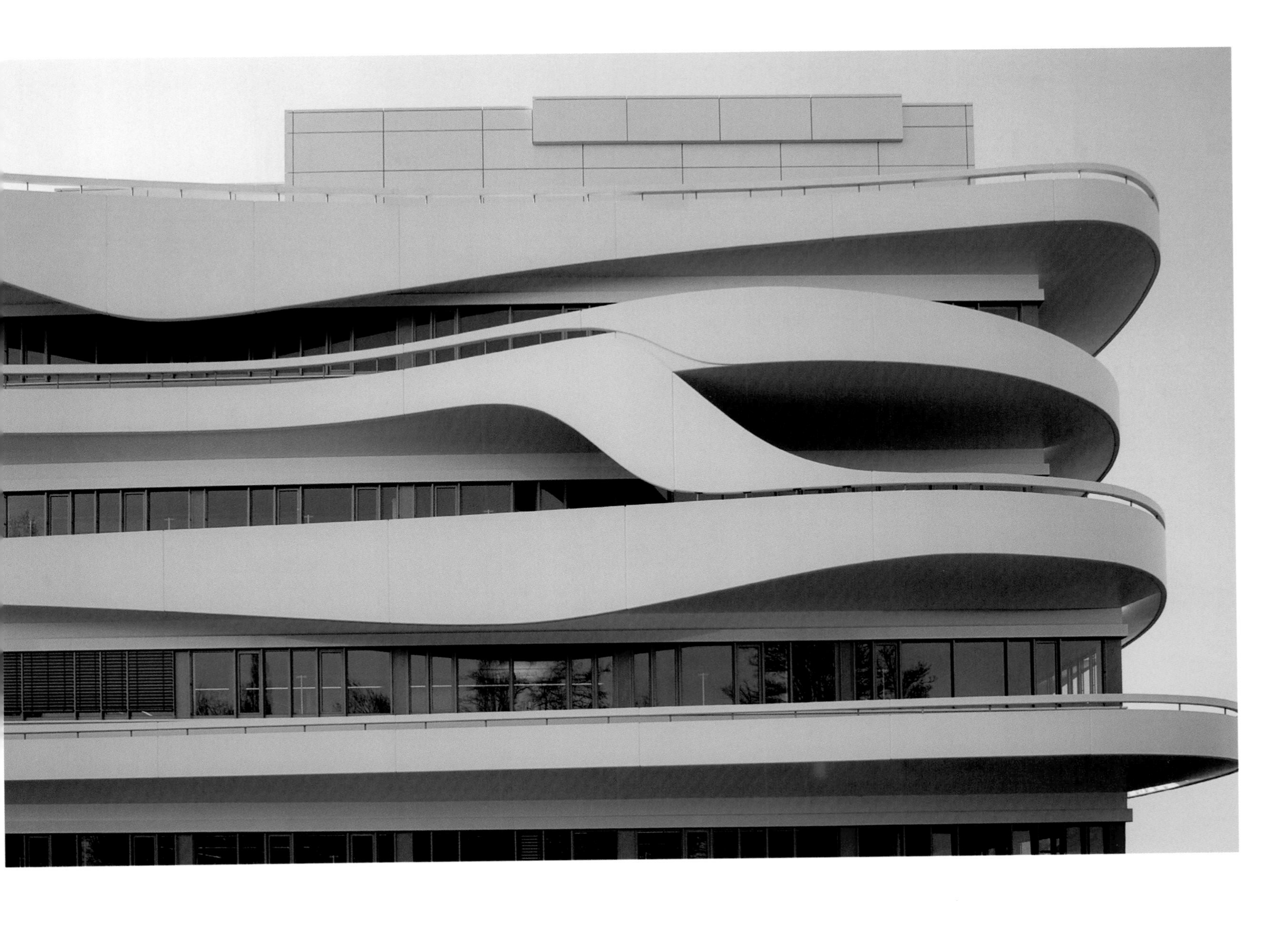

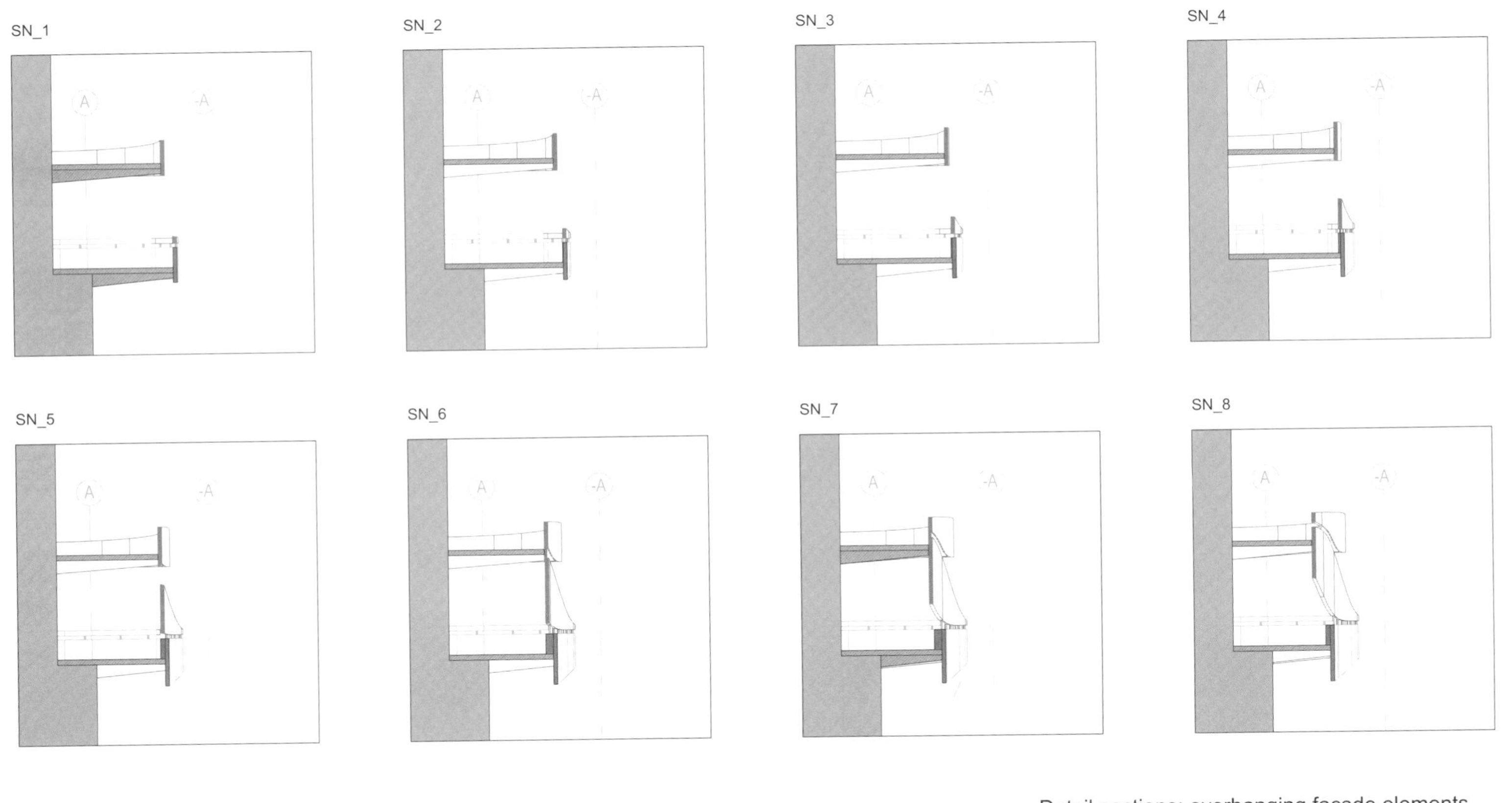

Detail sections: overhanging façade elements

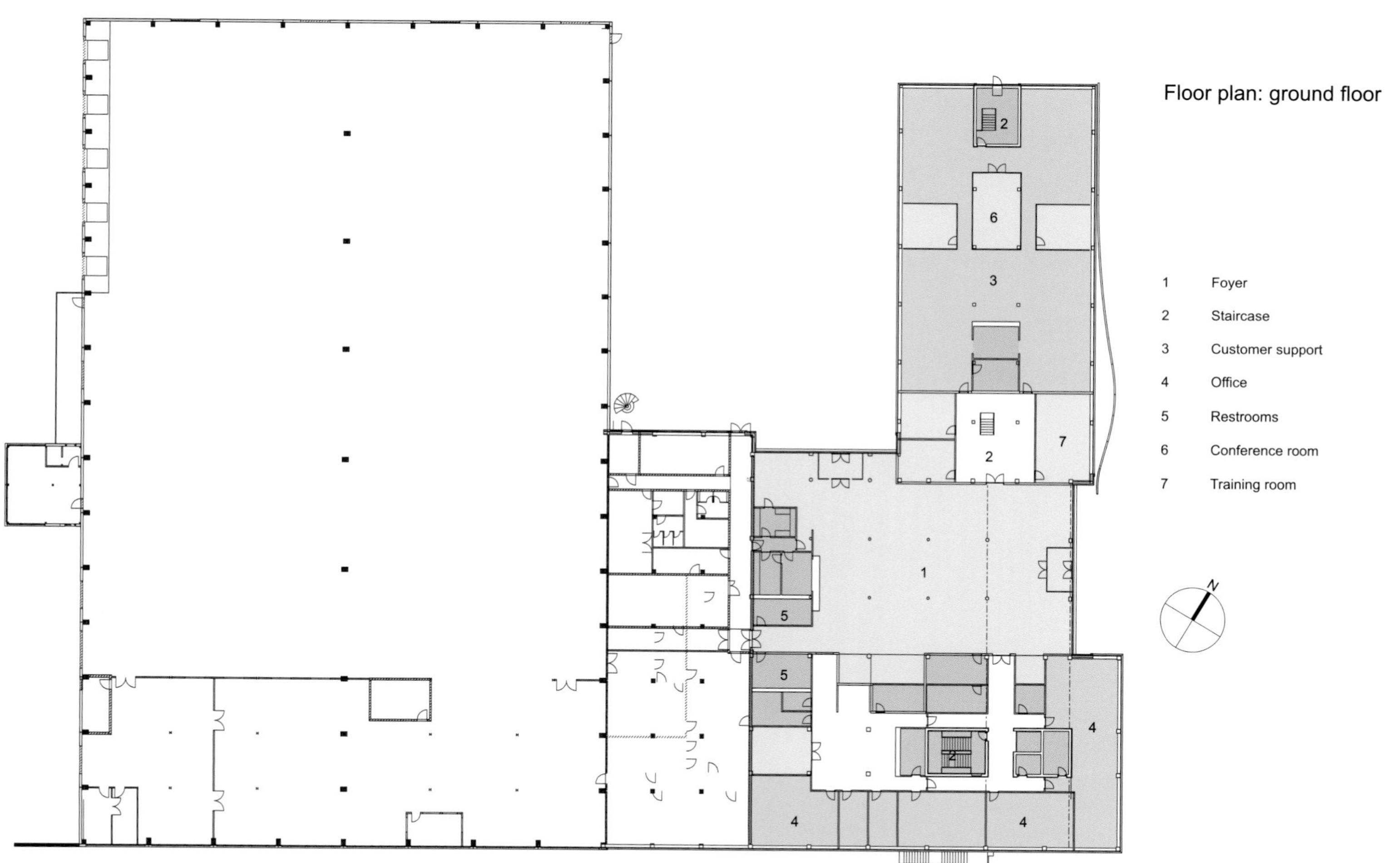

Floor plan: ground floor

1 Foyer

2 Staircase

3 Customer support

4 Office

5 Restrooms

6 Conference room

7 Training room

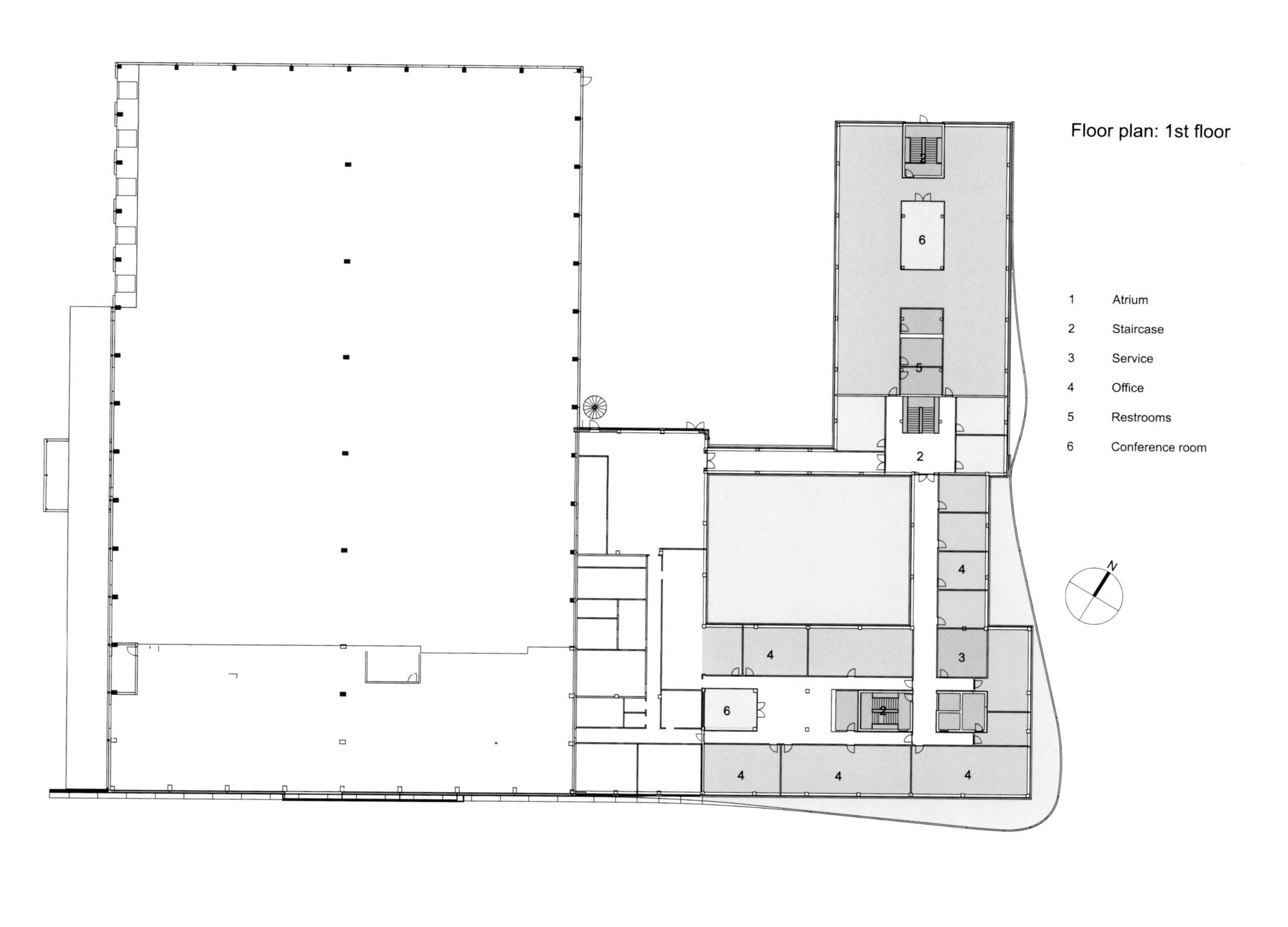

Floor plan: 1st floor
1 Atrium
2 Staircase
3 Service
4 Office
5 Restrooms
6 Conference room
N

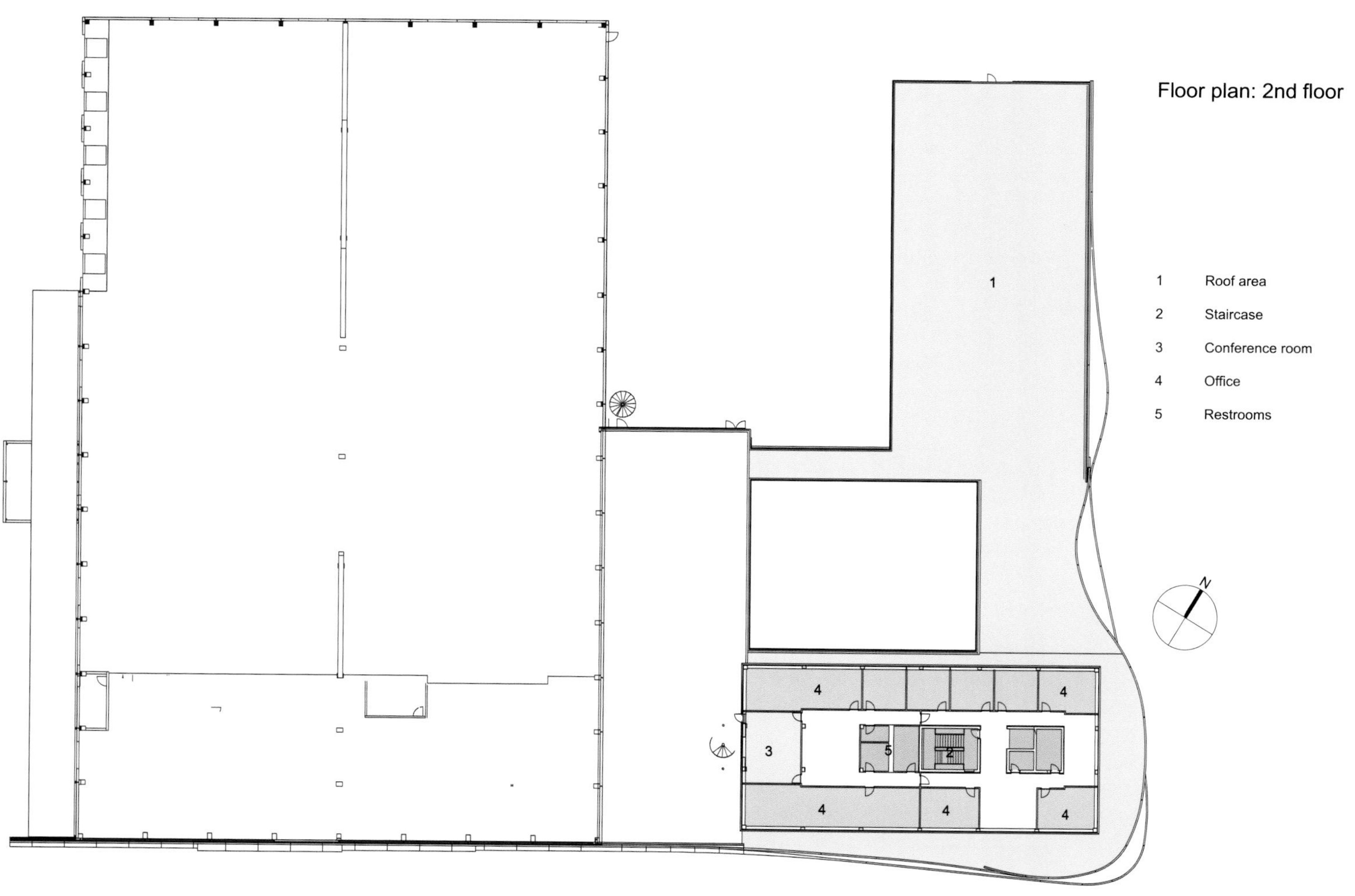

Floor plan: 2nd floor
1 Roof area
2 Staircase
3 Conference room
4 Office
5 Restrooms
N

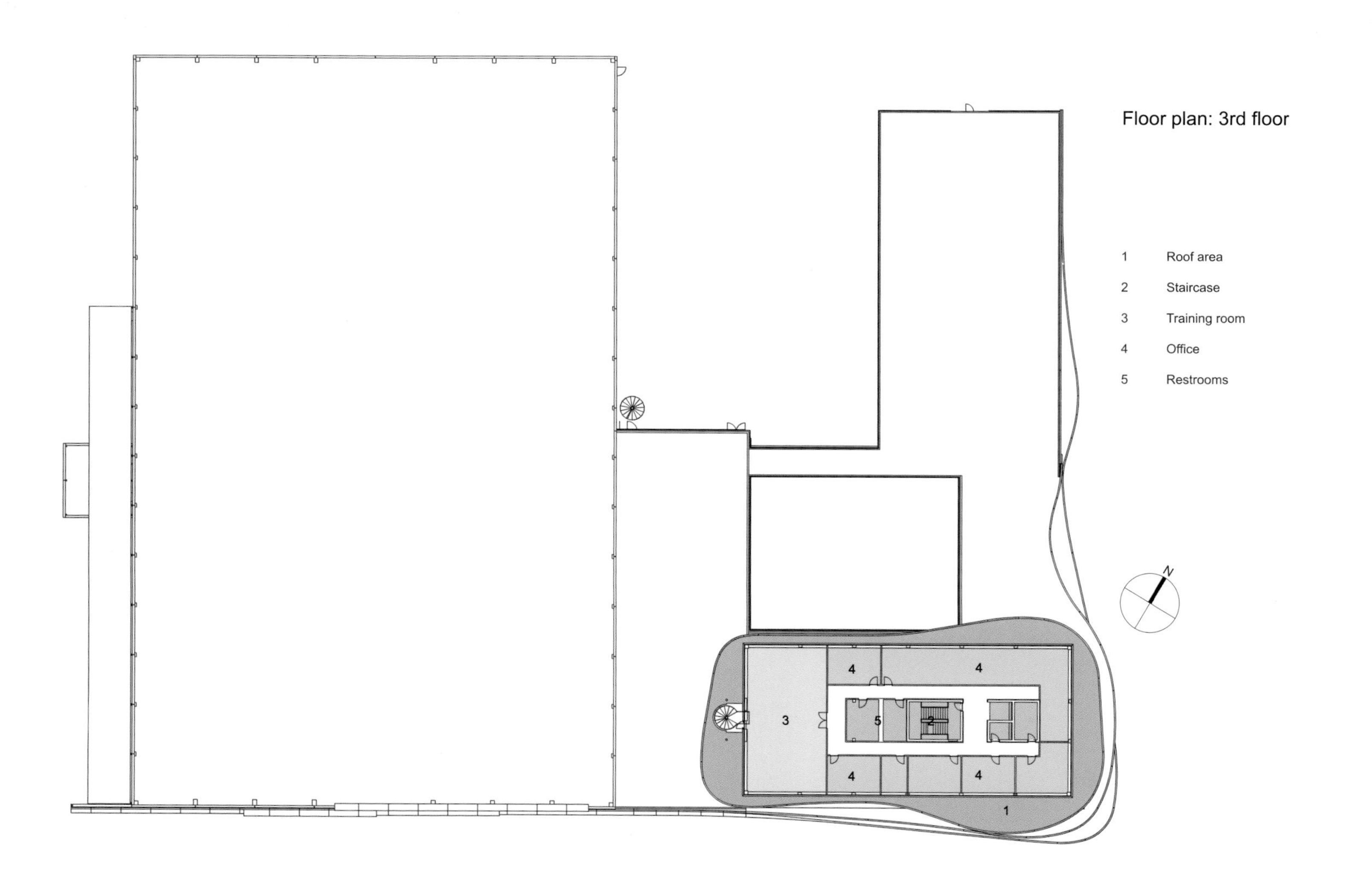
Floor plan: 3rd floor
1 Roof area
2 Staircase
3 Training room
4 Office
5 Restrooms
N
4
4
3
5
2
4
4
1

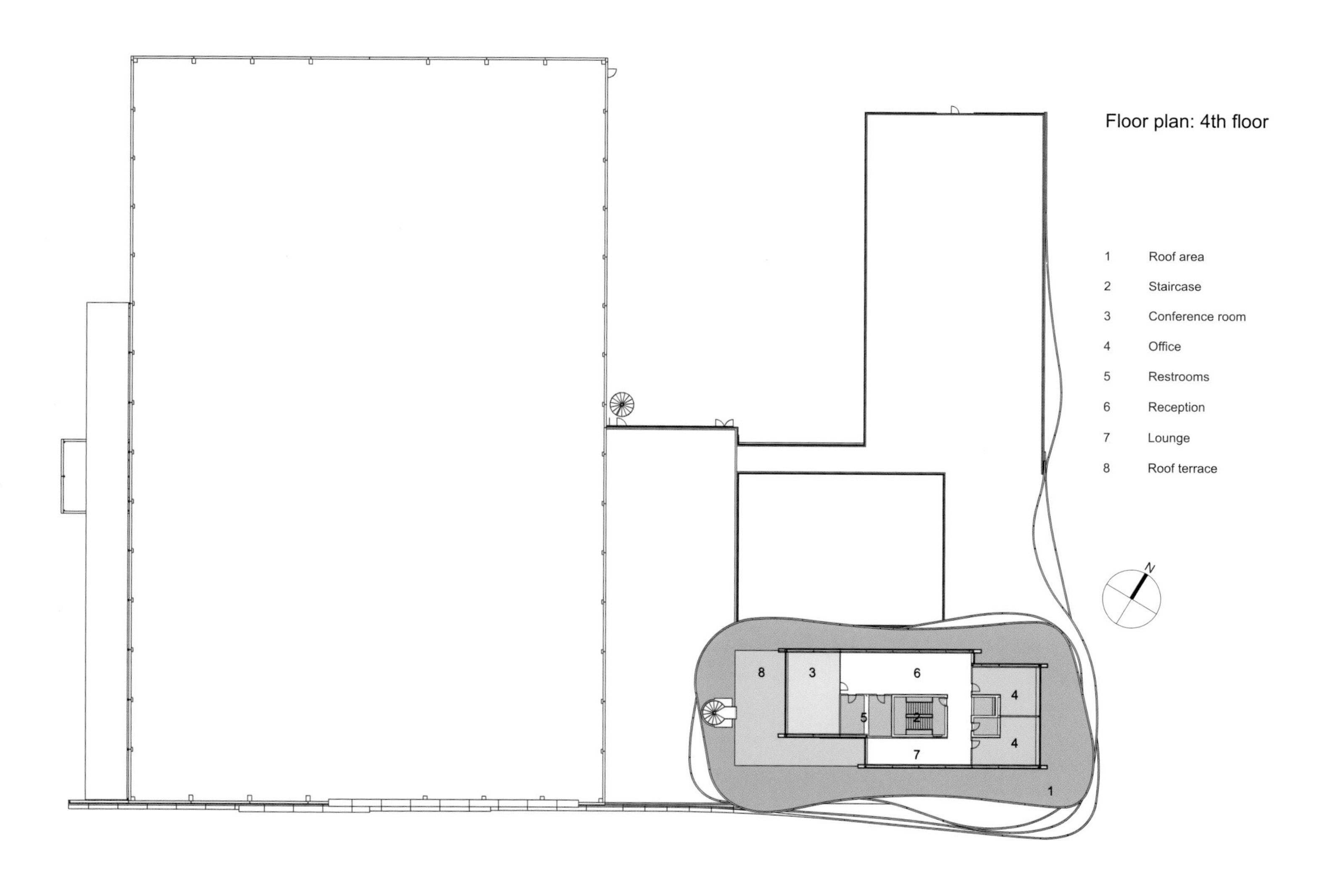
Floor plan: 4th floor
1 Roof area
2 Staircase
3 Conference room
4 Office
5 Restrooms
6 Reception
7 Lounge
8 Roof terrace
N
8
3
6
5
2
4
4
7
1

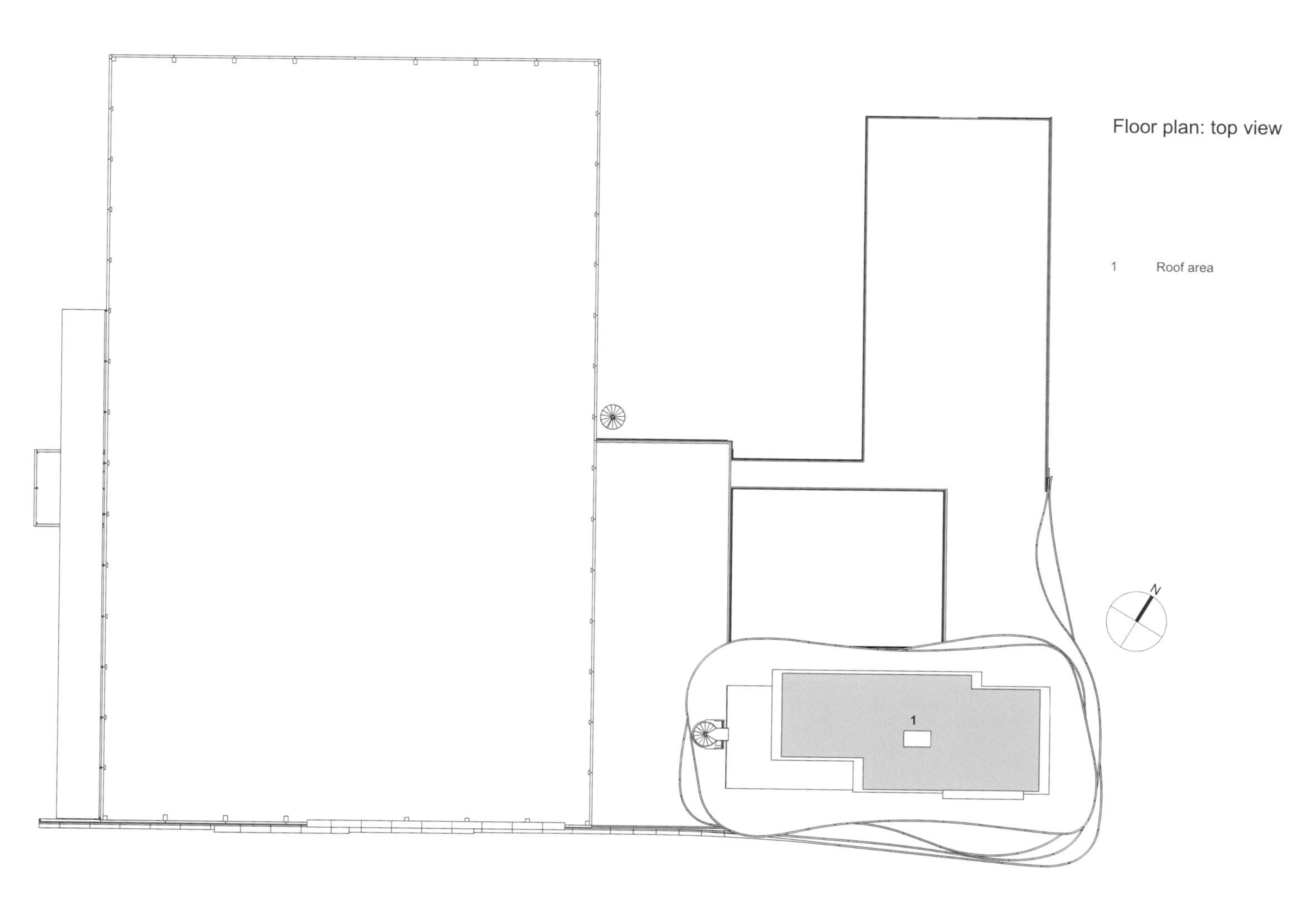

Floor plan: top view

1 Roof area

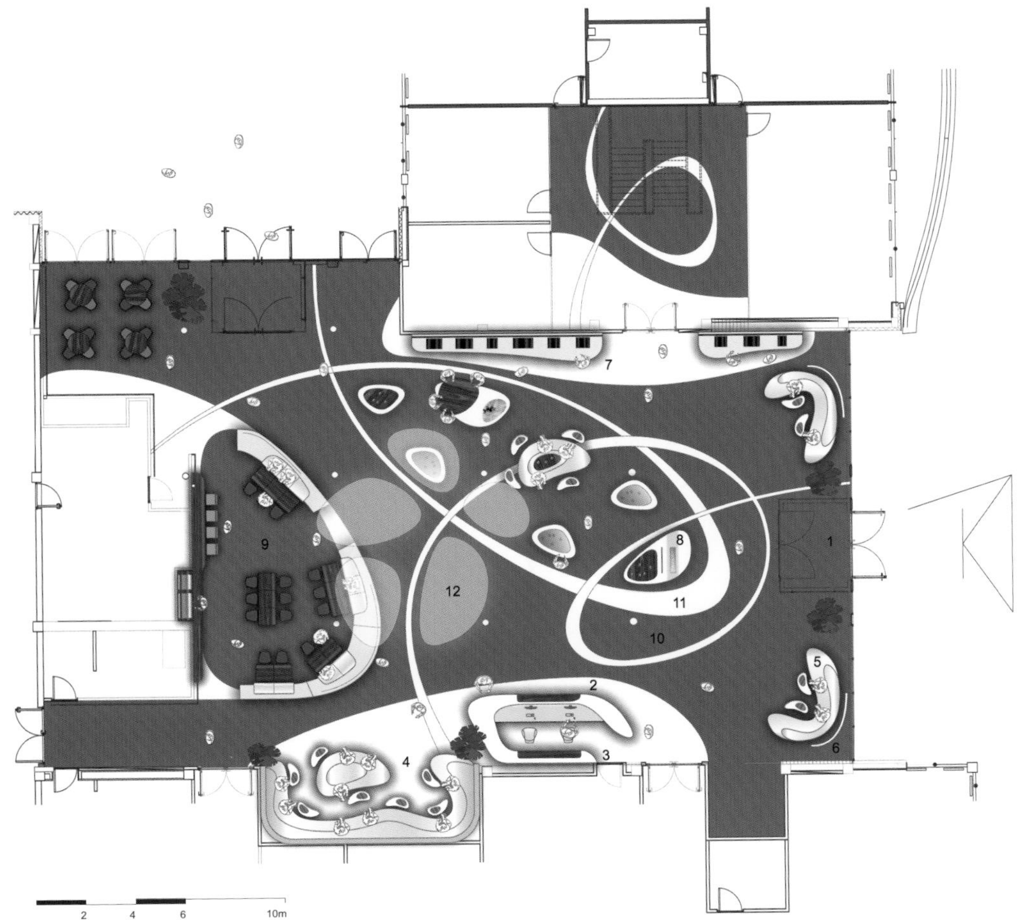

Floor plan: forum

1 Main entrance
2 Reception
3 Welcome board
4 Lounge
5 Lounge sofa
6 Room divider
7 Product presentation of coffee machines
8 Exhibition area
9 Cafeteria
10 Oak parquet
11 Floor inlay, engineered quartz stone
12 Luminous ceiling

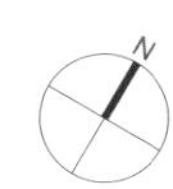

继 3deluxe 设计的莱昂纳多玻璃立方体获得国际赞誉之后，凭借宏伟的设计和富有表现力的形式语言，3deluxe 再次完成了一项企业建筑的设计。

在所有楼层上，外立面的不对称带状结构将三座大厦相连，它们之间的相互连接构成了 U 形综合体。它们巧妙地隐藏了混凝土框架结构的正交网格。这座大楼由 150 块独特的预制混凝土结构组装而成，每个结构重达 6 吨，面向道路的建筑外立面呈带状结构，构建起公司总部大楼 100 米的高度，从而产生一种深度感和动态感。而行政部由生动的、垂直交错的丝带组合构成，带状结构在此处达到了极致：带状结构再加上每个楼层不同造型和悬挑的天花板，这些使得综合体的拐角大楼成为了一处引人注目的美景。动态外观的设计不仅要与建筑的周围环境协调一致，还要考虑到人们能够在眼平视的视域内和短距离的范围内观察到大楼外观。从行人的角度来看，他们能看到下方表面的效果，即通过重叠的层板和立面的丝带所呈现出各种形状、光线和阴影的相互作用。有时，光线反射铜色的窗框，给白色的表面增添了柔和的金属光泽。

这座设计出色的角落建筑不仅栩栩如生，而且成为了公司总部的功能性体现的中心环节。除了个人办公室和会议室以外，还包括非正式交谈的公共休息区。董事会成员的办公室位于顶层大厦，覆盖着白色合金板，其形状采用建筑综合体的基本矩形结构。商务套房配置了定制的家具和高端配件，而在室内设计中再次应用了建筑元素。

快速轨道办公与研发大楼

FAST TRACK OFFICES AND LABS

ARCHITECT
Jan van Iersel, Renze Evenhuis, Cees Schott AvB

PROJECT MANAGEMENT
Willeke van de Groep

CLIENT
Synthon Bv Nijmegen

LOCATION
Nijmegen, the Netherlands

AREA
6,850 m²

PHOTOGRAPHER
Thea van den Heuvel /DAPh

Synthon Nijmegen is growing fast, very fast. Additional laboratories and offices proved necessary within a year. The so-called "Fast Track Offices and Labs", also known as "FTOL", were completed over a period of exactly twelve months – from design to completion. This took place at an unprecedented rate and the outcome is a high-quality, safe and friendly building. Construction team and high quality standards – it is possible. The key to this success? Optimal cooperation between the parties involved. Trusting each other and sharing knowledge and expertise.

FTOL's design is clear and transparent. The building is made up of five layers. A warehouse is located under the deck and on top of the green deck is a glass office. Floating above this is a horizontal volume of three layers comprising laboratories, offices and technical installations. Two atriums create an open connection between the spacious entrance hall and the first floor. Tall glass fibres spanning floor to ceiling strengthen the interaction with the landscape. The overhang of the upper building volume keeps the sun's heat at bay in a natural way. The facade is built up from two layers. A wind and water tight inner layer of bright lime-yellow facade elements is protected by an outer layer of expanded metal mesh. From the motorway the volume appears quite closed, but as one approaches the building it opens up and the lime-yellow color of the inner layer becomes increasingly visible and clear.

Broekbakema has been working on the Synthon master plan since 2008. FTOL is the first building to be completed. Other designs, from landscape to interior, are currently being worked on.

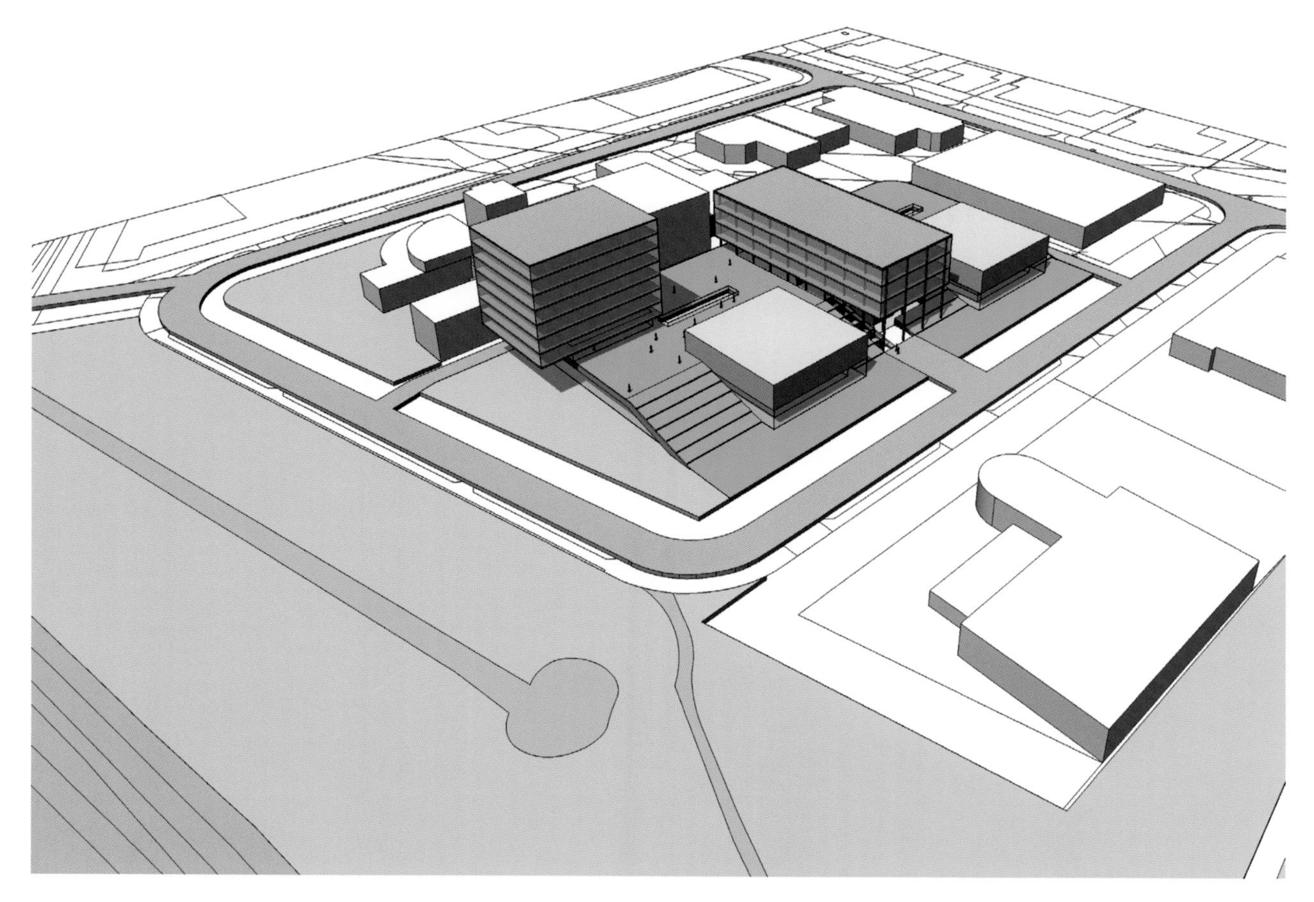

Groene Omgeving

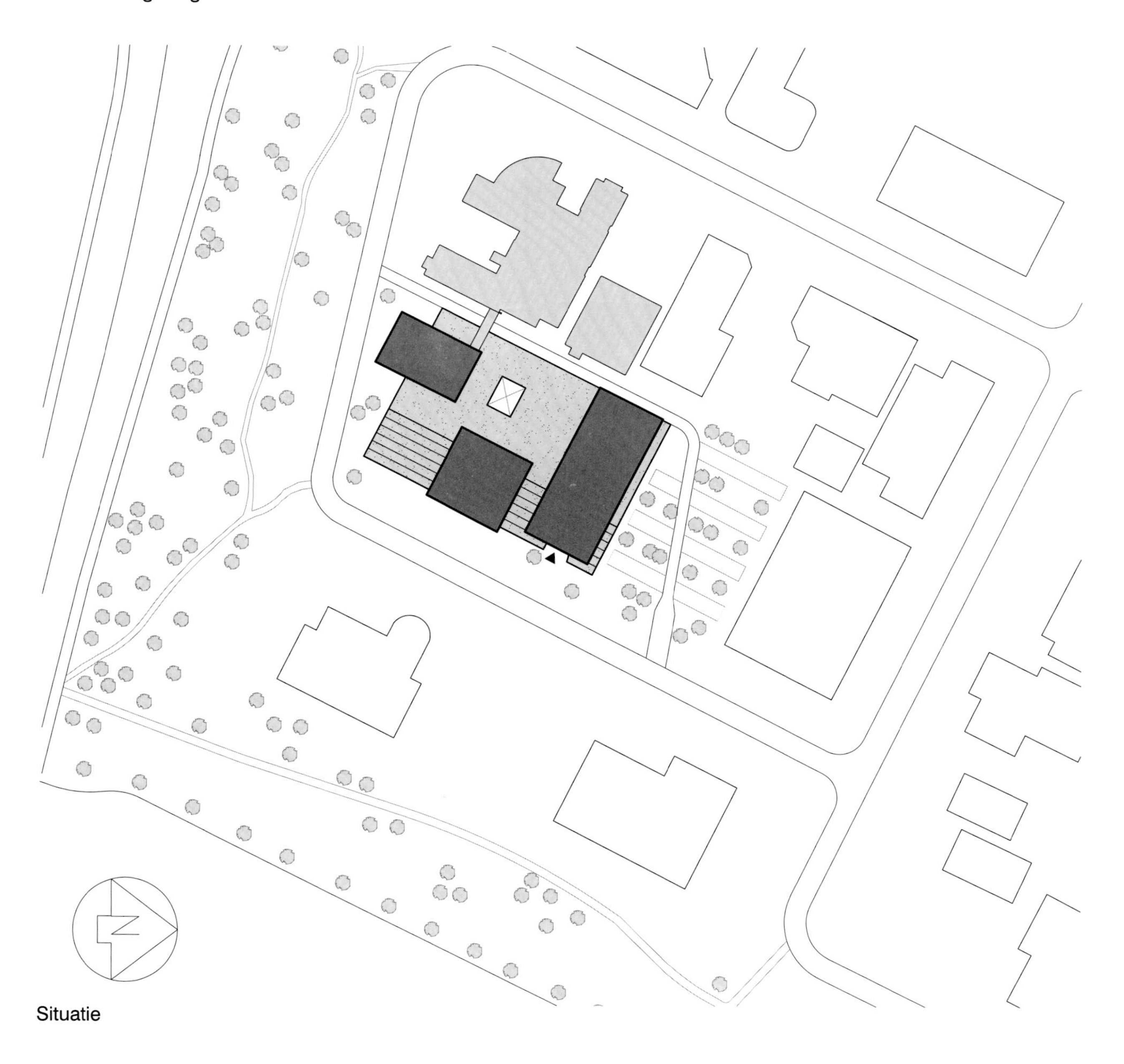

Situatie

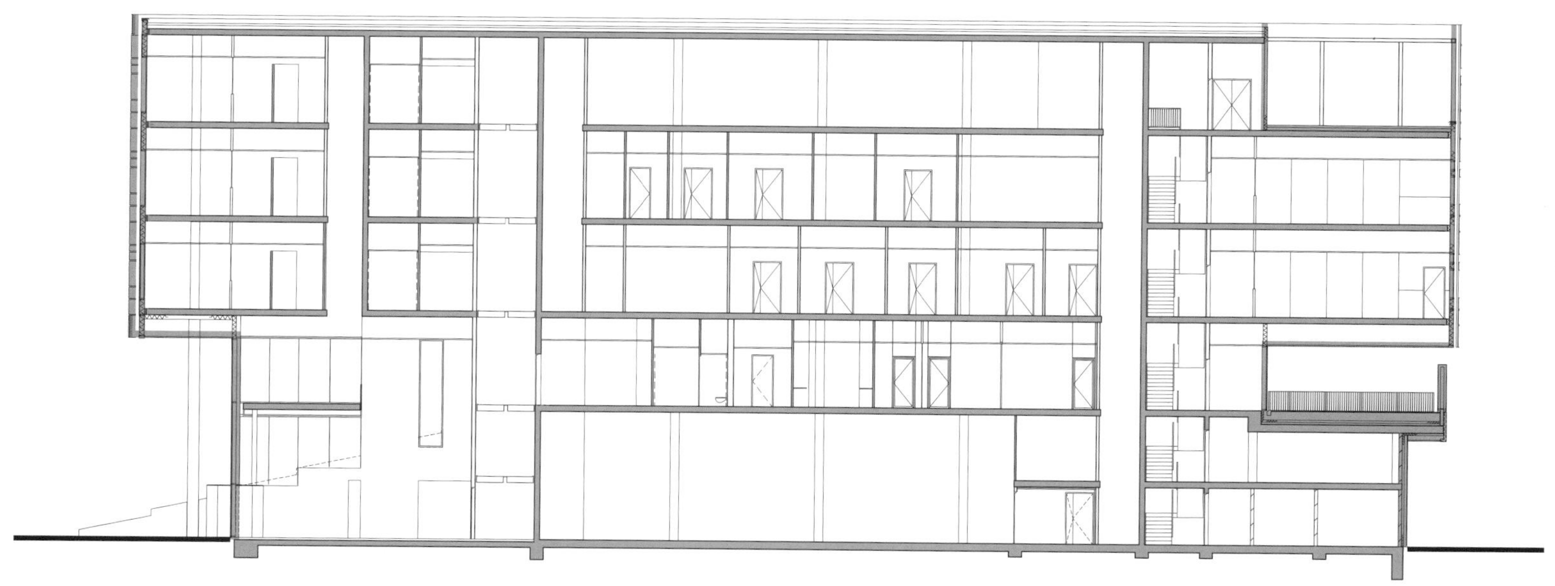

奈梅根西恩松公司近年来发展得十分迅猛，急需在一年内建立更多的研发实验室和办公室。快速轨道办公与研发大楼(FTOL)就是在这样的背景下，仅用十二个月，就完成了从设计到完工的整个过程。设计师和建筑工人们创造了前所未有的建设速度，建造了一栋集高品质、安全性以及友好性于一体的宏伟建筑。建设团队是如何成功解决高质量的建设标准和紧张的时间之间的矛盾呢？其中最重要的就是优化团队的合作，互相信任，共享资源和专家团队。

快速轨道办公与研发大楼的设计风格清晰而透明。整栋大楼分为五个层次的建筑模块。地面基座底层是仓库区，上面是一间玻璃办公室。基座上方是水平方向延伸的三层模块，包括研发区、办公室和技术装置区。宽敞的门厅和一楼办公室由两个中庭连接。高大的玻璃幕墙从地面一直延伸到建筑顶端，通透的视野效果使得大楼内外环境有机地融为一体。位于上方的建筑模块部分向墙外突出，能够充分保证自然光线的射入。整个大楼的外墙体分为两层，内层是明黄色的防水层，外层是金属网保护层。从高速公路向大楼望去，大楼看起来十分封闭，而实际上当人们走近时，能感觉到建筑整体的开放式效果，而外墙体上明黄色的内层更加显得分外醒目。

布罗贝克玛公司自2008年起就一直在从事建设西恩松公司的规划项目工作。快速轨道办公与研发大楼是第一个完工的建筑项目，其他的项目方案，包括景观和内部装饰工程，还在逐步地进行中。

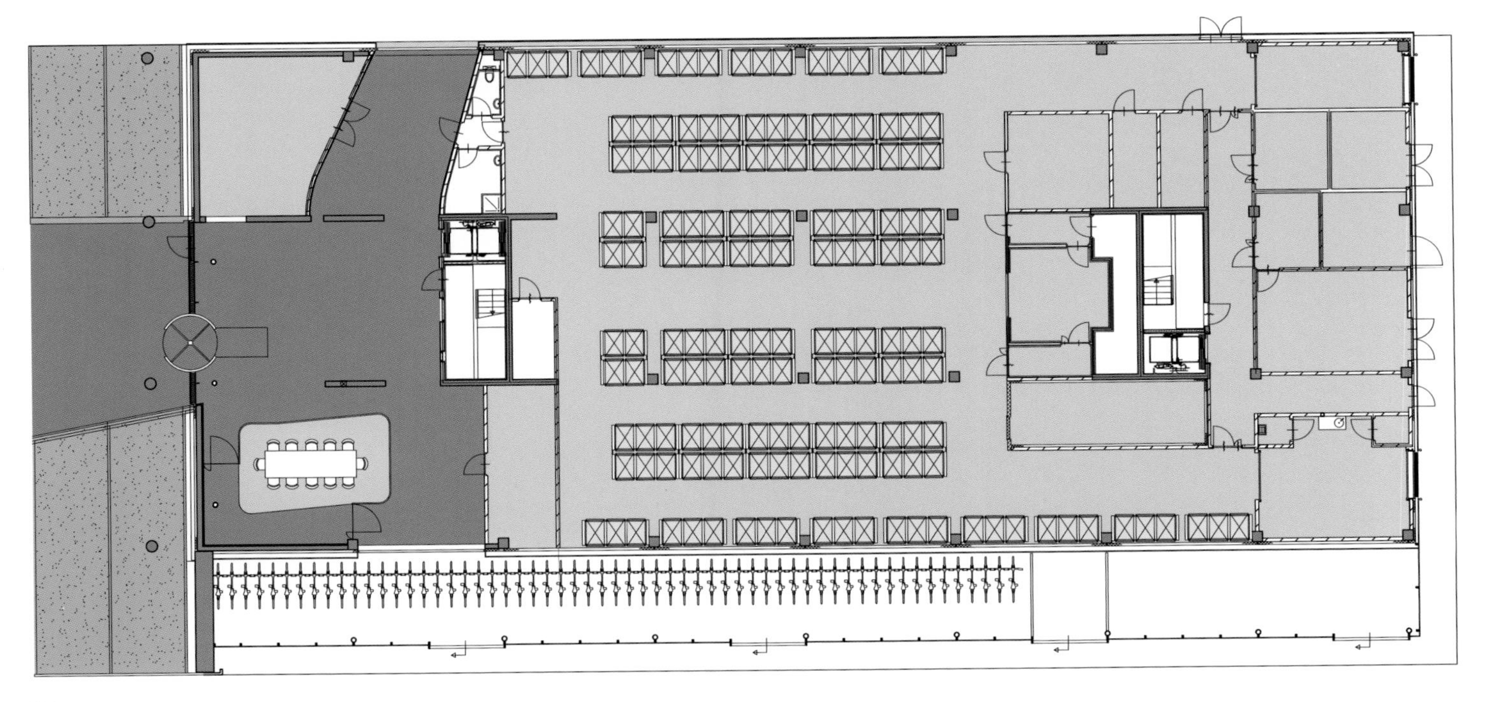

L0

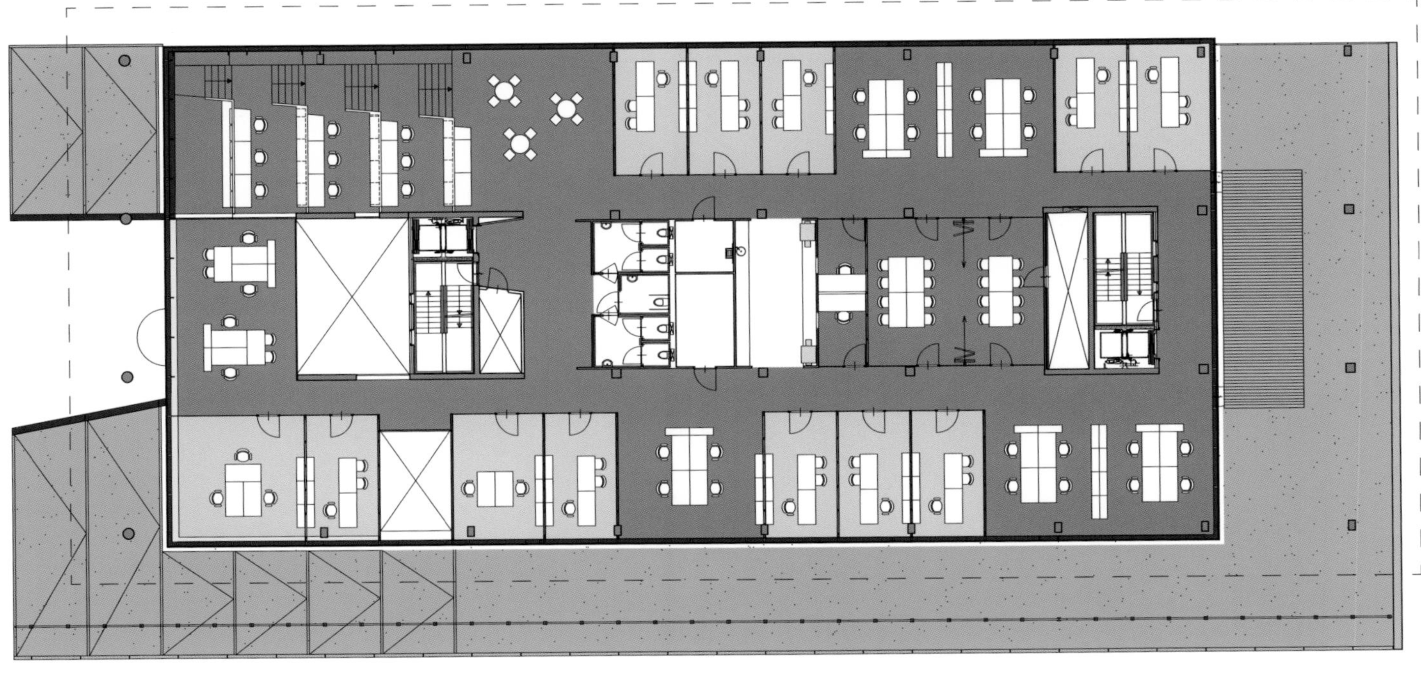

L1

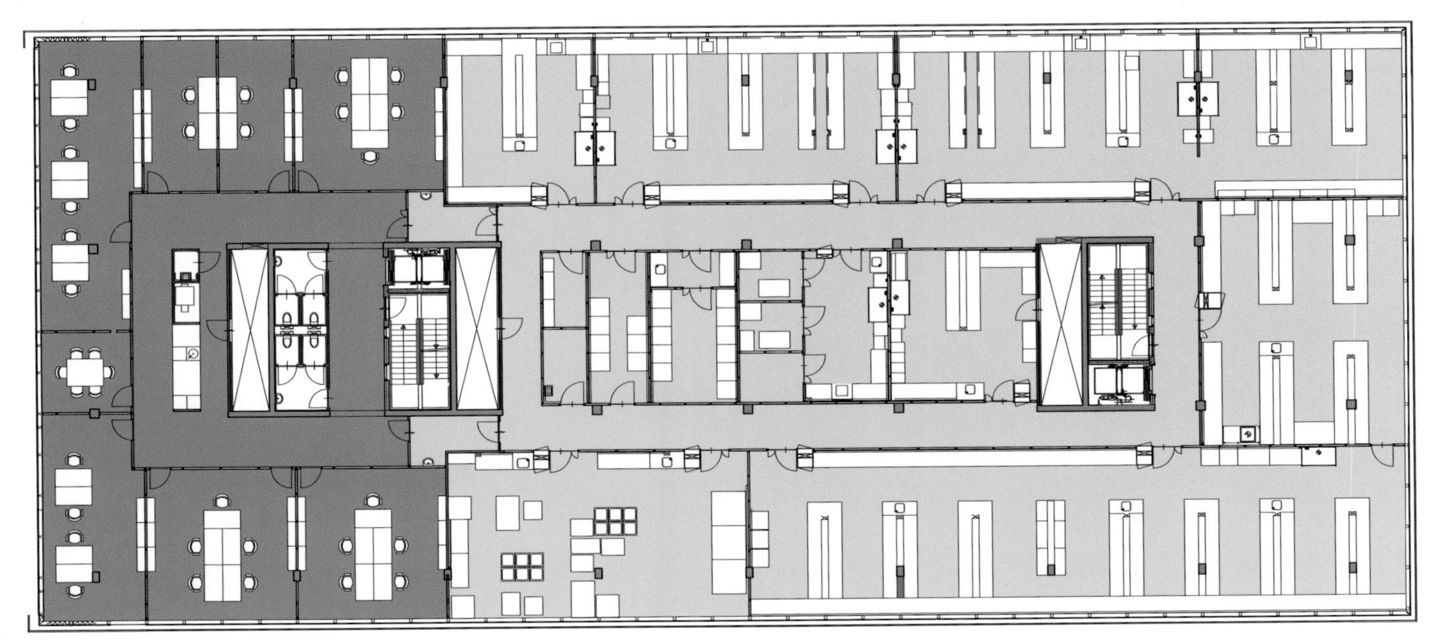

L2

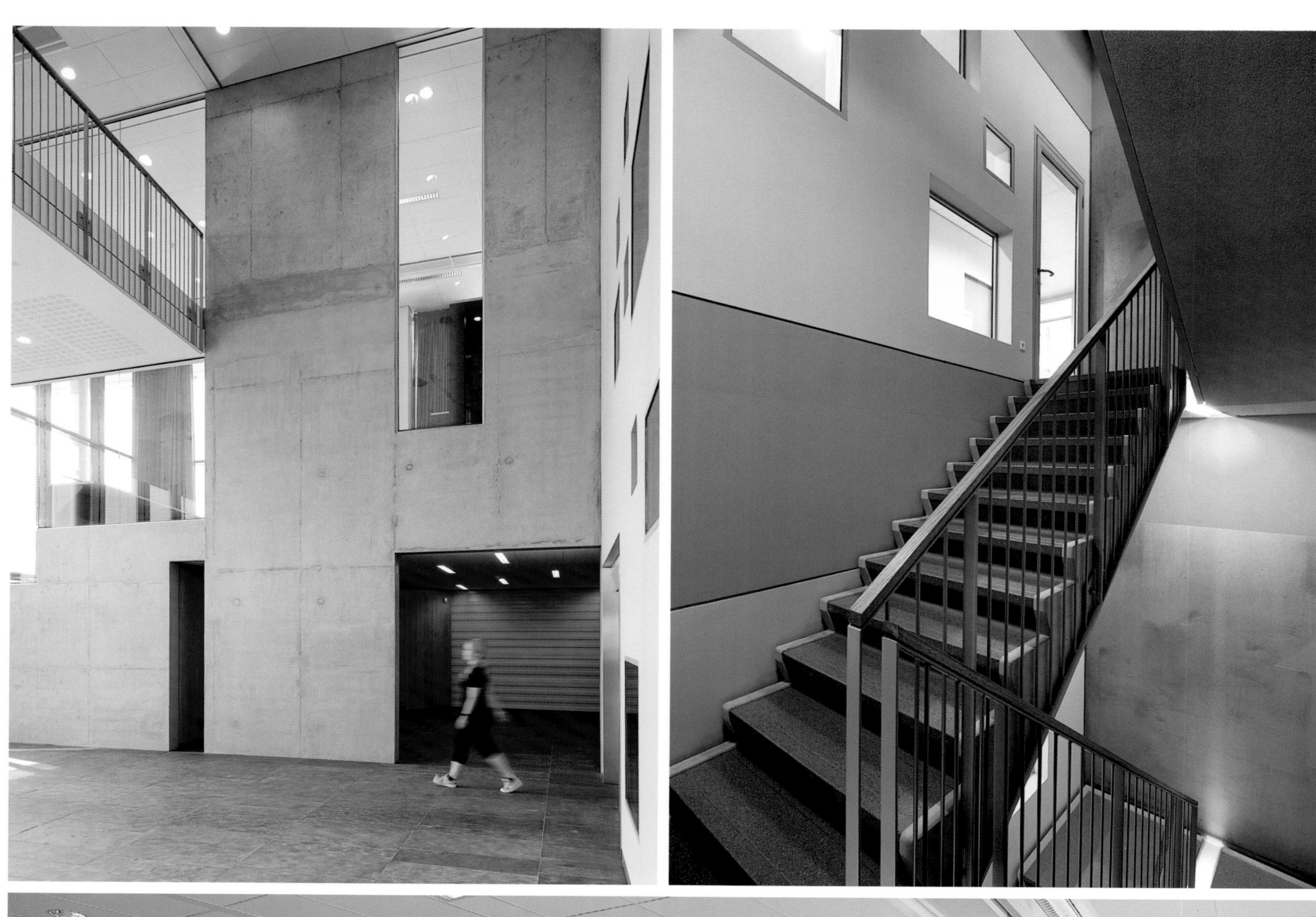

于家堡工程指挥中心

YJP ADMINISTRATIVE CENTER

ARCHITECT
HHD FUN

CLIENT
The city of Binhai New Area in Tianjin

LOCATION
Tianjin, China

PHOTOGRAPHER
Wei Gang, Wang Zhenfei

BD
天津滨海新区中心商务区于家堡金融区
YUJIAPU FINANCIAL DISTRICT OF CBD IN TIANJIN BINHAI NEW AREA

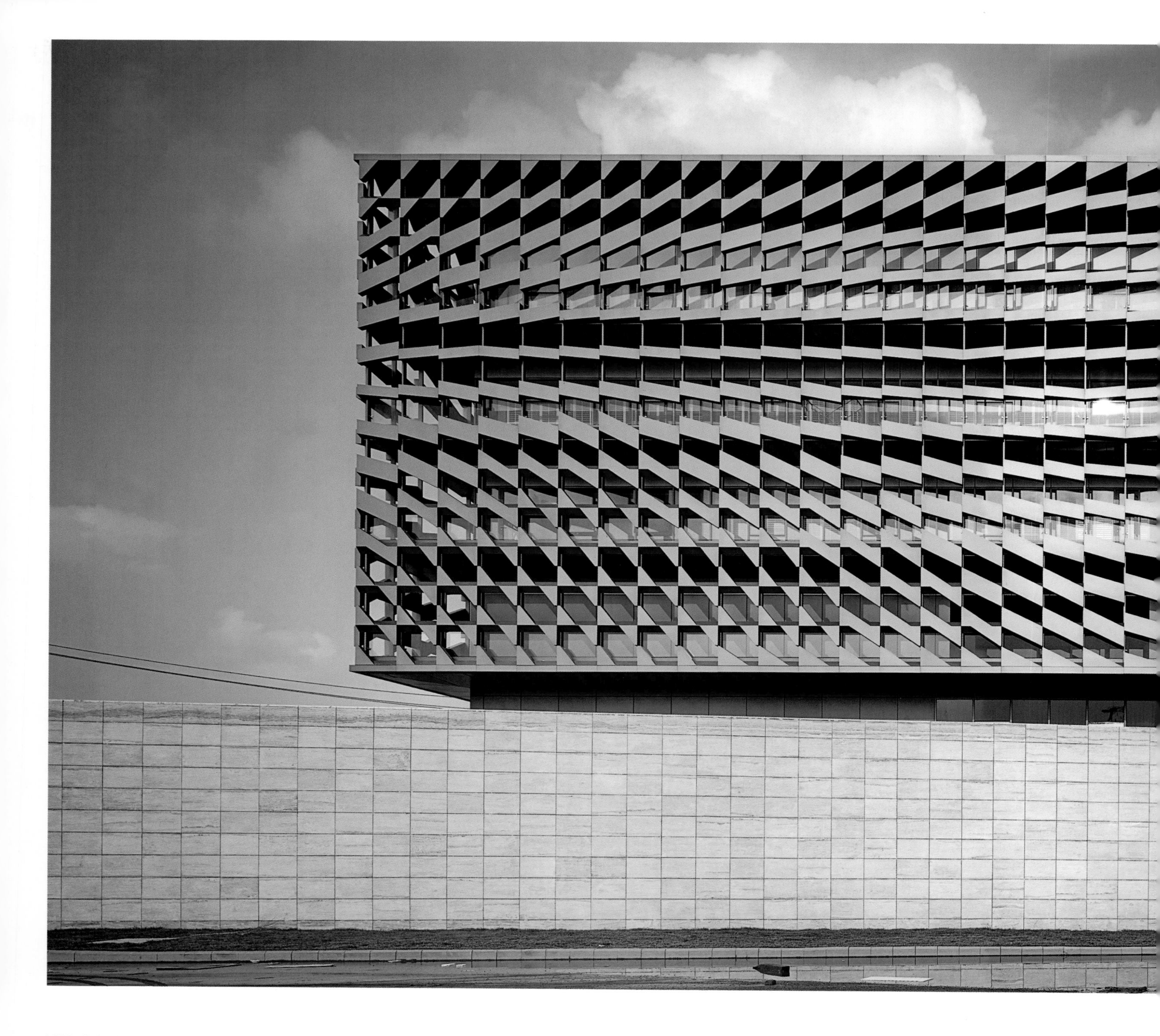

YJP Administrative Center, as the office center for all the projects to be developed in YJP financial district in Tianjin Binhai New Area in the next five years, needs to be designed and completed in a very short time to adapt to the construction schedule of the financial district. In China where it is widely thought that construction is in a state of low technology, parametric design has undisputedly shown its advantages in construction schedule and construction process.

As the project command center of the financial district, circular corridors are designed on the outside of all the rooms above the second floor for the observation of the entire site. Due to their different functions as well as their different requirements of lighting, the rooms can be divided into offices, lounges, halls, warehouses and elevator halls, etc. Through a script, facade lighting information can be converted to a curved surface, on which the height of each point corresponds to the day lighting rate of this very point, so the facade components shall meet the requirements of the lighting of its location. Based on this principle, different opening rate is used on the treatment of the facade of the verandah to adapt to the different requirements of internal functions and to generate different view windows as well, making the viewing process more interesting. Facade components are designed to be a series of geometric components which can change day lighting rate through its own rotation; there are continuous topological equivalence changes as well as continuous topological unequivalence changes among components, which mean that opened linear view windows of high day lighting rate change continuously to closed quadrilateral and hexagonal view windows of low day lighting rate. The design indicates that simple geometric relationship can produce complicated geometric transform relation after a deep research on its characteristics and a simple list. Through the operation of another script, the curved surface is used to control the installation and the arrangement of different modules.

In order to be able to control the time limit for the project, the facade solution is finally restricted within six kinds of different modules, six kinds of continuous variation modules represents six different day lighting rates; after the calculation of curved surface by the script, the construction drawings generate automatically with different modules marked in different positions.

十四局集团三公司

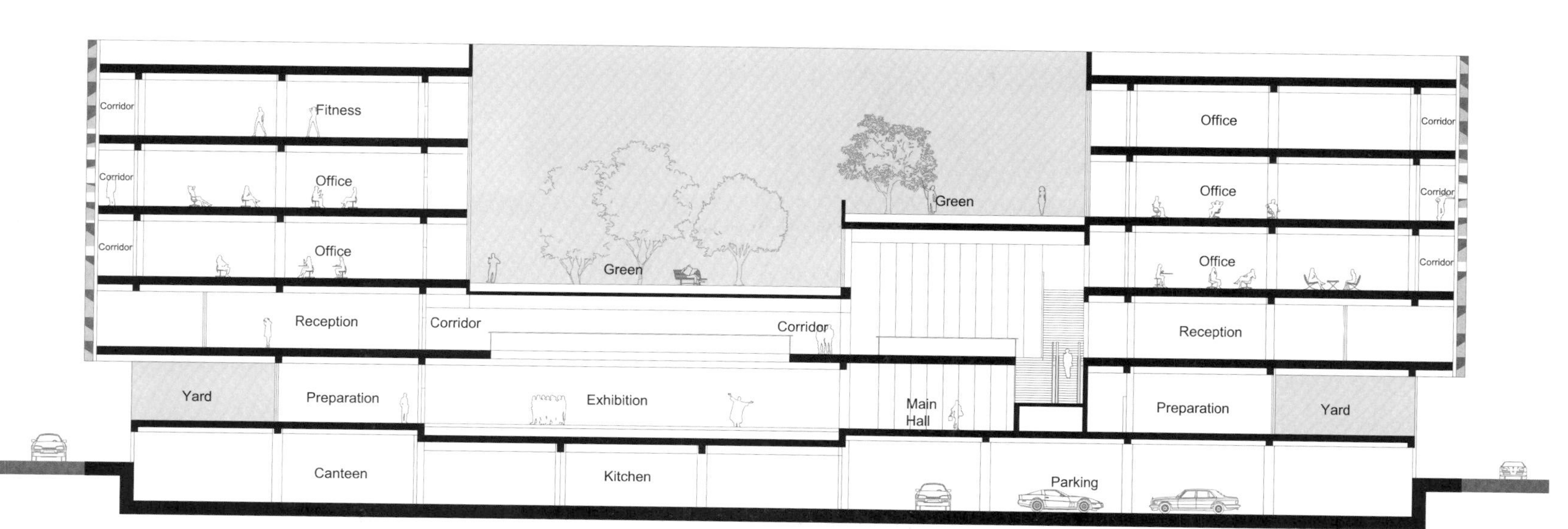
Corridor
Fitness
Corridor
Office
Corridor
Office
Reception
Corridor
Corridor
Green
Green
Office
Corridor
Office
Corridor
Office
Corridor
Reception
Yard
Preparation
Exhibition
Main
Hall
Preparation
Yard
Canteen
Kitchen
Parking

Section

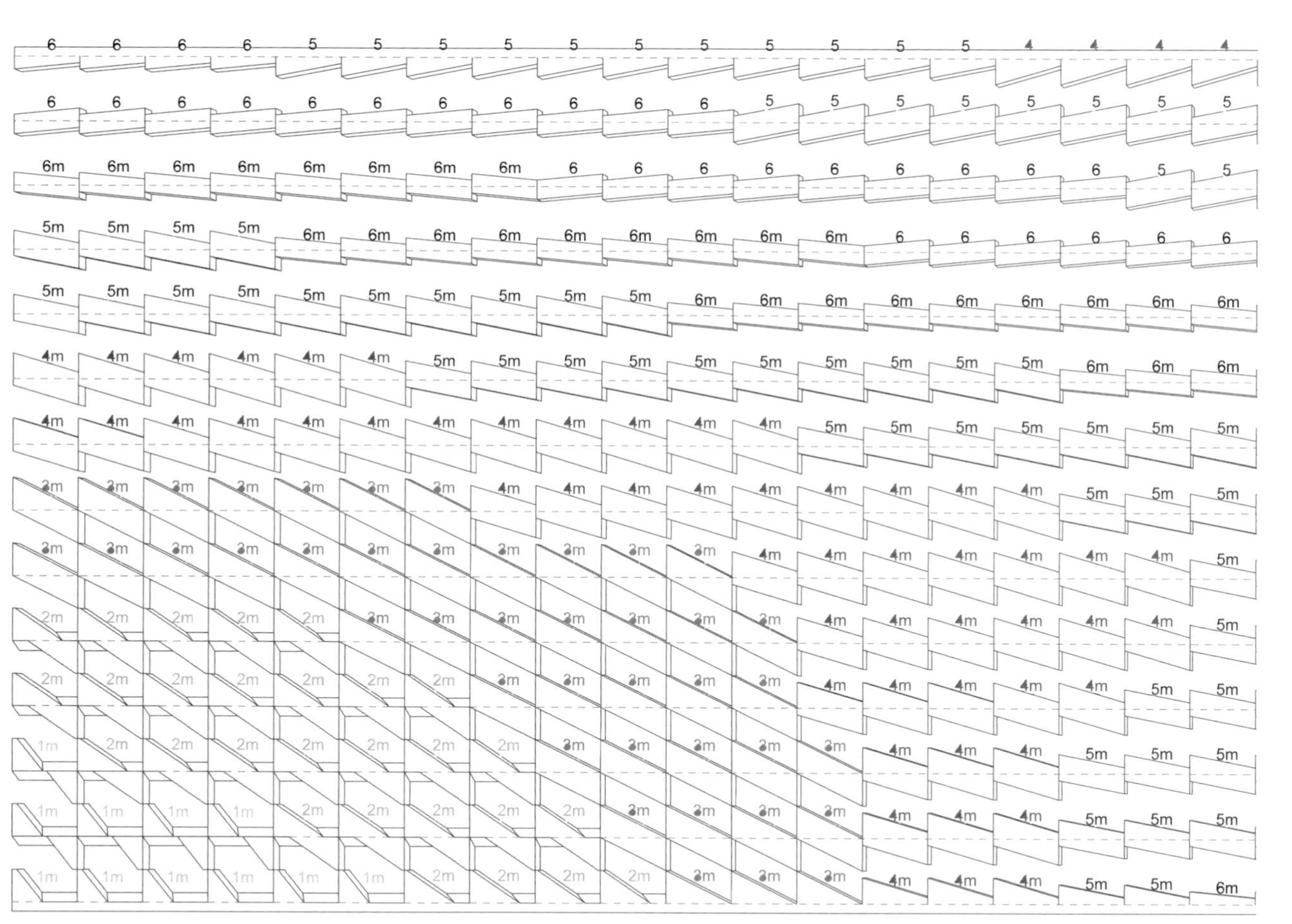

参考案例

Reference Projects

路易威登旗舰店　LV flagship store,UNstudio

三菱日联大厦　MUFG Nagoya, Neil M. Denari

马德里市民法院　Civil Courts of Justice, Zaha Hadid

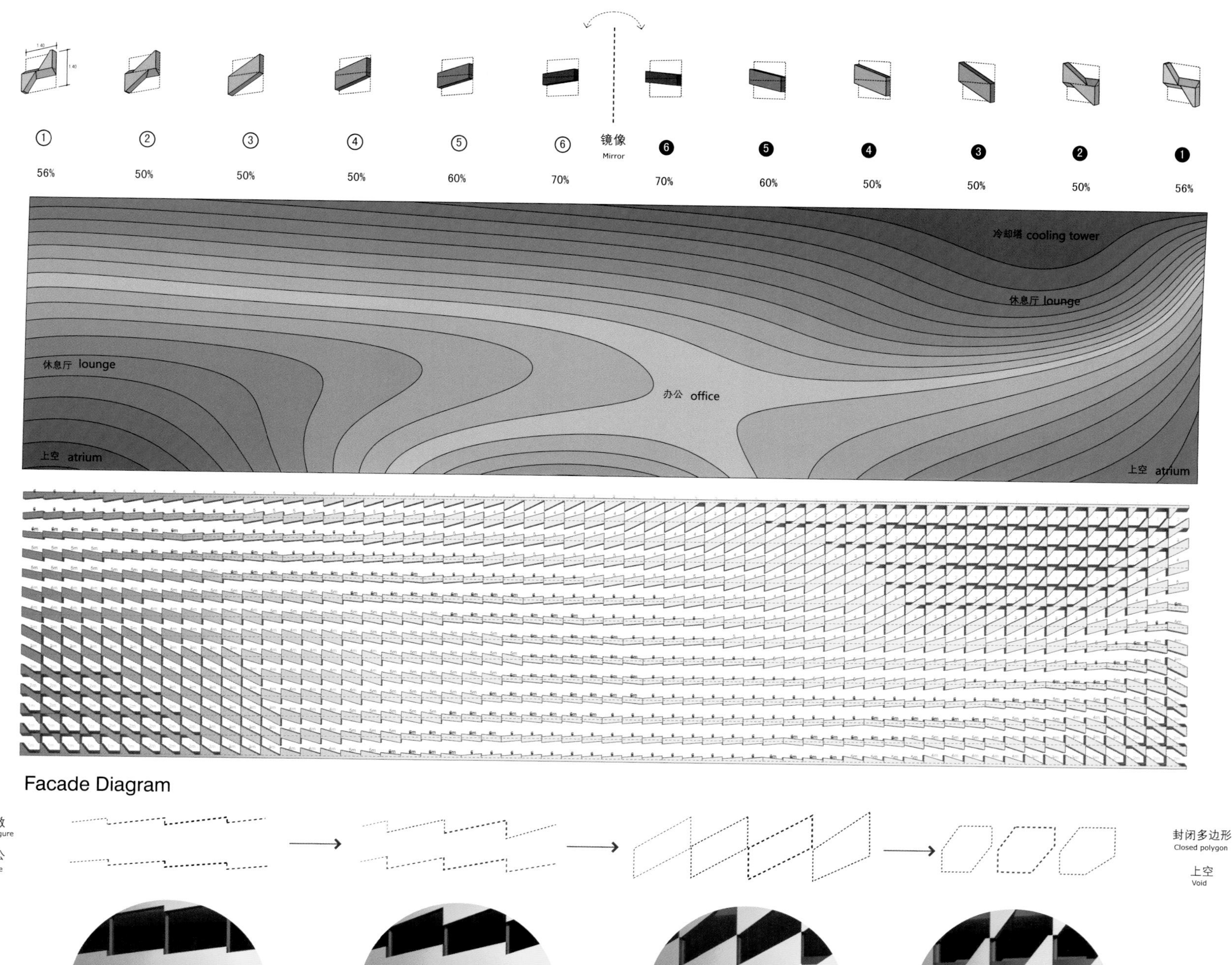

Facade Diagram

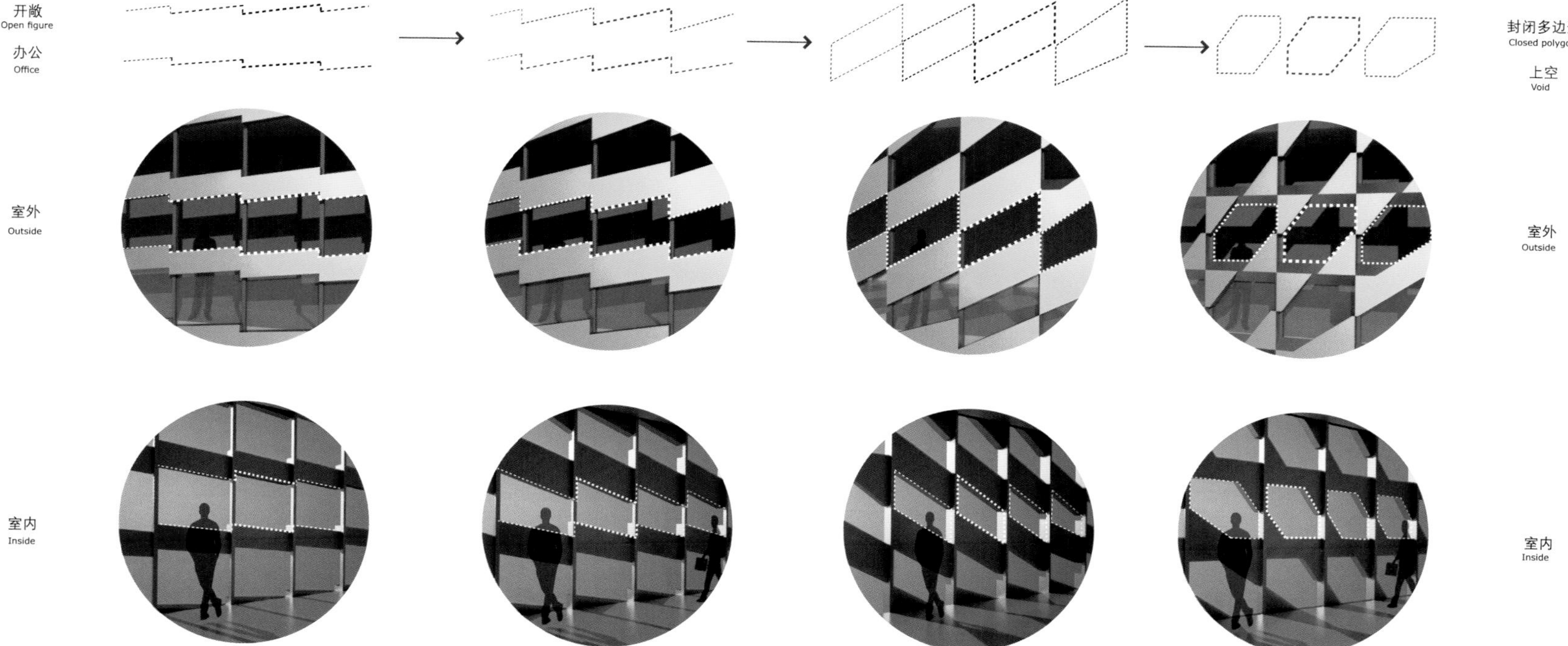

不同功能决定不同的开窗方式，产生不同的内部感受。
Different patterns based on different function. Therefore generate different feelings.

于家堡工程指挥中心作为天津滨海新区于家堡金融区未来五年内的所有工程办公中心，需要在极短的时间内完成设计施工以适应整个金融区的工程进度。参数化设计在工程进度及施工过程中的优势被毫无疑问地体现出来，即使是在被普遍认为施工处于低技状态的中国。

作为整个金融区的工程管理中心，二层以上所有房间的外侧均设有环通的走廊以便于对整个工地进行观察。由于各房间功能的不同，按采光要求高低可分为办公区、休息区、大厅、库房和电梯厅等。通过一段脚本可以将立面的采光信息转化成一个曲面，这个曲面上每一个点的高低对应着该点采光率的高低，因此立面构件应满足其所在位置的采光要求。外廊的外立面处理以此为基础形成不同的开孔率以适应内部功能的不同需求，同时还为外廊生成了不同的景窗，使观赏过程变得更加有趣。立面构件被设计成一系列的以旋转来改变采光率的几何构件，构件间既有连续的拓扑等价变化，又有连续的拓扑不等价变化，即由开放式的线性高采光率的景窗连续变化到封闭的四边形及六边形的低采光率景窗。设计充分体现了即使是极为简单的几何关联关系在对其特性进行充分研究以后只经过简单的罗列也可以产生丰富的几何变换关系。将这个曲面通过另外一段脚本的运算用于控制不同模块的安装排列。

为了能够控制工期，立面方案最后被限制在六种不同的模块内，六种连续变化的模块代表六种不同的采光率，经过脚本对控制曲面的计算，标有不同模块的排列位置的施工图纸自动生成。

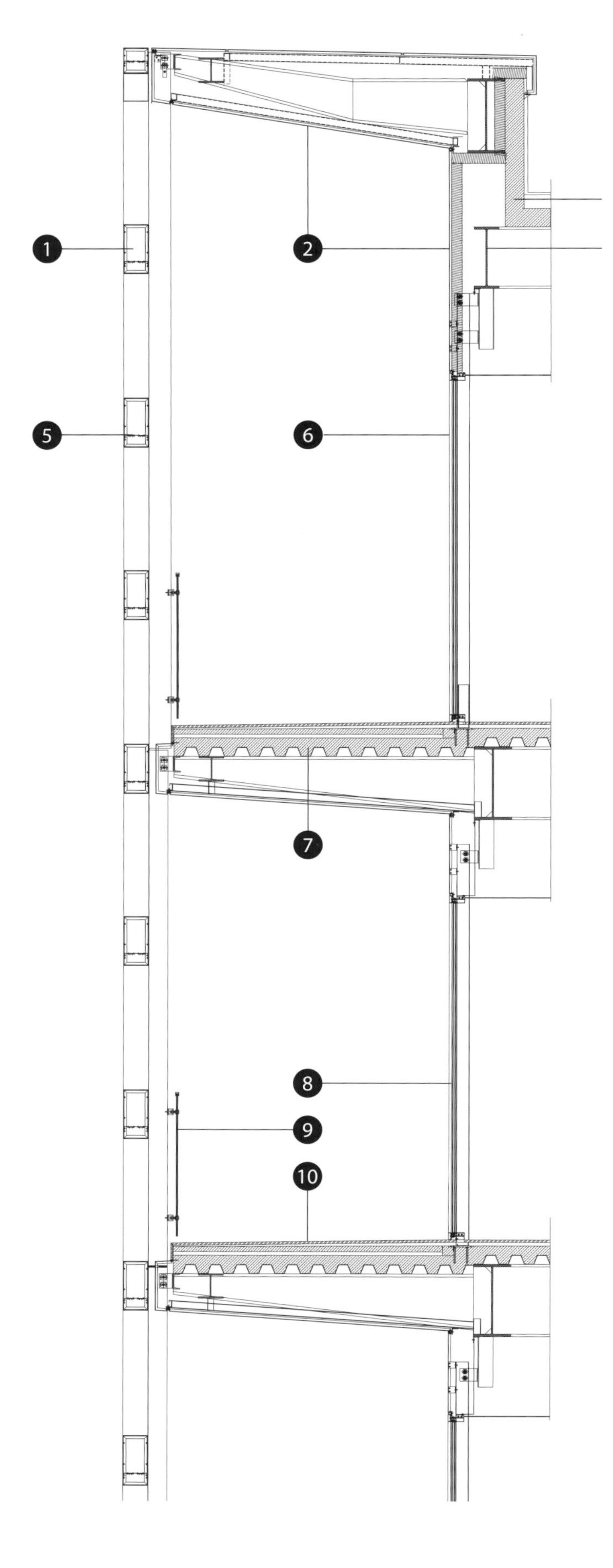

1	Prefabricated aluminum facade component
2	Aluminum panel
3	150mm XPS insulation
4	Steel beam
5	Facade components bolted together at joint
6	Aluminum frame glass curtain wall
7	Steel reinforced concrete slab
8	Double-glazing
9	Tempered safety glass
10	Stone floor finish

TJBINHAI
天津滨海新区中心商务区于家堡
YUJIAPU FINANCIAL DISTRICT OF CBD IN TIANJIN

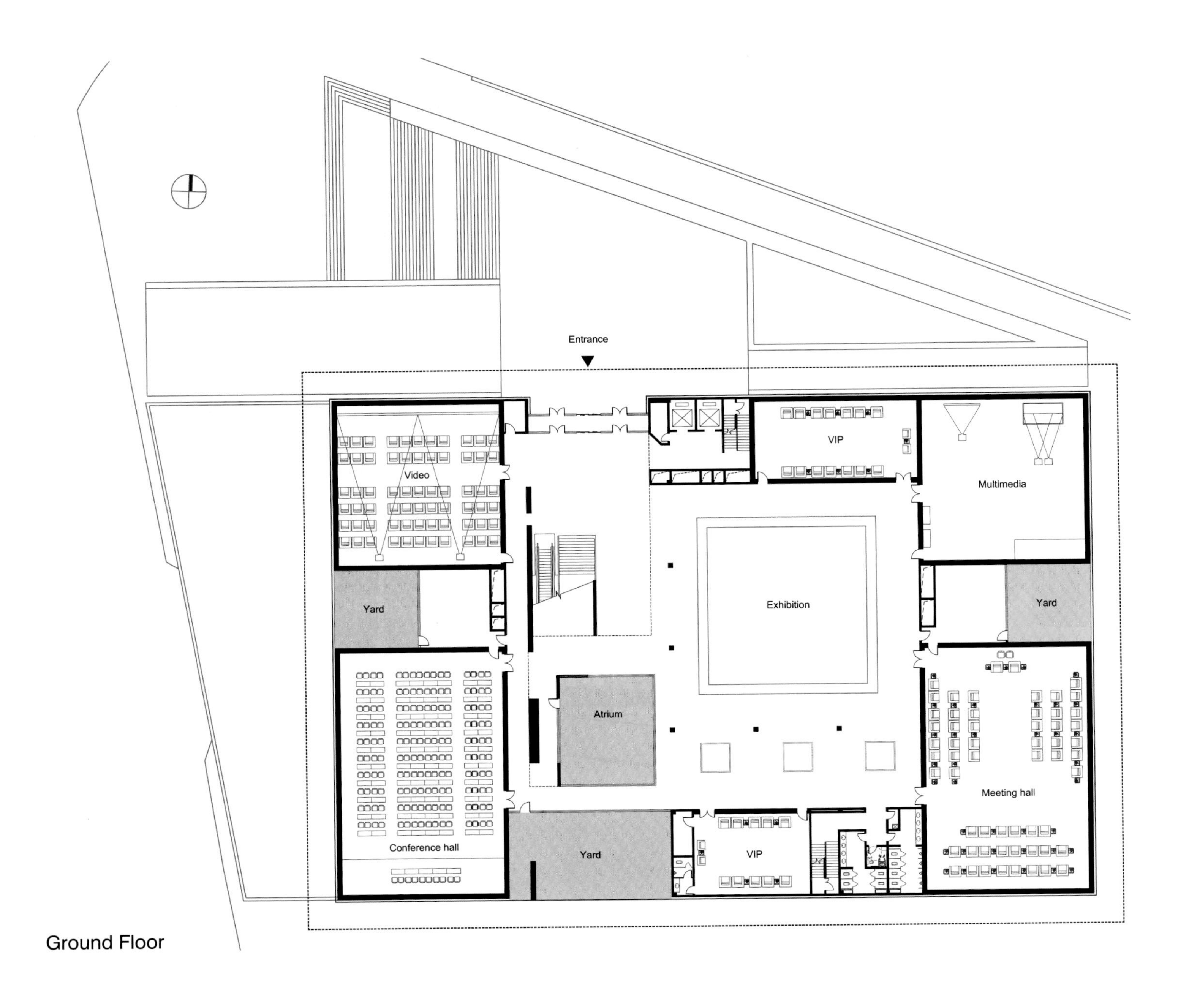

Ground Floor

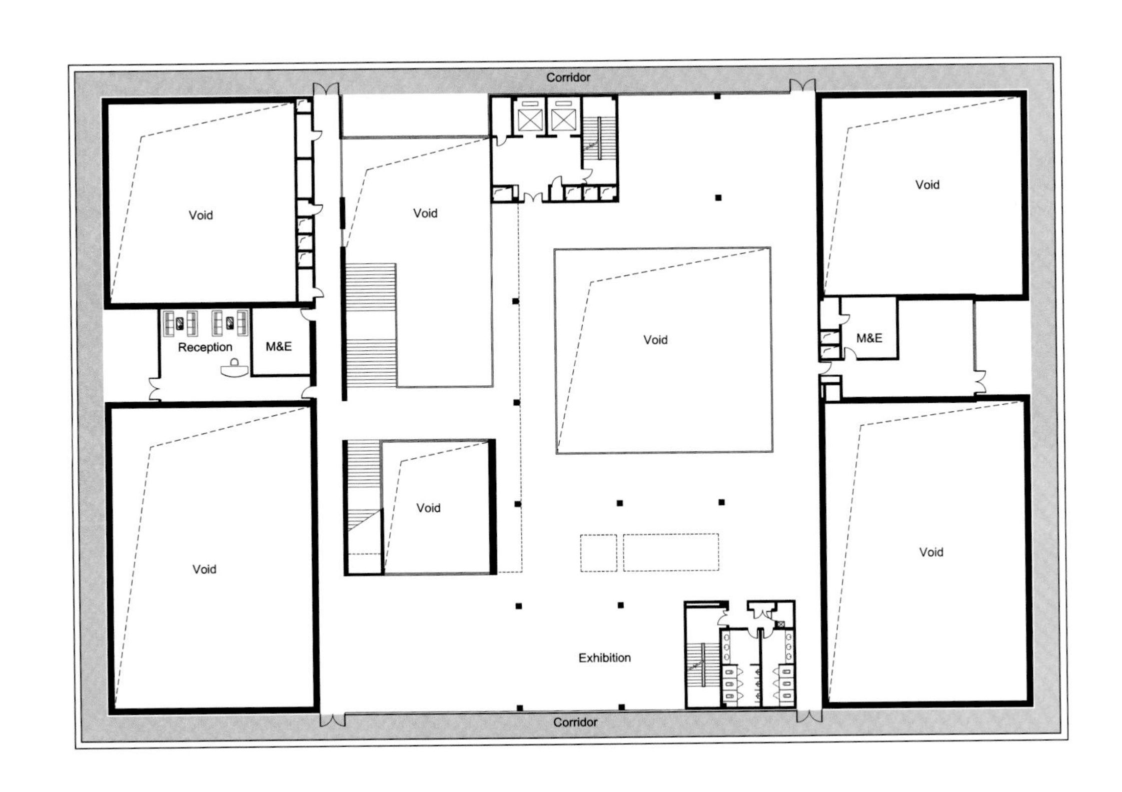

Second Floor

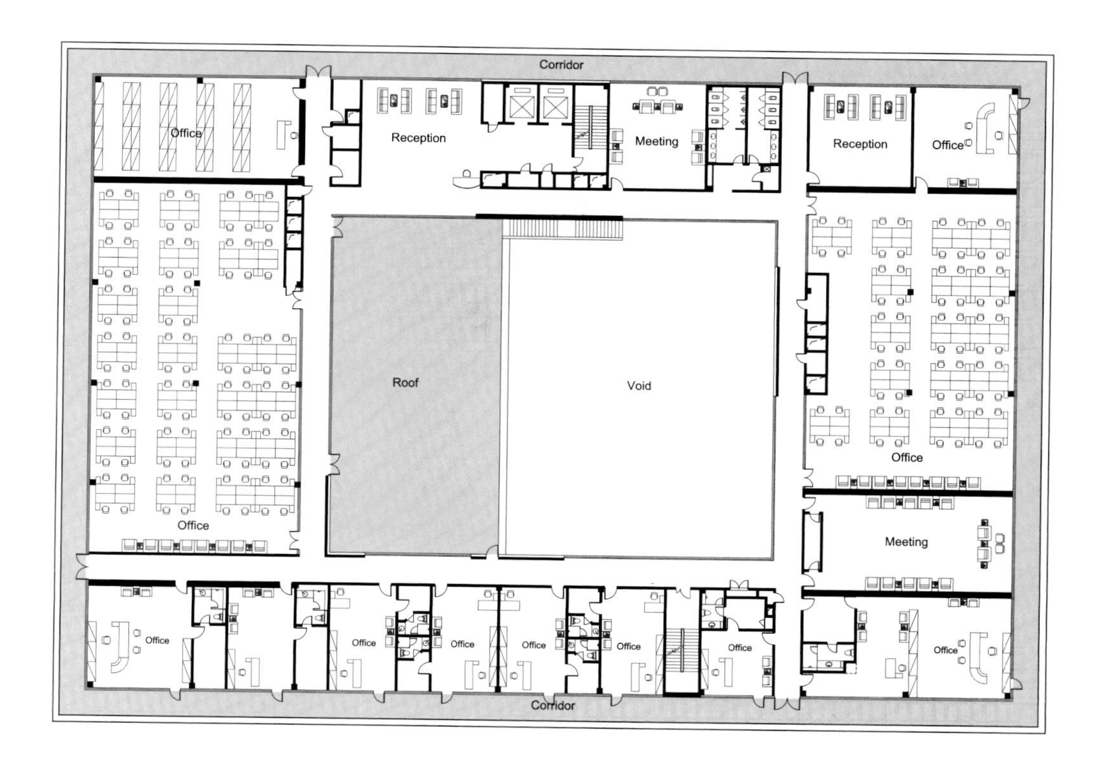

Third Floor

Fourth Floor

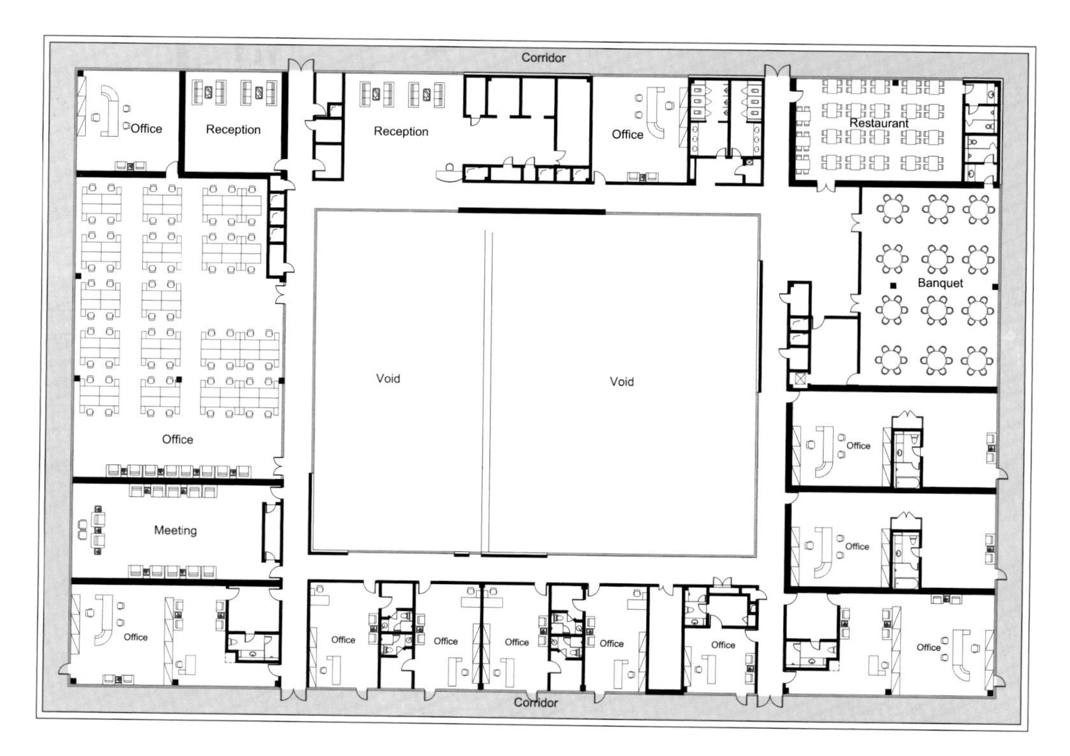

Fifth Floor

PLATINIUM 商业园区第 4 栋办公楼

PLATINIUM 4 OFFICE BUILDING, WARSAW

ARCHITECT
Roman Dziedziejko, Mikolaj Kadlubowski, Krzysztof Mycielski

Firm
Grupa 5 Architects

PHOTOGRAPHER
Marcin Czechowicz

LOCATION
Warsaw, Poland

AREA
22,600 m²

This 12-storey A-class office building is the second of four within the Platinium Bis Office Park in the post-industrial area of Mokotów SłuŻew, which has become a bustling white collar hub within the last 15 years. Grupa 5 Architects obtained a planning decision for 12 storeys, 5 storeys higher than the previous planning permission allowed.

This building has been fully rented. The four planned buildings designed by Grupa 5 Architects complete a complex of six buildings situated around a lush green courtyard with a water pool, wooden deck terraces looking over the water adjacent to a cafeteria for the employees of the complex.

The character of the architecture harmonises with the two existing stainless steel clad buildings with, while keeping the overall budget of the building envelope within a very economical standard.

AVIVA

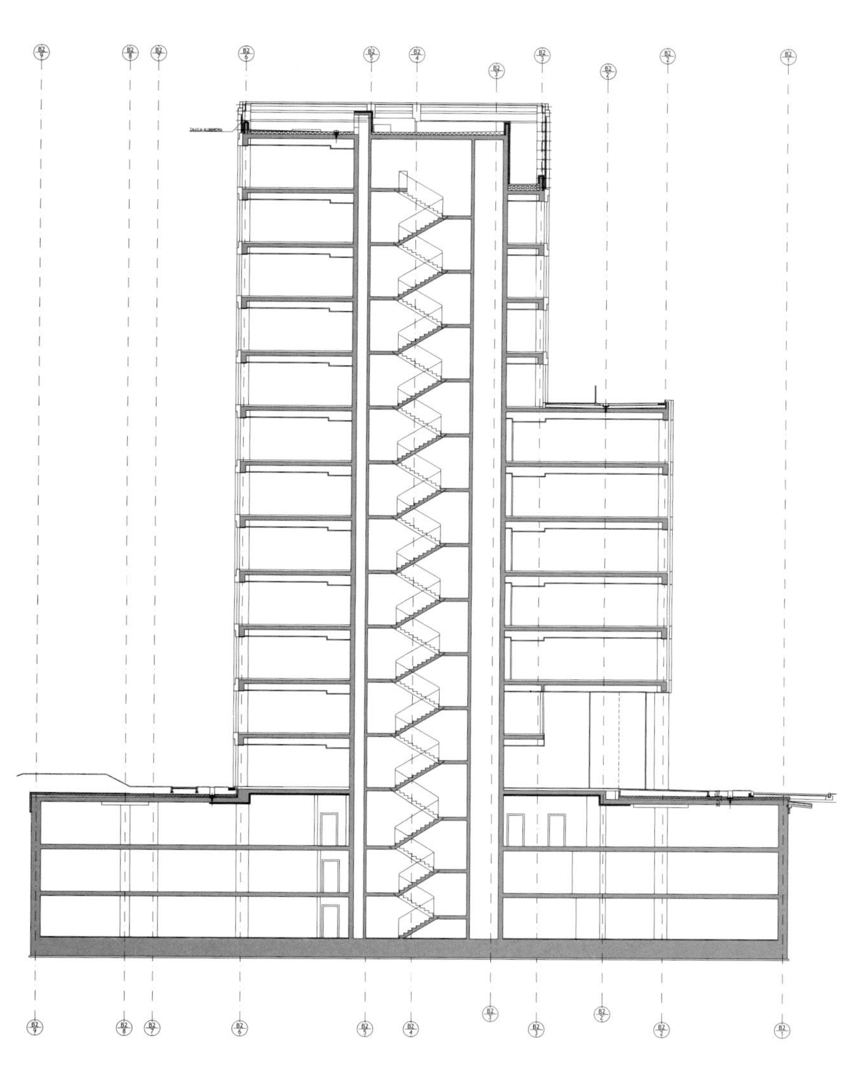

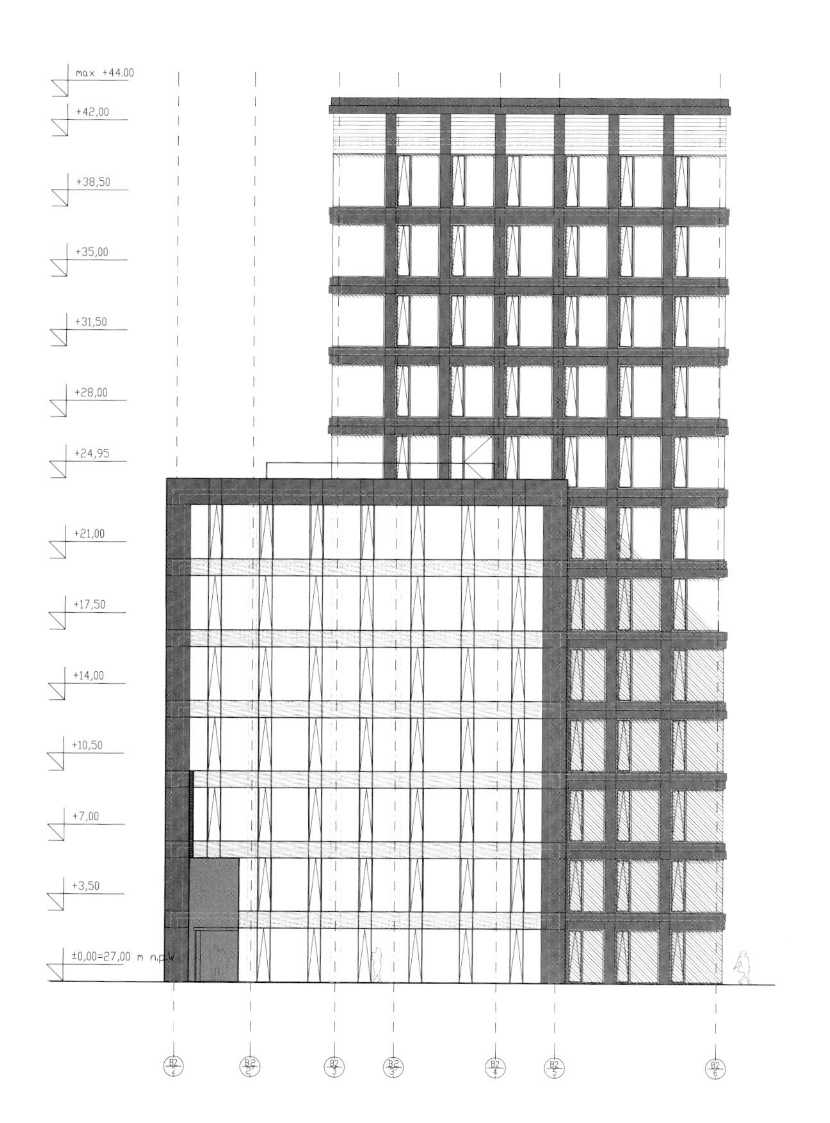
max +44.00
+42,00
+38,50
+35,00
+31,50
+28,00
+24,95
+21,00
+17,50
+14,00
+10,50
+7,00
+3,50
±0,00=27,00 m n.p.W

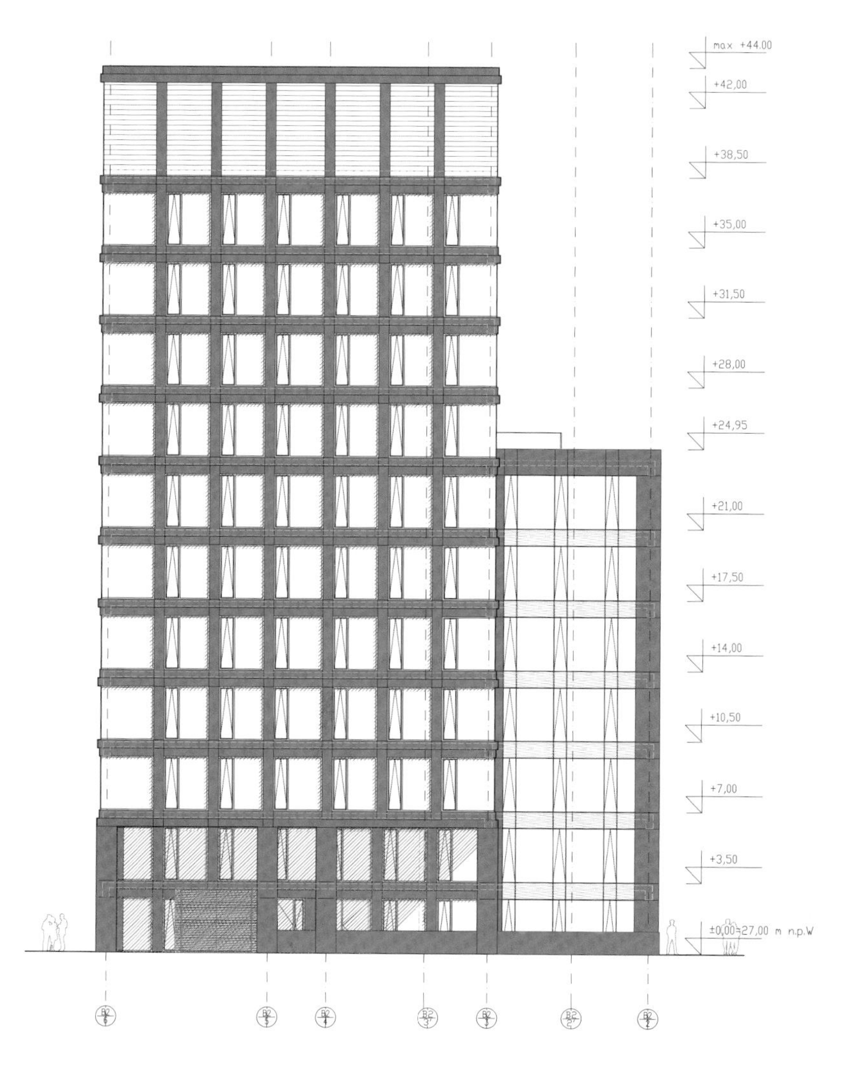
max +44.00
+42,00
+38,50
+35,00
+31,50
+28,00
+24,95
+21,00
+17,50
+14,00
+10,50
+7,00
+3,50
±0,00=27,00 m n.p.W

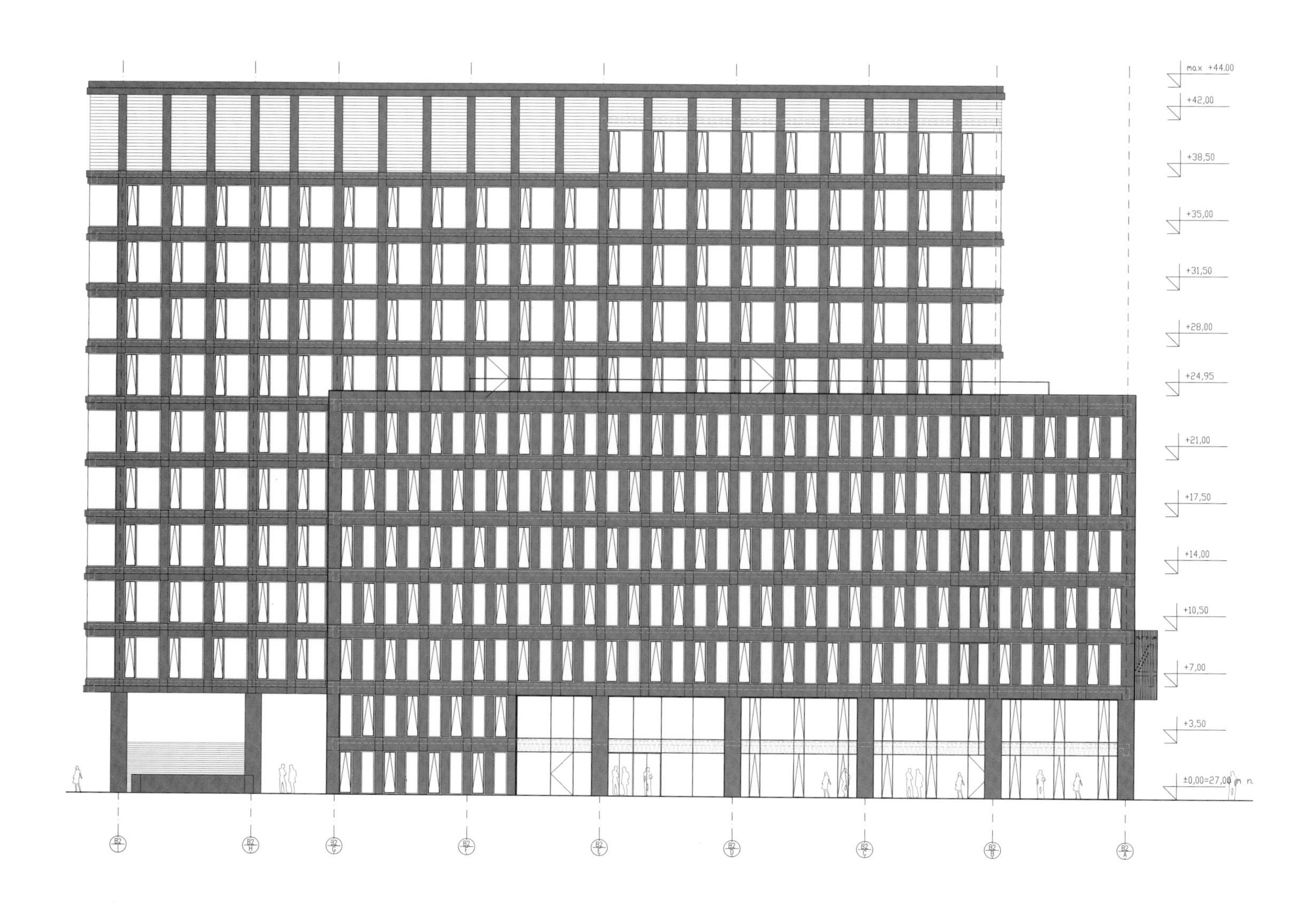
max +44,00
+42,00
+38,50
+35,00
+31,50
+28,00
+24,95
+21,00
+17,50
+14,00
+10,50
+7,00
+3,50
±0,00=27,00 m n.

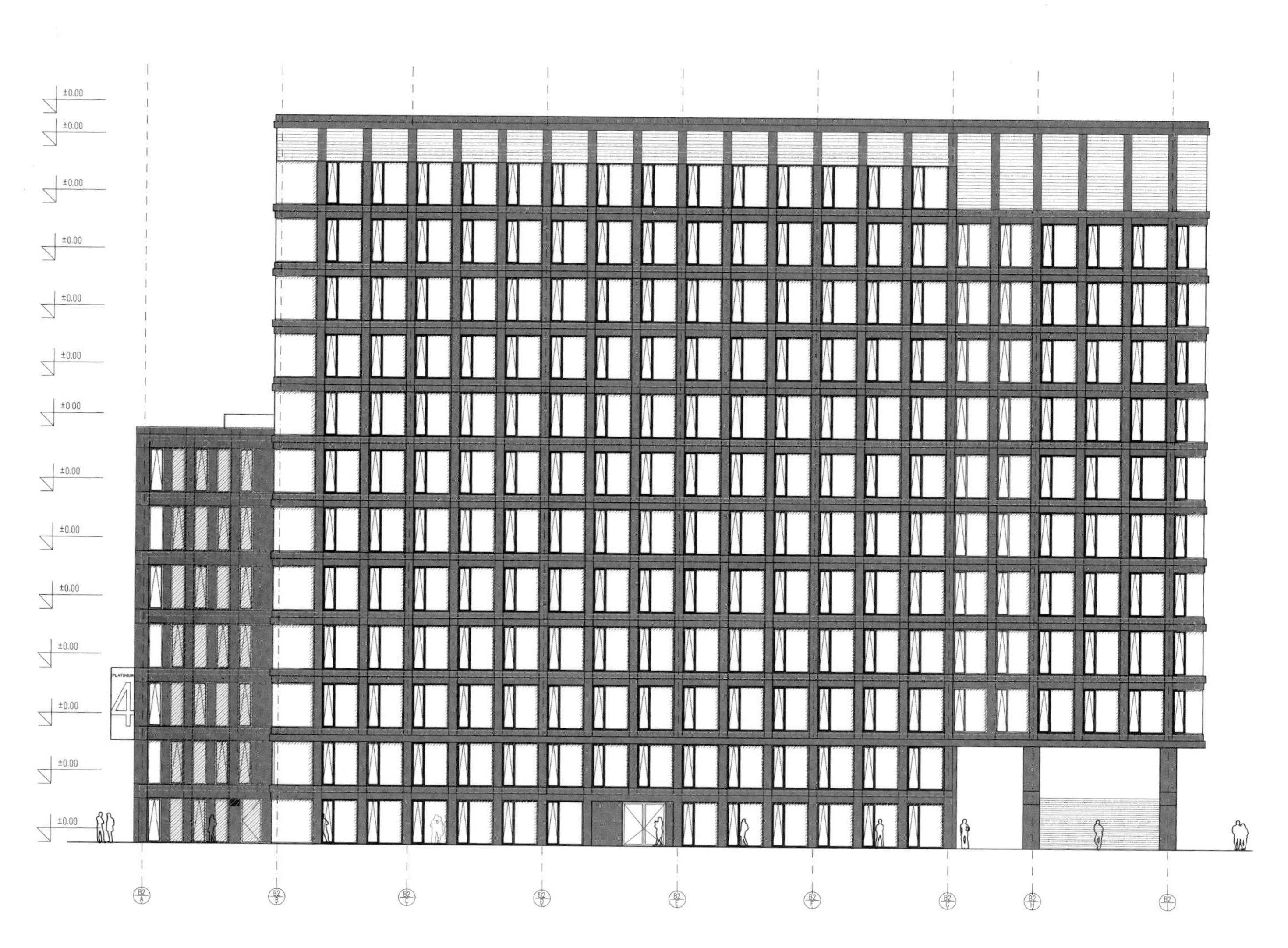
±0.00
±0.00
±0.00
±0.00
±0.00
±0.00
±0.00
±0.00
±0.00
±0.00
±0.00
±0.00
±0.00
±0.00

PLATINIUM 商业园区第4栋办公楼是一栋十二层高的顶级办公大楼，属于 Platinium Bis 办公园区中新建四栋大楼中的第二栋。这个园区是在早年华沙莫克托夫・苏泽夫工业区的基础上建造的，近十五年内已经逐渐成为白领的工作聚集地。格鲁帕公司获得了该项目的规划设计授权，计划建造一栋十二层高的建筑，比原计划多了五层。

这栋办公大楼完全对外出租。包括格鲁帕公司设计规划的四栋大楼在内，一共六栋大楼分布在一个植被茂盛的绿色中央庭院广场的周围，广场上还设有水系景观，旁边是木质台阶，临近还设有自助餐厅，为办公大楼中的人们提供服务。

经过设计师们的精心设计，这四栋大楼的建筑风格与原有的两栋不锈钢结构大厦相互协调，包括大楼四周的护栏在内，总体预算控制在一个十分经济的标准之内。

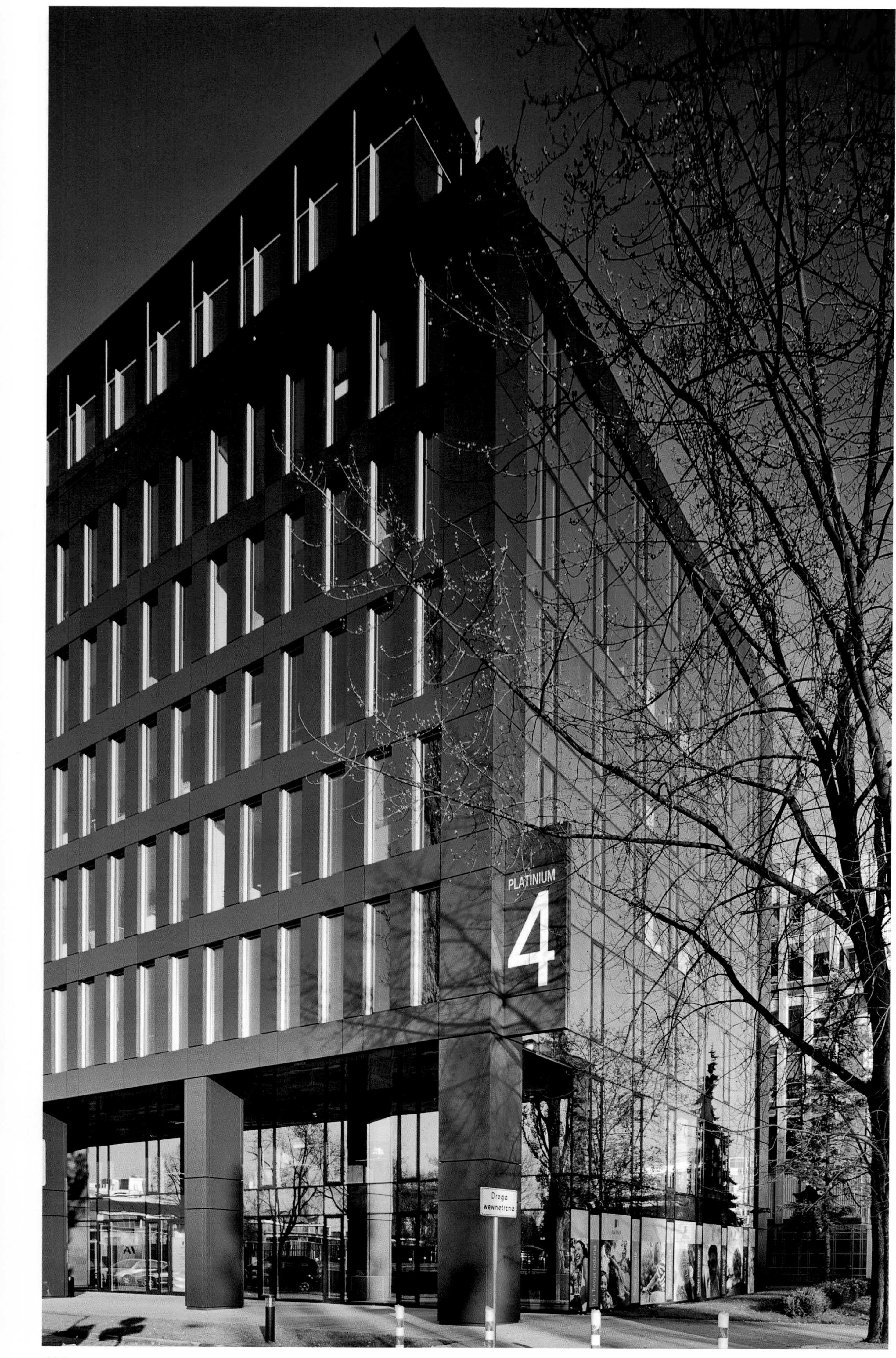
PLATINIUM
4
Droga
wewnętrzna

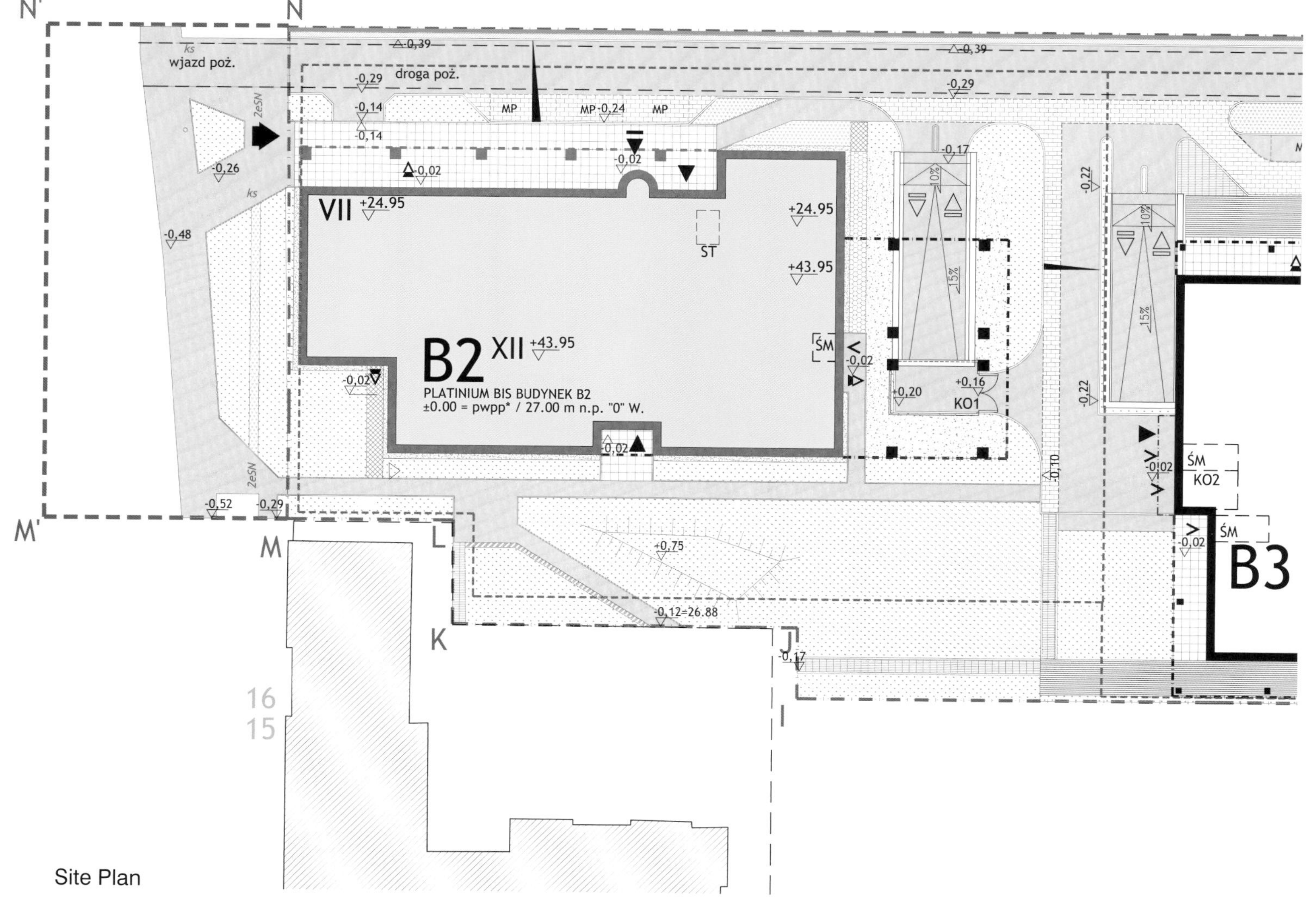
N'
N
wjazd poż.
droga poż.
B2
PLATINIUM BIS BUDYNEK B2
±0.00 = pwpp* / 27.00 m n.p. "0" W.
KO1
KO2
B3
M'
M
L
K
J
I
16
15

Site Plan

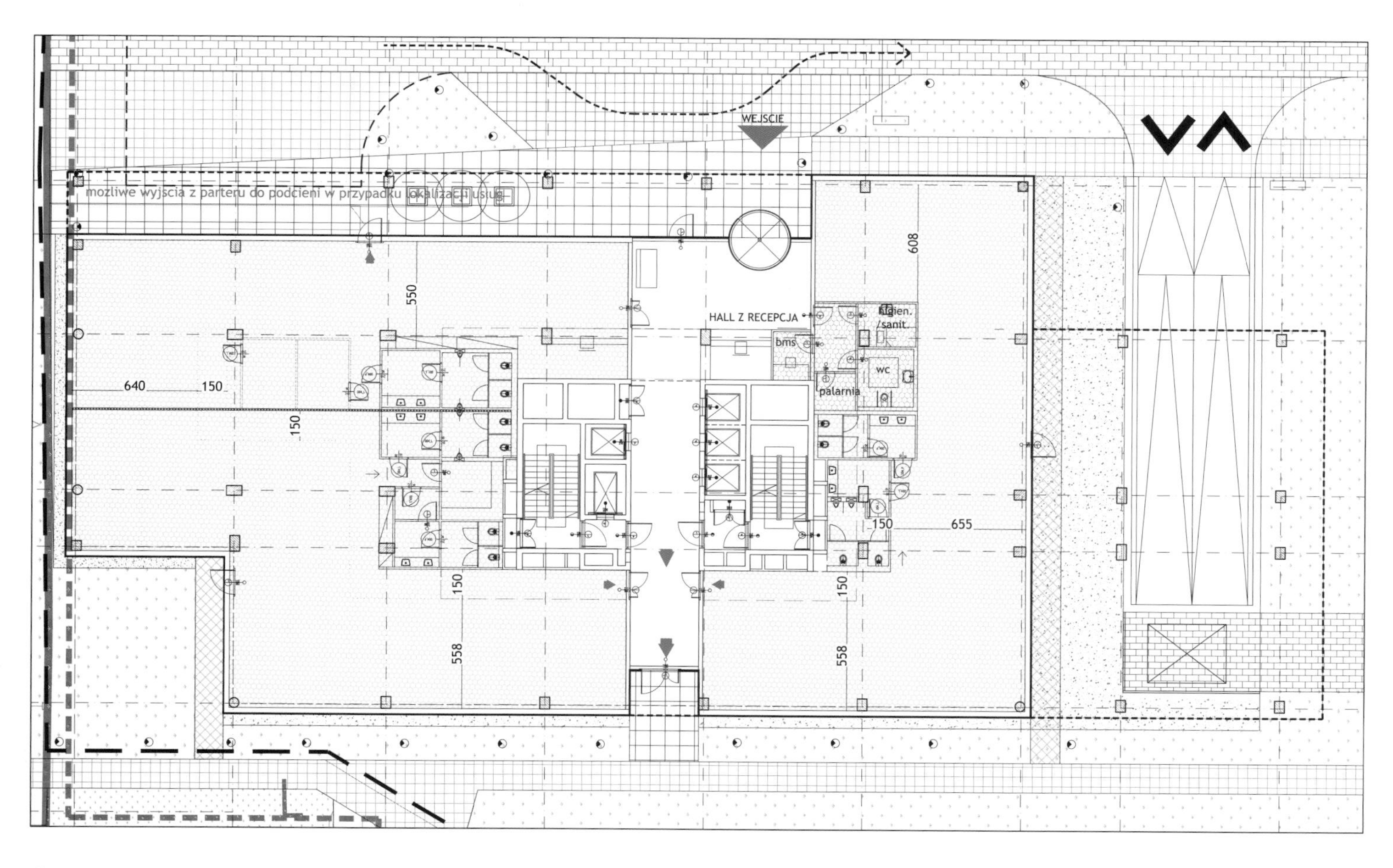

Ground Floor

AVIVA

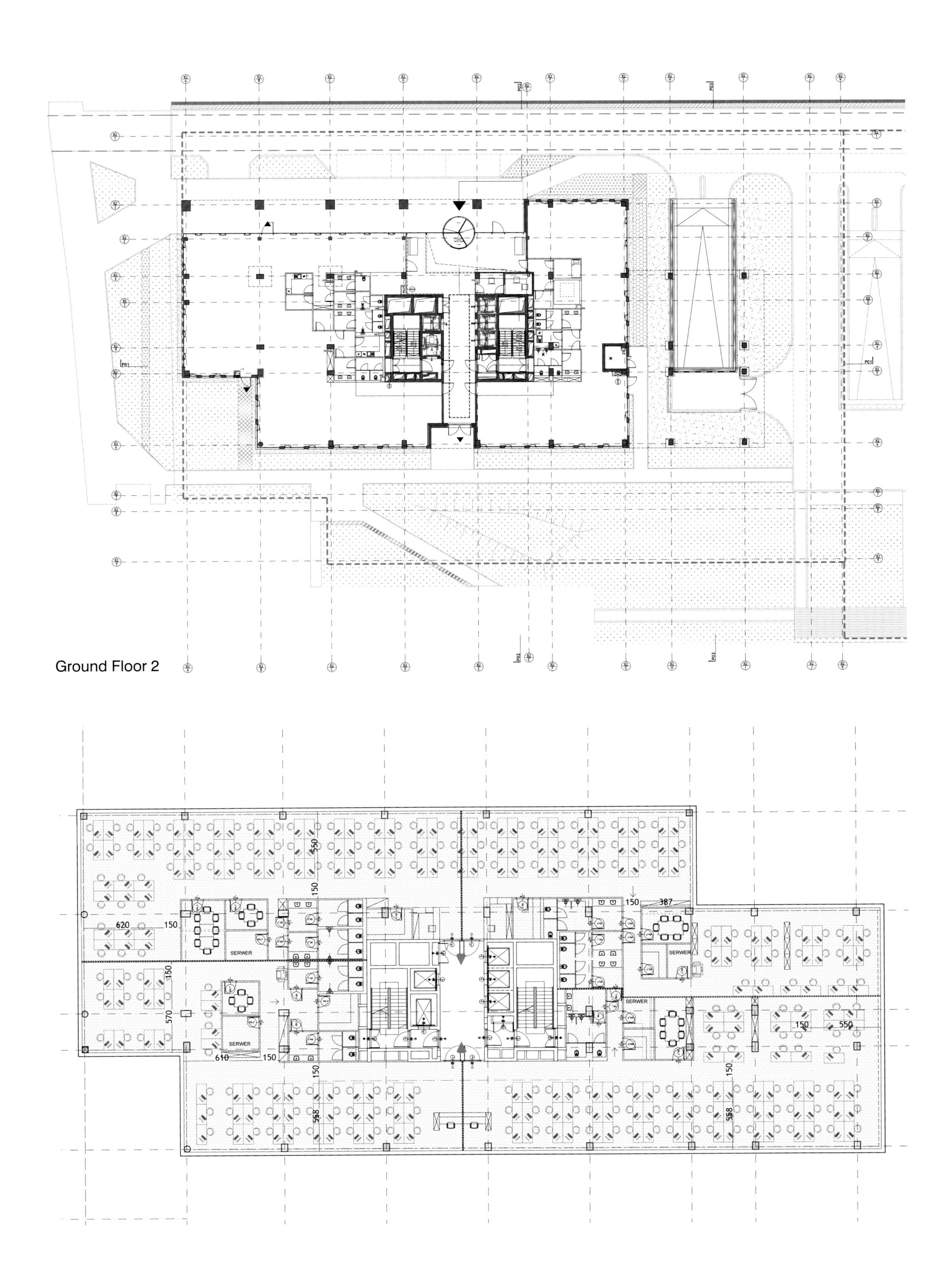

Ground Floor 2

Typical Floor

THE NVE BUILDING

ARCHITECT
Hege Thorvaldsen, Vivian Steinsåker, Simona Ferrari

FIRM
Dark Arkitekter AS.

INTERIOR DESIGN
Dark Arkitekter AS(Hege Thorvaldsen and Vivian Steinsåker), Zinc AS(Nina Sperling and Therese Jonassen), Andersen & Flåte AS(Anne Ruth Flåte)

CLIENT
Entra Eiendom AS

LOCATION
Oslo, Norway

AREA
21,000 m²

PHOTOGRAPHER
Nils Petter Dale

NVE

NVE

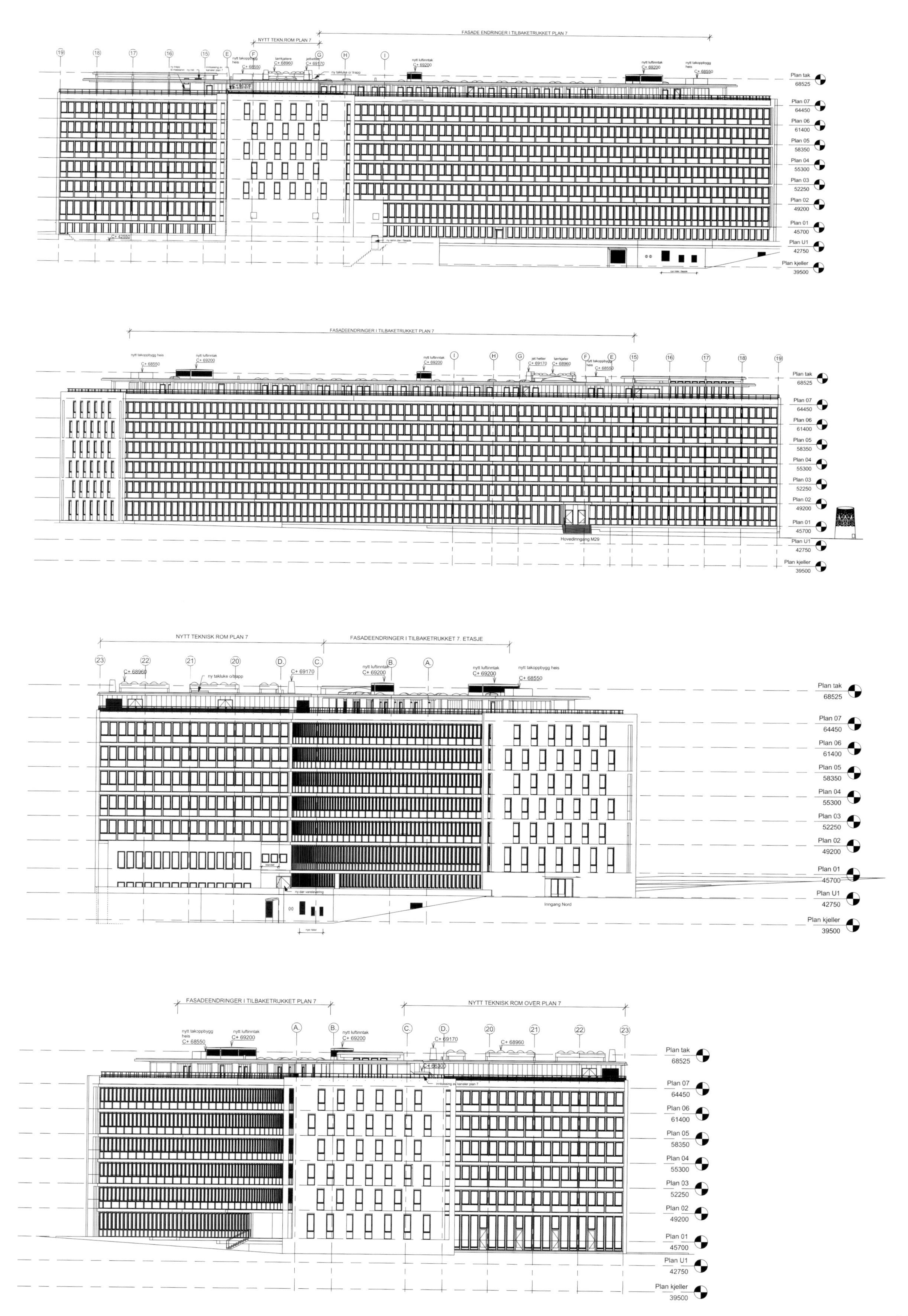

FASADE ENDRINGER I TILBAKETRUKKET PLAN 7
NYTT TEKN.ROM PLAN 7
Plan tak
68525
Plan 07
64450
Plan 06
61400
Plan 05
58350
Plan 04
55300
Plan 03
52250
Plan 02
49200
Plan 01
45700
Plan U1
42750
Plan kjeller
39500
FASADEENDRINGER I TILBAKETRUKKET PLAN 7
Hovedinngang M29
NYTT TEKNISK ROM PLAN 7
FASADEENDRINGER I TILBAKETRUKKET 7. ETASJE
nytt luftinntak
C+ 69200
nytt takoppbygg heis
C+ 68550
C+ 68960
C+ 69170
ny takluke o/trapp
Inngang Nord
FASADEENDRINGER I TILBAKETRUKKET PLAN 7
NYTT TEKNISK ROM OVER PLAN 7

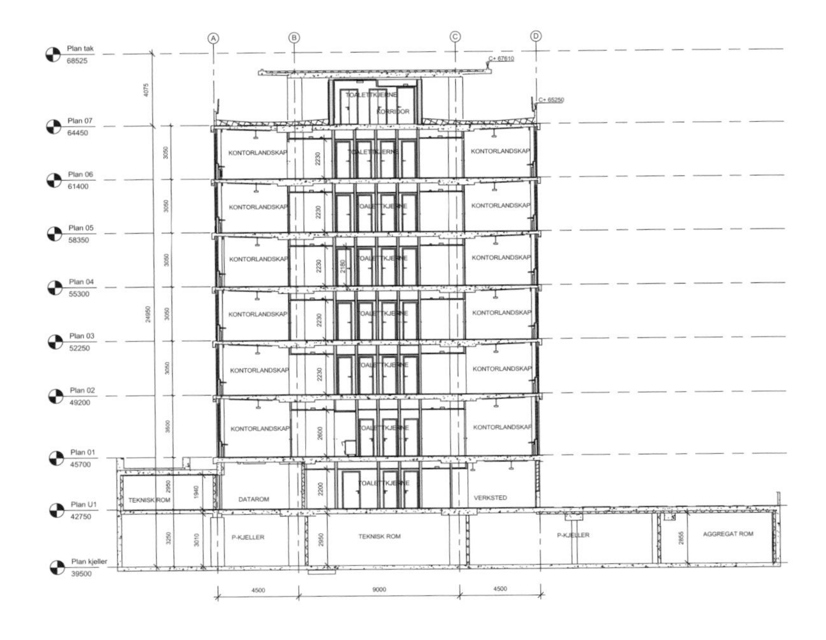

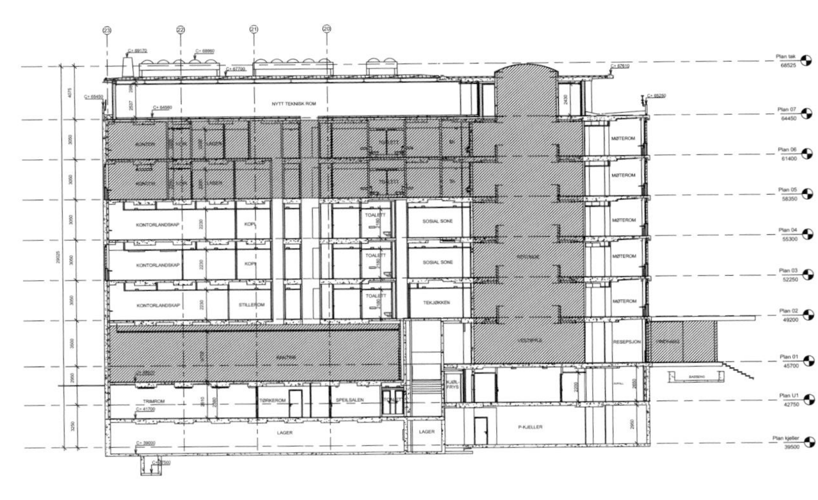

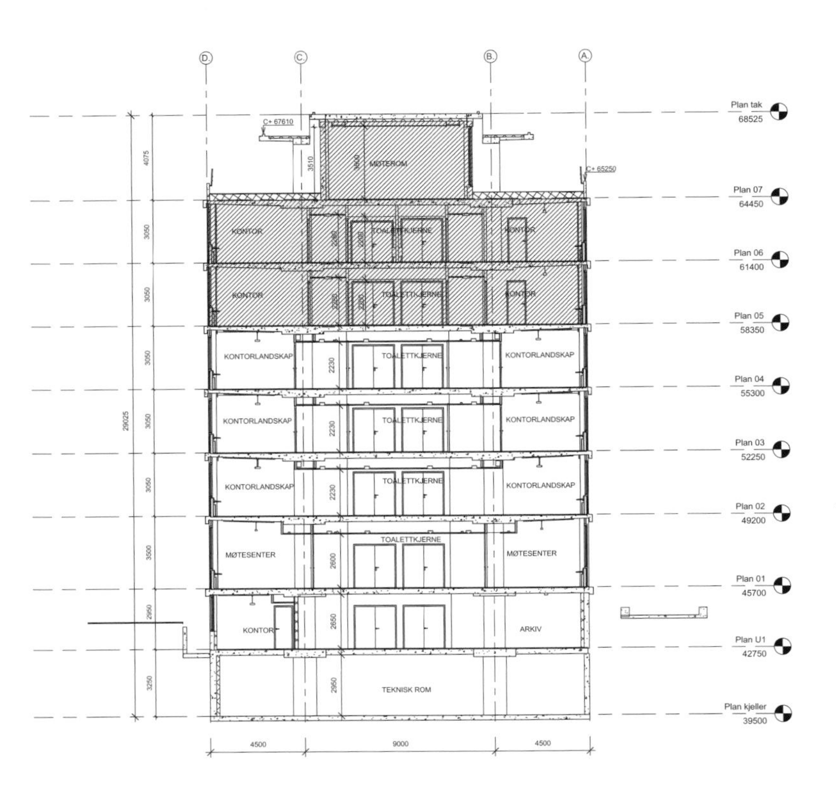

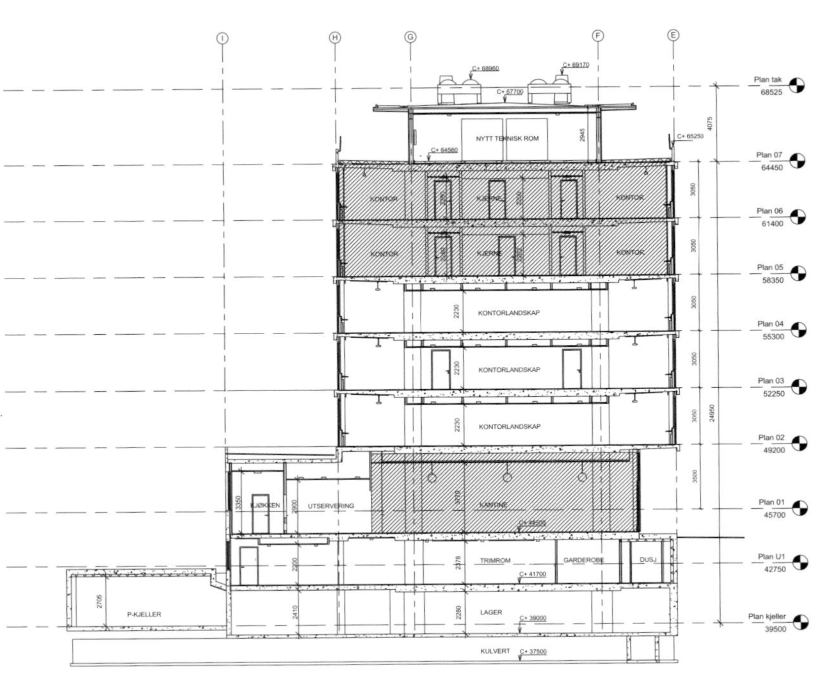

Middelthunsgate 29 was built 1962-1964 as the Administration Building for the Norwegian Water Resources and Energy Directorate, designed by the brothers Fr. Lykke-Enger and Knut Enger, with artistic decoration by Odd Tandberg and Nic Schiøll. The building is elongated and curved with a transverse rear wing in modernistic style. The curved shape gives the building a monumental character, and it stands out among similar buildings from the same period. The building has a high level of architectural quality, represented with exclusive details and solid materials.

Parts of the building are currently under conservation. That includes the cafeteria, the main entrance, the central staircase and two office wings, in which preserve "the little bureaucrat's office". The facade, both exterior and interior, is also preserved. A close collaboration with Norwegian building preservation authorities has strengthened the rehabilitation process. Maintenance of the original qualities of the building has been emphasized, together with the ambition of creating a functional workplace for NVE by creating flexible work spaces customized for present and future requirements and needs.

The renovated facilities appear to be a pioneer building when it comes to environmental design and energy-use, and there has been a unique emphasis on environmental- and resource-friendly solutions. The added energy of the building will not exceed 120 kWh/m^2 per year (ie Class B). Reaching class B, considering the high degree of conservation, is highly extraordinary. The rehabilitation is carried out with a full attention on universal design.

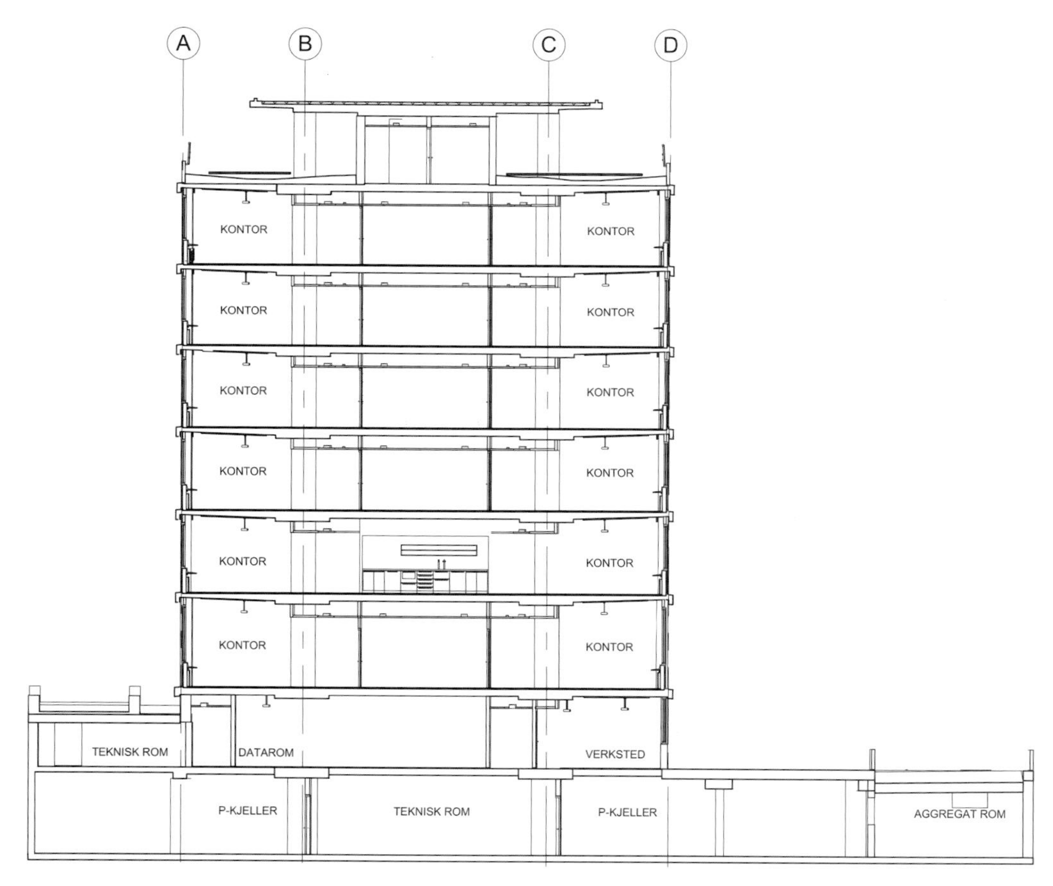

Middelthunsgate 29号建于1962—1964年间，是挪威水资源和能源理事会的行政大楼所在地，由Fr. Lykke-Enger和Knut Enger兄弟设计，艺术装饰由Odd Tandberg和Nic Schiøll共同完成。整座建筑呈横尾翼状，显得狭长又带有弧度，具有浓郁的现代主义风格。曲面造型是大楼的一个显著特点，独特的造型使大楼从同一时期的类似建筑物中脱颖而出。这座大楼建筑质量上乘，细部设计独特，使用材料坚固，不愧是挪威建筑中的典范之作。

目前，大楼的部分设施已被保留下来。这包括餐厅、主入口、中央楼梯及两个办公室翼部，其中“行政管理人员”办公室也在保留之列。大楼的立面，无论是外立面还是内立面，也均被保留。与挪威建筑保护当局的密切合作加快了改造的进程。原建筑的优质品质也得到了高度重视，同时还要为NVE创建一个功能性的工作场所，而这个灵活的工作空间必须按照现在和未来的需求与需要量身打造。

翻修后的大楼在环境设计和能源利用方面将会处于领先地位，因为改造方案无处不传达着环境友好型、资源节约型的新理念。大楼新增的能源将不会超过每年每平方米120千瓦时（即B级）。由于大楼的大部分设施被保留，能耗达到B级已实属不易。改造工程在实施过程中将重点关注通用设计。

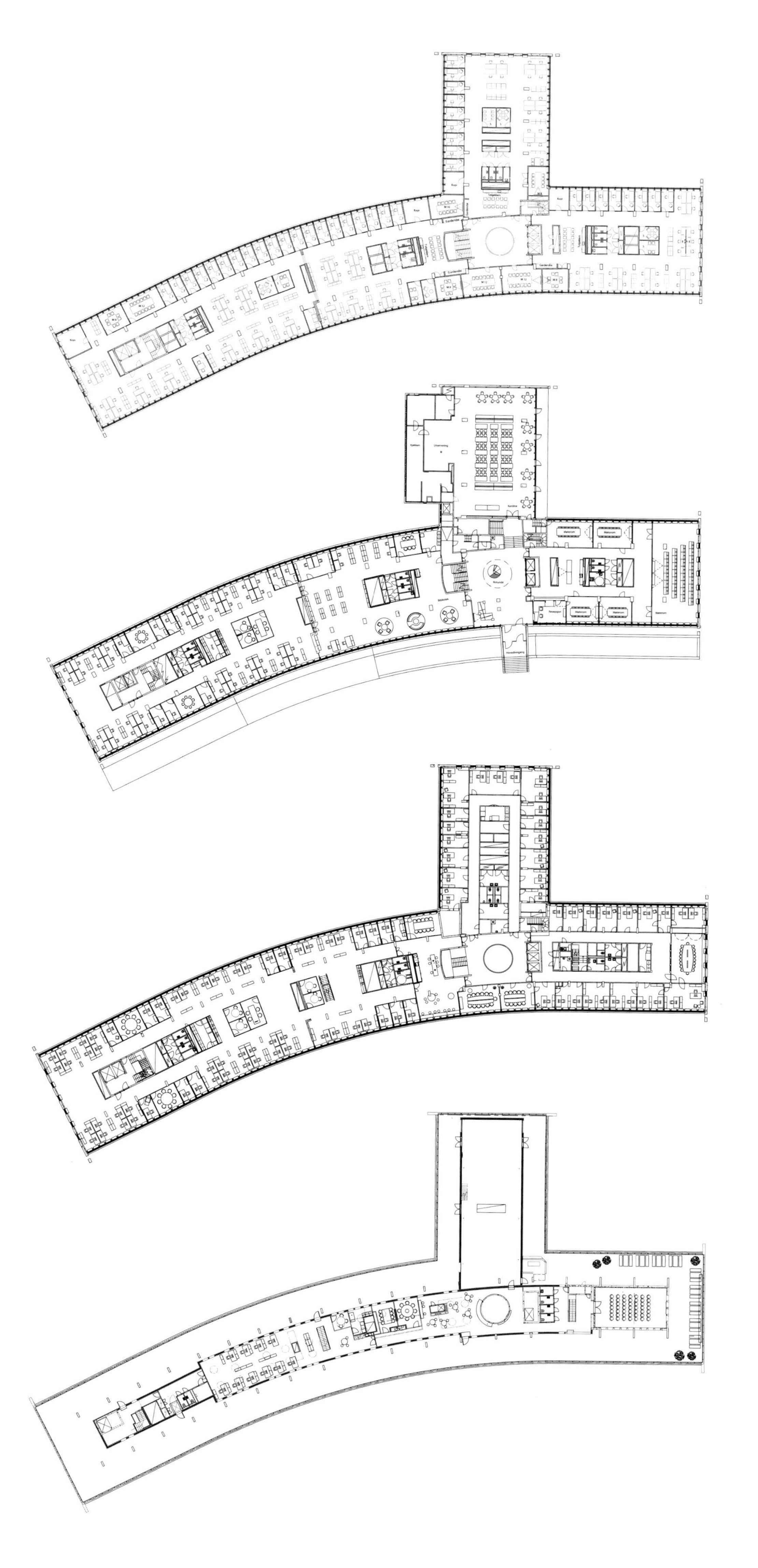

阿德莱迪湾大厦

BAY ADELAIDE CENTER

ARCHITECT
WZMH Architects

OWNER
Brookfield Properties Ltd.

GENERAL CONTRACTOR
EllisDon

LANDSCAPE ARCHITECT
Dillon Consulting

CIVIL ENGINEER
Dillon Consulting

ELECTRICAL ENGINEER
Genivar

MECHANICAL ENGINEER
The Mitchell Partnership Inc.

STRUCTURAL ENGINEER
Halcrow Yolles

COMMISSIONING AGENT
The Mitchell Partnership Inc.

LEED CONSULTANT
Enermodal Engineering Ltd. & Mulvey + Banani International Inc.Building Envelope Consultant: BVDA Facade Engineering Ltd.

LOCATION
Toronto, Ontario, Canada

PHOTOGRAPHER
Tom Arban

The Bay Adelaide Center is a signature 51-storey tower in downtown Toronto distinguished by its elemental, modernist form – a refined rectangular plan with notched corners – and a prism-like skin of clear and fritted glass that make it one of the downtown core's most transparent towers. At the top of the tower, the extension of the glass skin beyond the rooftop gives the building profile a distinctive identity.

The highly transparent tower base seamlessly incorporates the historic facade of the National Building on Bay Street (designed by Chapman and Oxley, 1926) and the lobby features a major integrated public art project by the world-renowned artist James Turrell. The project also has a half-acre outdoor urban plaza landscaped with ginkgo trees and ornamental grasses that frame benches and an open seating area, contributing a much-needed public open space to the central business district.

Certified to an LEED Gold standard, the project is among Canada's largest sustainable buildings. The tower contains more than 100,840m² of rentable Class-AAA office space, as well as over 3,700m² of below-grade retail space linked to the extensive underground PATH network.

The project was awarded the 2012 Ontario Association of Architects Design Excellence Award.

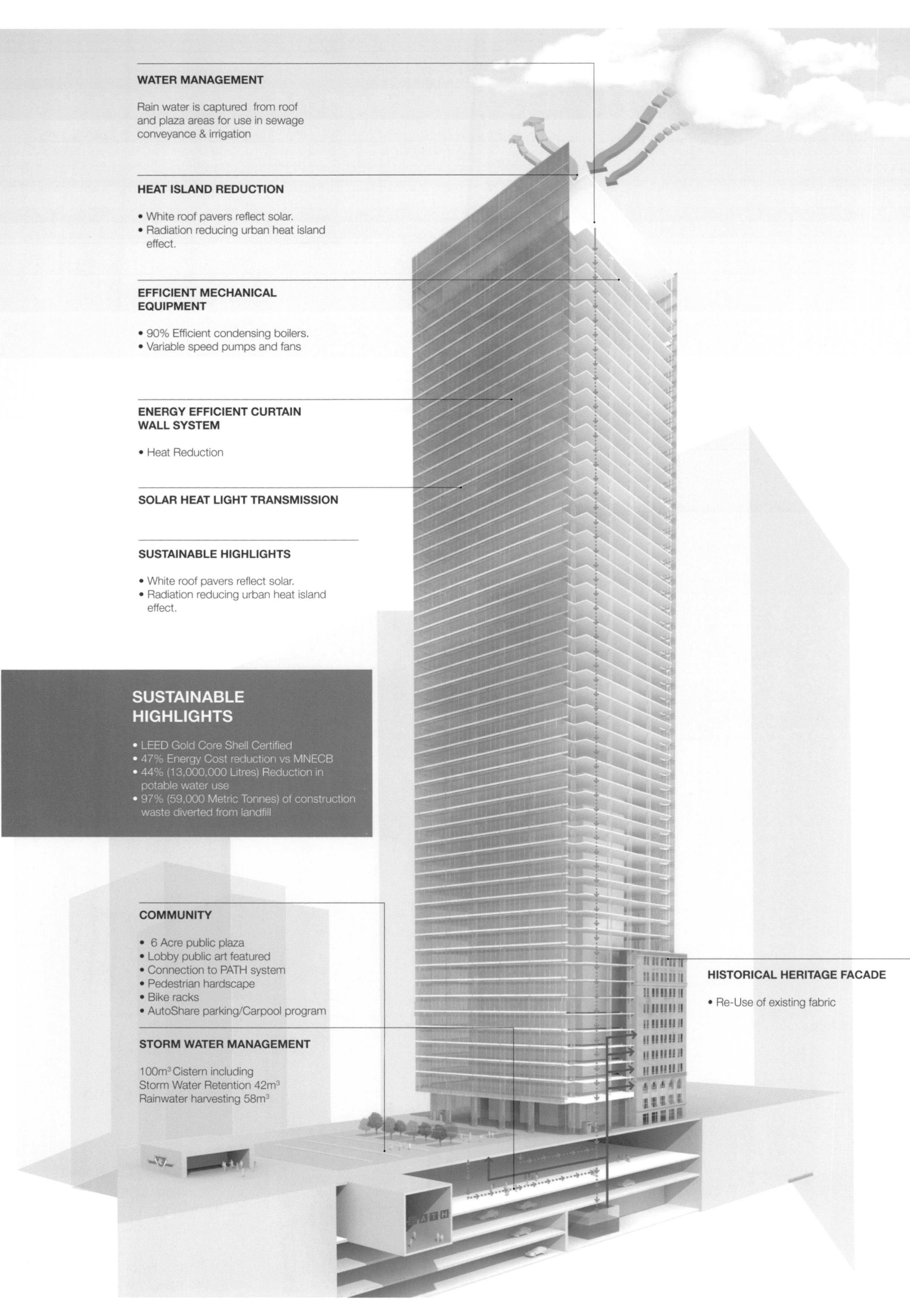
WATER MANAGEMENT
Rain water is captured from roof and plaza areas for use in sewage conveyance & irrigation
HEAT ISLAND REDUCTION
• White roof pavers reflect solar.
• Radiation reducing urban heat island effect.
EFFICIENT MECHANICAL EQUIPMENT
• 90% Efficient condensing boilers.
• Variable speed pumps and fans
ENERGY EFFICIENT CURTAIN WALL SYSTEM
• Heat Reduction
SOLAR HEAT LIGHT TRANSMISSION
SUSTAINABLE HIGHLIGHTS
• White roof pavers reflect solar.
• Radiation reducing urban heat island effect.
SUSTAINABLE HIGHLIGHTS
• LEED Gold Core Shell Certified
• 47% Energy Cost reduction vs MNECB
• 44% (13,000,000 Litres) Reduction in potable water use
• 97% (59,000 Metric Tonnes) of construction waste diverted from landfill
COMMUNITY
• 6 Acre public plaza
• Lobby public art featured
• Connection to PATH system
• Pedestrian hardscape
• Bike racks
• AutoShare parking/Carpool program
STORM WATER MANAGEMENT
100m³ Cistern including
Storm Water Retention 42m³
Rainwater harvesting 58m³
HISTORICAL HERITAGE FACADE
• Re-Use of existing fabric
PATH

HEAT REDUCTION

1. White roof pavers reflect solid radiation, reducing urban heat island effect

2. Efficient mechanical equipment includes 90%+ efficient condensing boilers, variable speed pumps & fans

LIGHT/AIR/HEAT

1. CO_2 controlled ventilation system ensures optimum balance of air quality & energy savings

2. High efficiency light fixtures controlled via occupancy sensors

3. Visible light transmittance (TVIS)-0.58

4. Energy efficient curtain wall system clear, around. Solar heat gain co-efficient (SHGC) 0.31

   Total Thermal conductivity - R 3.7 (1.55w/m^2)

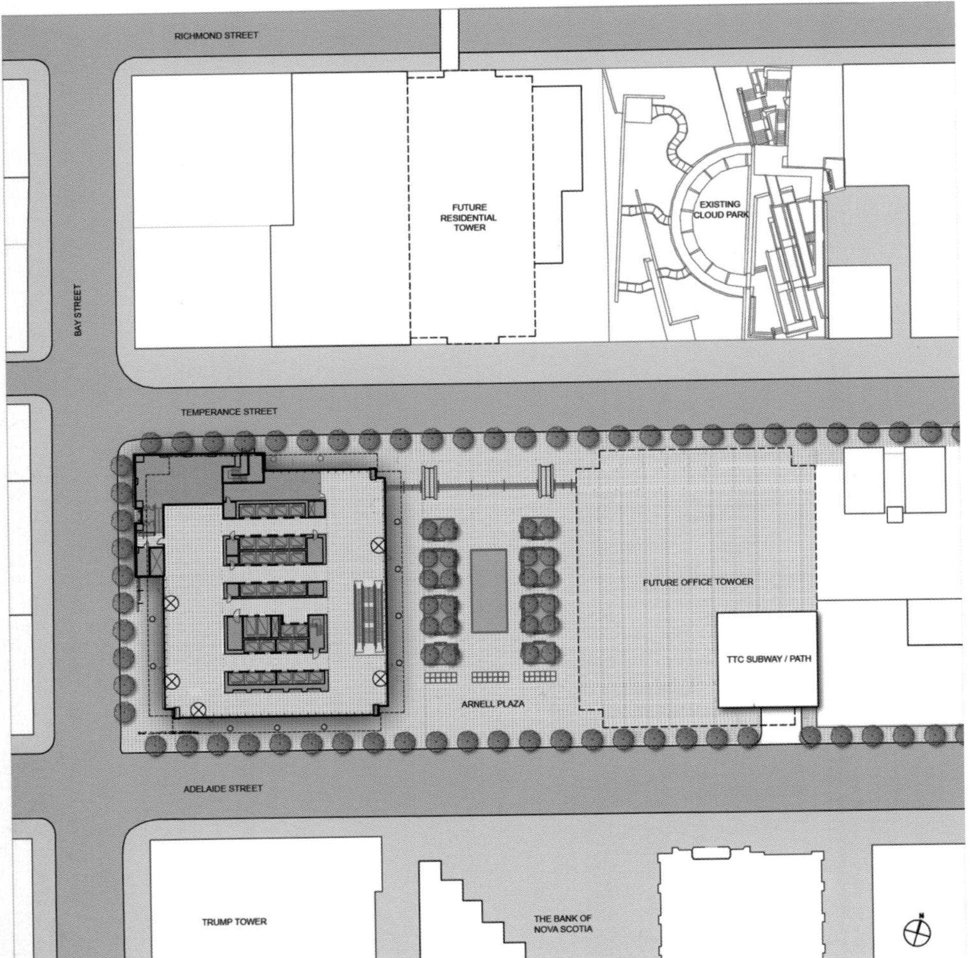

RICHMOND STREET
BAY STREET
FUTURE RESIDENTIAL TOWER
EXISTING CLOUD PARK
TEMPERANCE STREET
FUTURE OFFICE TOWOER
TTC SUBWAY / PATH
ARNELL PLAZA
ADELAIDE STREET
TRUMP TOWER
THE BANK OF NOVA SCOTIA

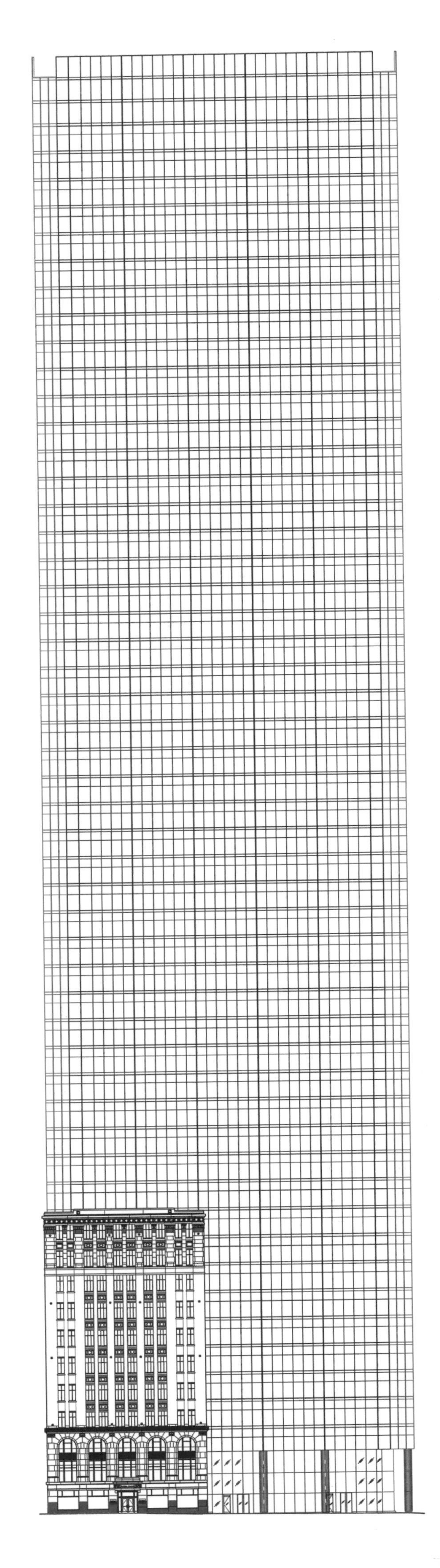

BMO
TRUMP

BAY ADELAIDE CENTRE
333

BAY ADELAIDE CENTRE
BAY ADELAIDE CENTRE
NO ACCESS PERMITTED
TENANT TRADE ENTRANCE ON P1 LEVEL

BAY ADELAIDE CENTRE

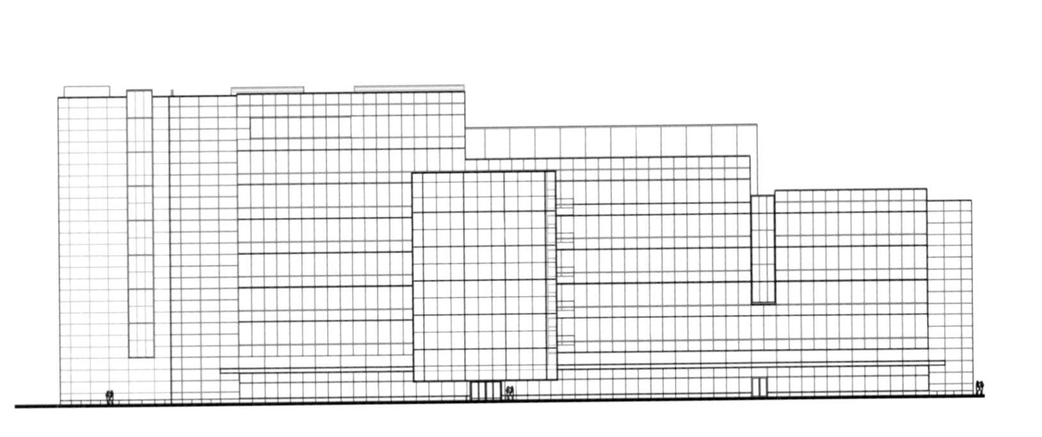

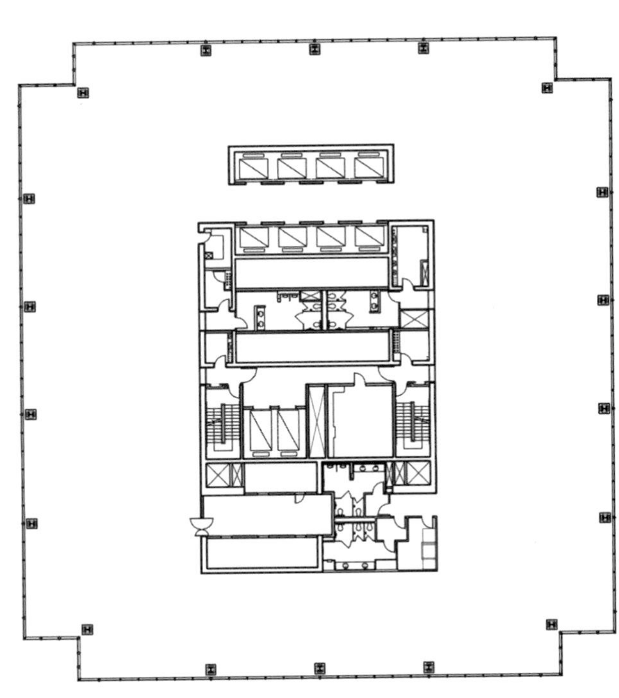

阿德莱迪湾大厦是一座具有鲜明特色的 51 层大厦，位于多伦多市中心，以其简单的现代主义造型——一个带凹口的精致的长方体——和一个棱柱状清晰的烧结玻璃立面，成为市中心最透明的大厦之一。在大厦顶端，玻璃立面延伸至天台，使大厦的侧面轮廓变得别具一格。

高度透明的大厦底座与海湾大街国立大厦著名的立面（由查普曼和奥克斯利于 1926 年设计）无缝化地连接在一起，而大堂则展示了由世界著名艺术家詹姆斯・特瑞尔设计的众多综合性的公共艺术品。本项目也包含一块 2025 平方米的室外城市广场，广场上将种植银杏树和观赏性草地，环绕着长凳和户外座位区，给中央商务区提供一个急需的公共露天场所。

此项目获得了 LEED 金级认证，并位于加拿大最大的可持续建筑之列。这座大厦有 100 840 多平方米的出租类 3A 级办公空间和 3700 多平方米的地下零售空间，并连接着市区众多的 PATH 线路。

这个项目被授予 2012 年安大略建筑师协会设计优秀奖。

DANIEL SWAROVSKI CORPORATION ZURICH

ARCHITECT AND GENERAL PLANNING
Ingenhoven architects international gmbh & co. kg

LANDSCAPE ARCHITECT
Ingenhoven architects

PROJECT TEAM
Christoph Ingenhoven, Thomas Höxtermann, Andrè Barton, Ingo Faulstich, Marion Heitplatz, Stefan Henfler, Peter Pistorius, Jörg Püschel, Alexander Thieme, Marc Oliver Wehner, Lutz Büsing

CLIENT
Swarovski Immobilien AG, Männedorf

AREA
19,100 m²

LOCATION
Zurich-Männedorf, Switzerland

PHOTOGRAPHER
R. Giesecke, H.G. Esch, A. Keller

This transparent office building for 500 employees of the Swarovski Corporation is situated on the eastern side of Lake Zurich, 19 kilometers south of the city of Zurich, in the village of Männedorf. The main focus of the design is the lake view, through the transparency of the facade as well as the arrangement of the work places.

The ground floor contains a lobby with a lounge, a restaurant, conference rooms and workshops. The upper floors contain mostly open office space for maximum flexibility. There are 170 parking spaces at the basement level.

The design and the sustainability concept follow the Swiss Minergie eco standard. Water from the lake is used both for heating and cooling.

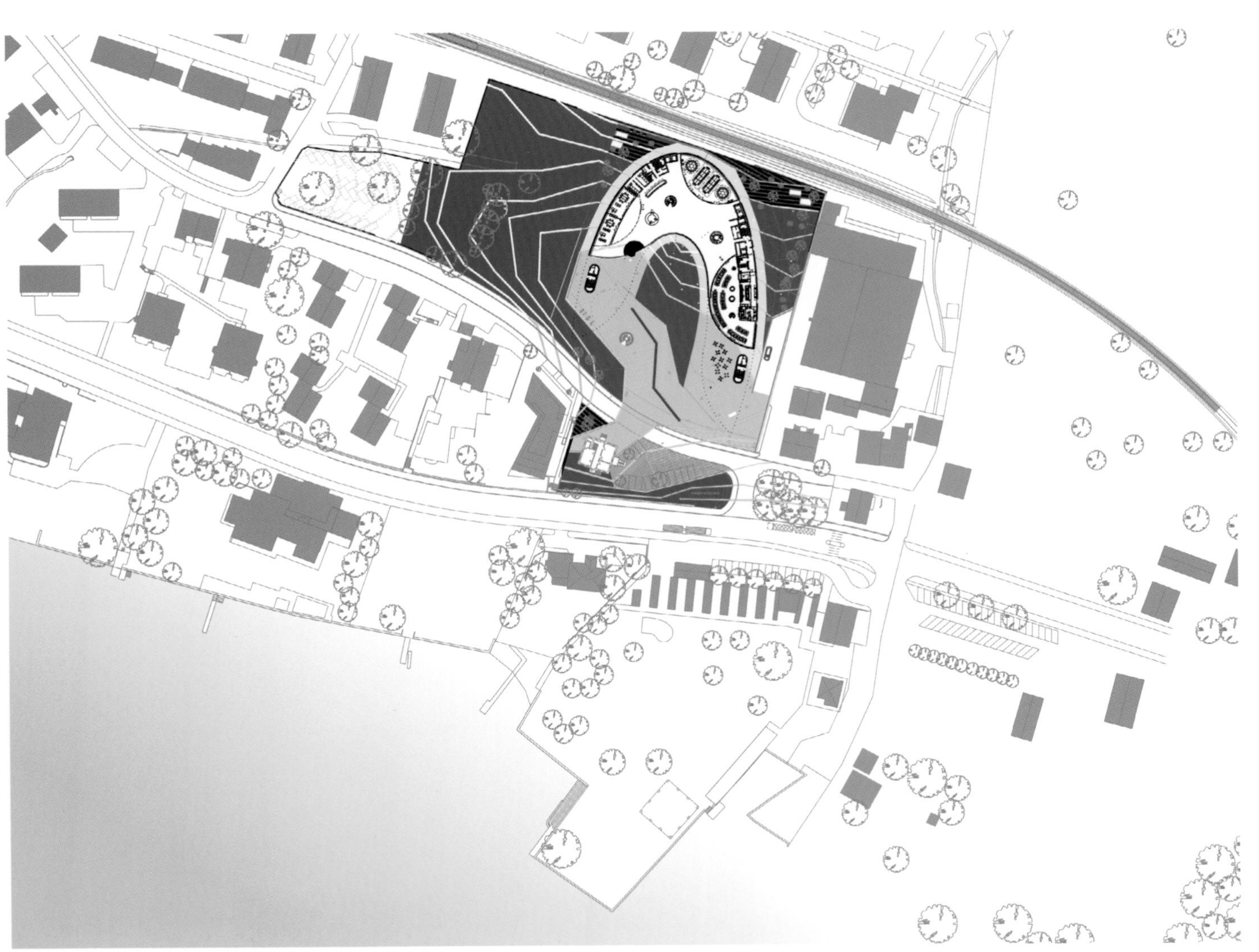

Site Plan

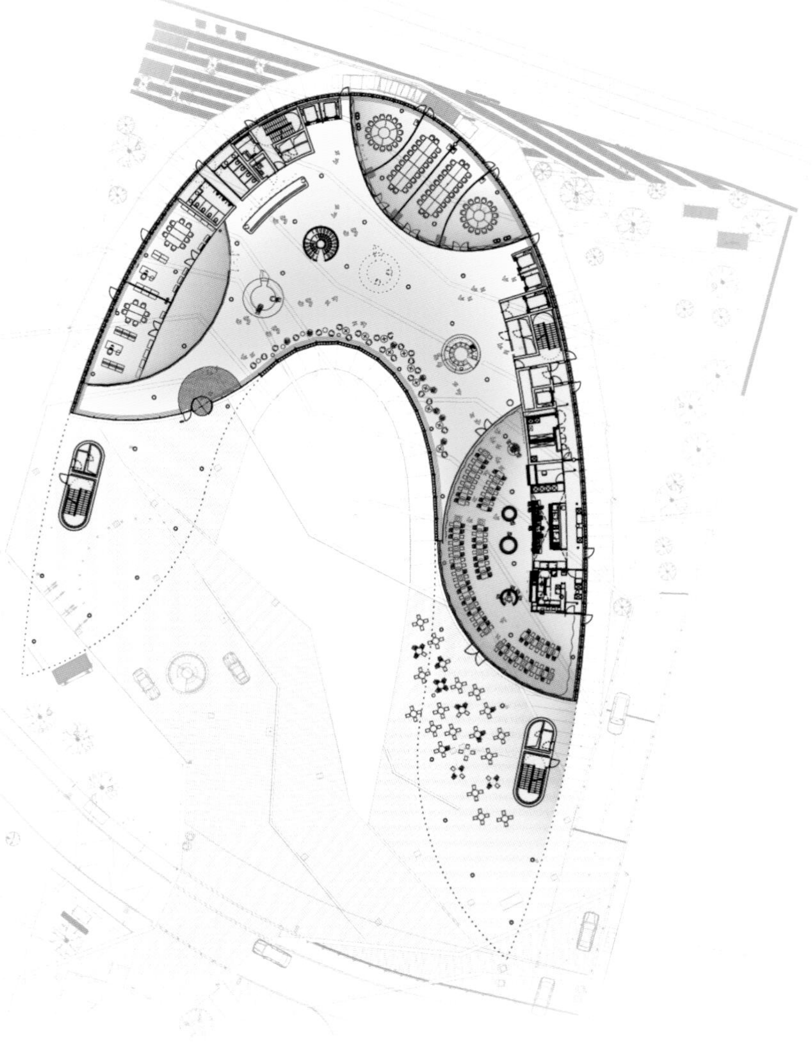

Ground Floor Plan

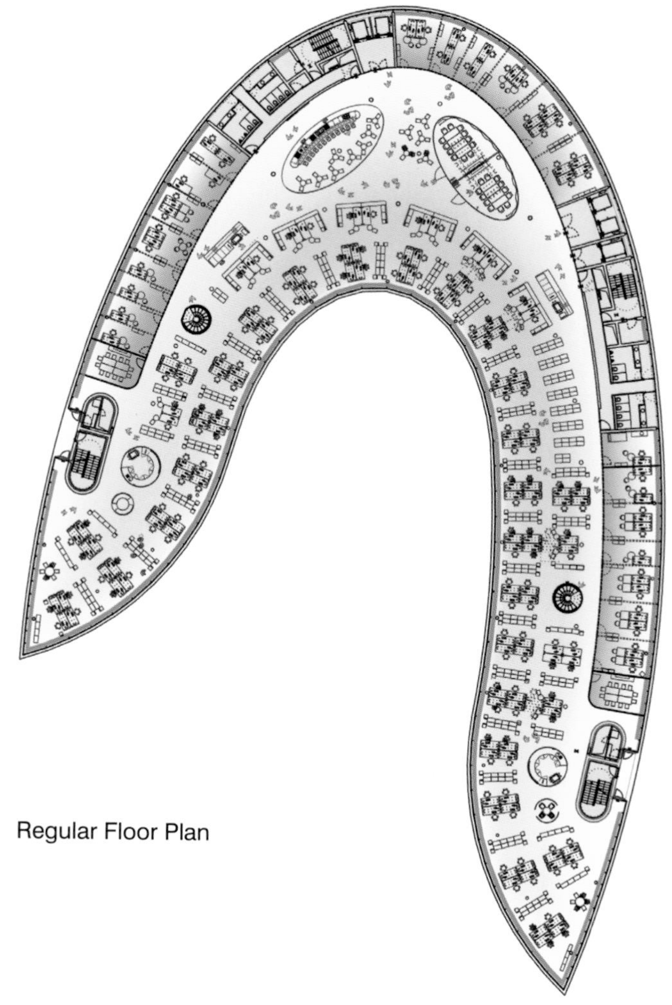

Regular Floor Plan

施华洛世奇公司大楼是一座透明的办公大楼，能容纳500名雇员，该大楼位于苏黎世湖的东部，距苏黎世城市南部19公里的Männedorf。通过透明的外观以及办公空间的布局，设计的主要关注点落在了湖景上。

一层设有大厅，包含休息室、餐厅、会议室和研讨室。高层主要用于开放的办公室空间以展示建筑的最大灵活性。地下停车场有170个停车位。

整栋建筑的设计和可持续性发展理念都遵循了瑞士迷你能源生态标准。湖中的水既可用于加热也可用于冷却。

TAI TOWER IN ISTANBUL

ARCHITECT
Dante O. Benini, Luca Gonzo

GRAPHICS AND BUILDING DIRECTORY
Massimo Vignelli

FIRM
Dante O. Benini & Partners Architects

LOCATION
Istanbul, Turkey

AREA
21,000 m²

PHOTOGRAPHER
Beppe Raso

Abdi Ibrahim pharmaceutical company's new headquarters skyscraper is in Istanbul office district. The tower of 21,000 square meters floor space distributed over 20 floors above ground and 5 below has a total height of 120m. Conceived as a single articulated volume, on its North and West facades it has hanging micro-perforated sheet metal screen panels masking service utilities, emergency stairs and at the top the air-conditioning plant.

On the upper floors, where the Presidential and executive offices are located, offsetting volumes create ample open terraces, protected by micro-perforated sheet metal sails. On the South facade, an inclined steel tubular superstructure, besides supporting the metallic shades, connects the lower stout volume to the top penthouse presidential gallery suspended at the tower's highest level. At intermediate level there is a restaurant-café, meeting point of the tower; which widens to the outside terrace indicating the fall between the lower and the higher volume. Open space standard offices are furnished with modular equipment suitable for installation according to the staff requirements. A gallery museum, meeting and waiting rooms and the reception, as well as the archives, overlook the three levels entrance hall, marked by a huge square glass window. Hanging ladders and connecting ramps enhance the perception of triple-height space extention. At first underground floor there is the 250 seatings auditorium with foyer. The last three underground levels provide car parking for up to 100 vehicles. The so well marked building's architectural volumes are exalted and enhanced by shade and light plays.

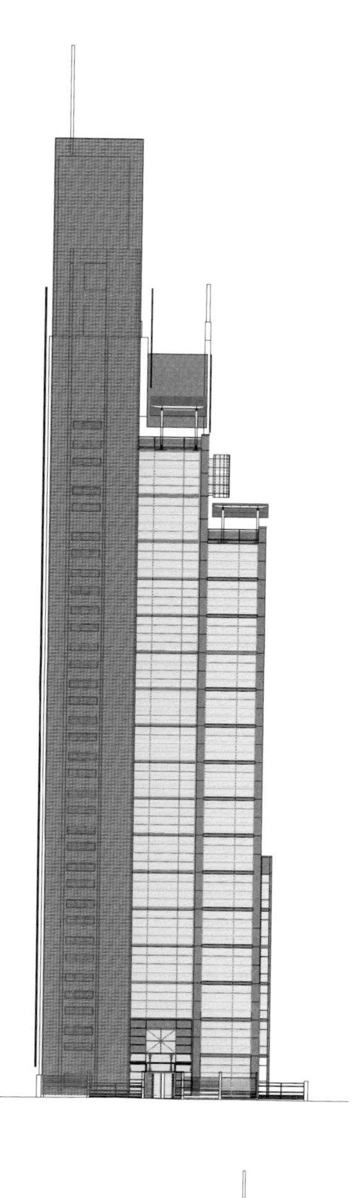

ABDI IBRAHIM

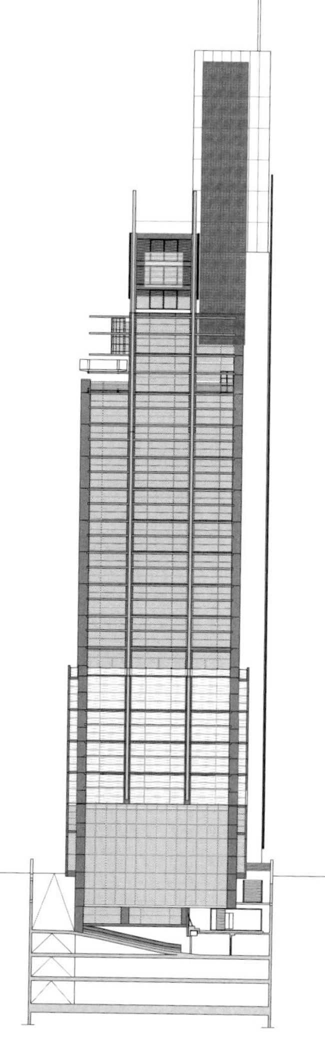

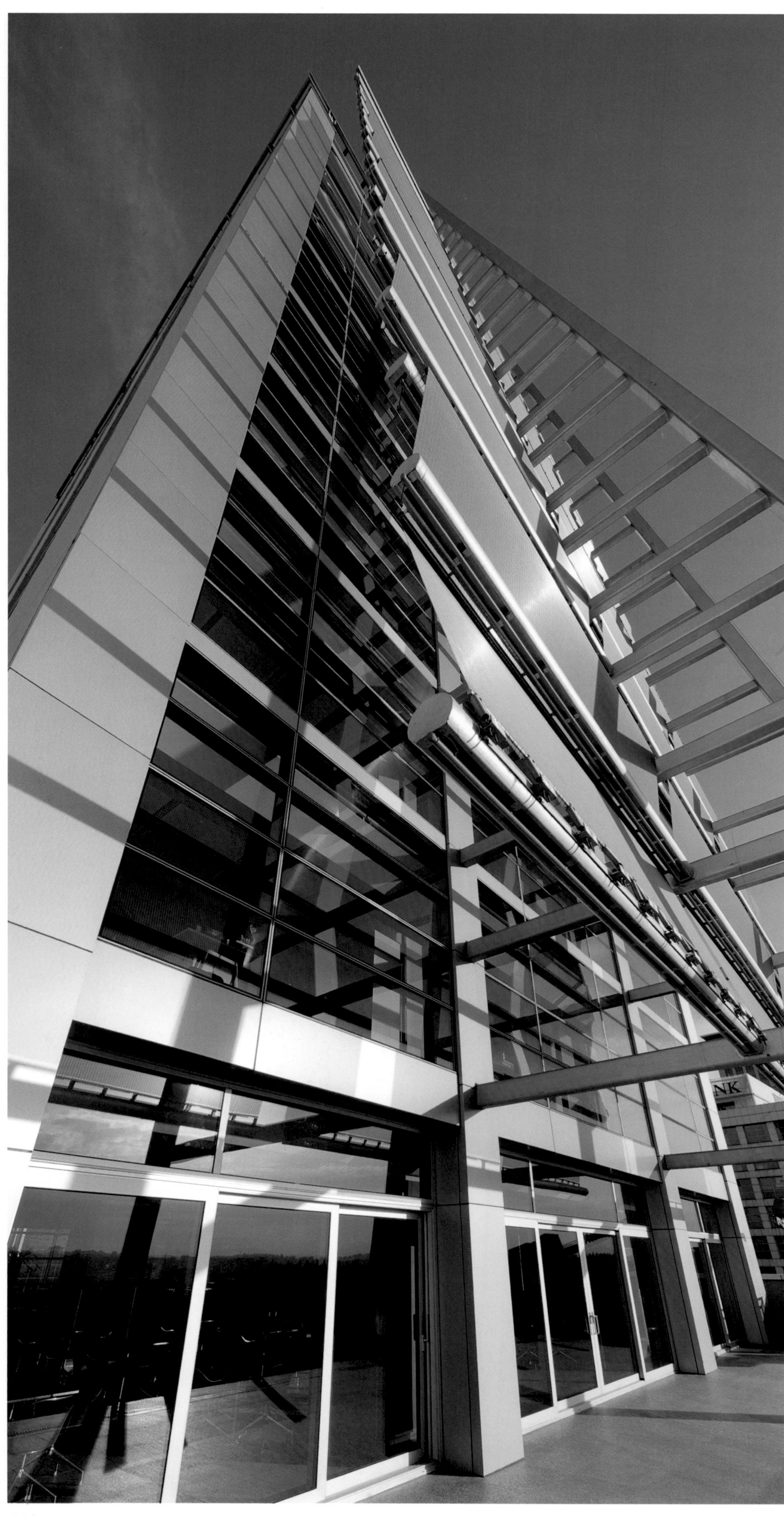

阿卜迪•易卜拉欣制药公司新总部摩天大楼位于伊斯坦布尔办公区。摩天大楼的建筑面积为21 000平方米，分布在地上20层和地下5层，总高度为120米。作为一个单一的铰接体，在其北立面和西立面均设计了悬挂式微穿孔金属片屏幕面板，用以掩蔽服务设施、逃生通道和顶部空调装置。

大厦的上层是总裁办公室和行政办公室的所在地；壁阶柱创建了宽敞的露天平台，平台顶部由帆状微穿孔金属板覆盖。在南立面，向大楼顶部倾斜的钢管状结构，除了作为金属百叶窗的支架，还支撑着整个南立面：下至下层结实的壁柱，上到悬浮于大楼顶层的总裁肖像画廊。在中间层有一个餐馆兼咖啡馆，是大楼的两节塔楼的交汇处；向外延伸的露台上，两节塔楼之间的落差显而易见。根据员工需求，开放空间标准办公室均配备适合安装的模块化设备。在画廊博物馆、会议室、候客厅、接待室以及档案室，都可以俯视三层高的入口大厅，大厅的特点是一个巨大的方形玻璃窗口。悬挂的梯子和连接的坡道设计增强了三重高度空间的延伸感。地下一层有一个250座带门厅的礼堂。最底下三层是停车场，有100多个停车位。这座著名的建筑因其别具一格的造型，丰富多彩的光影变化被世人赞扬和传颂。

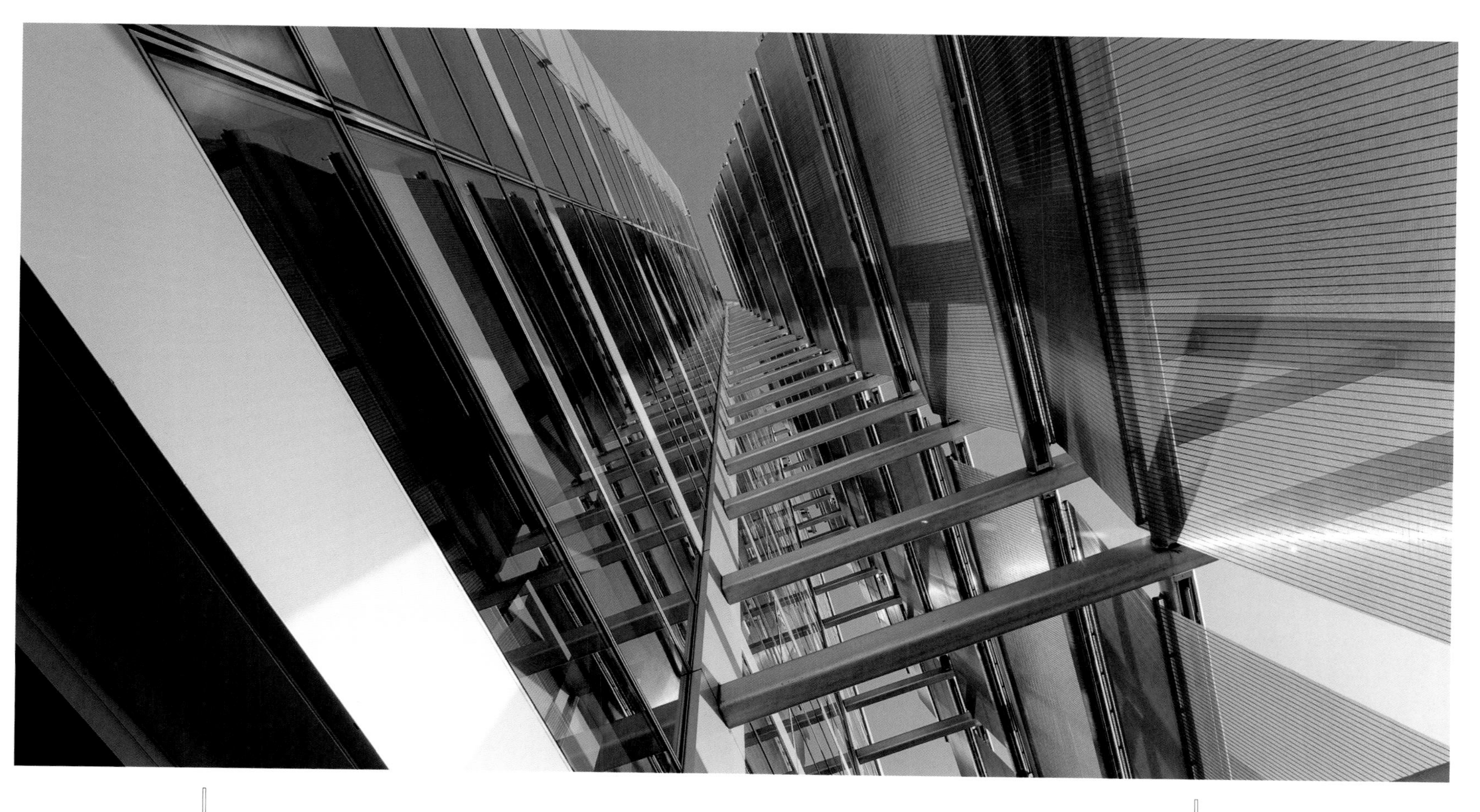

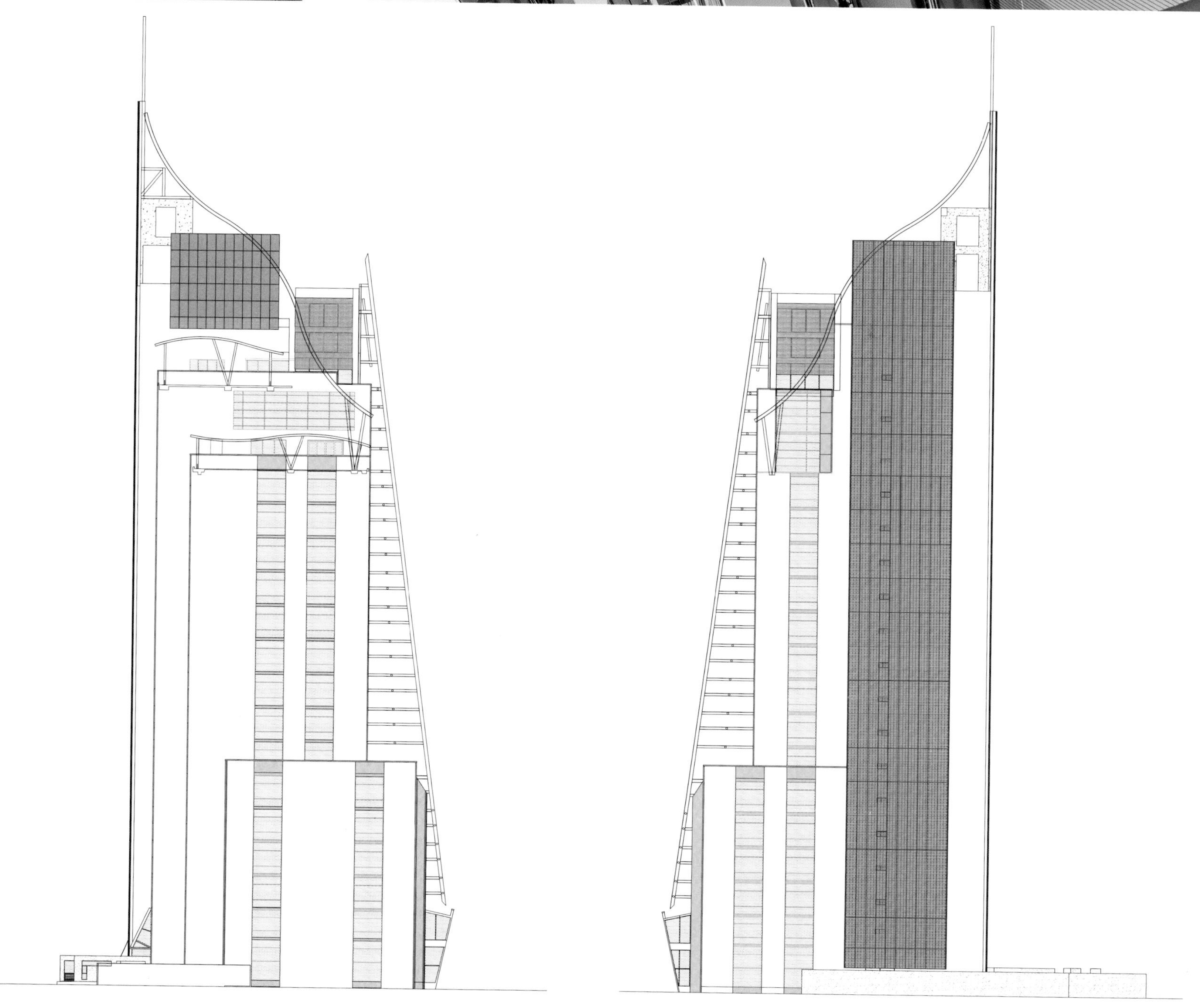

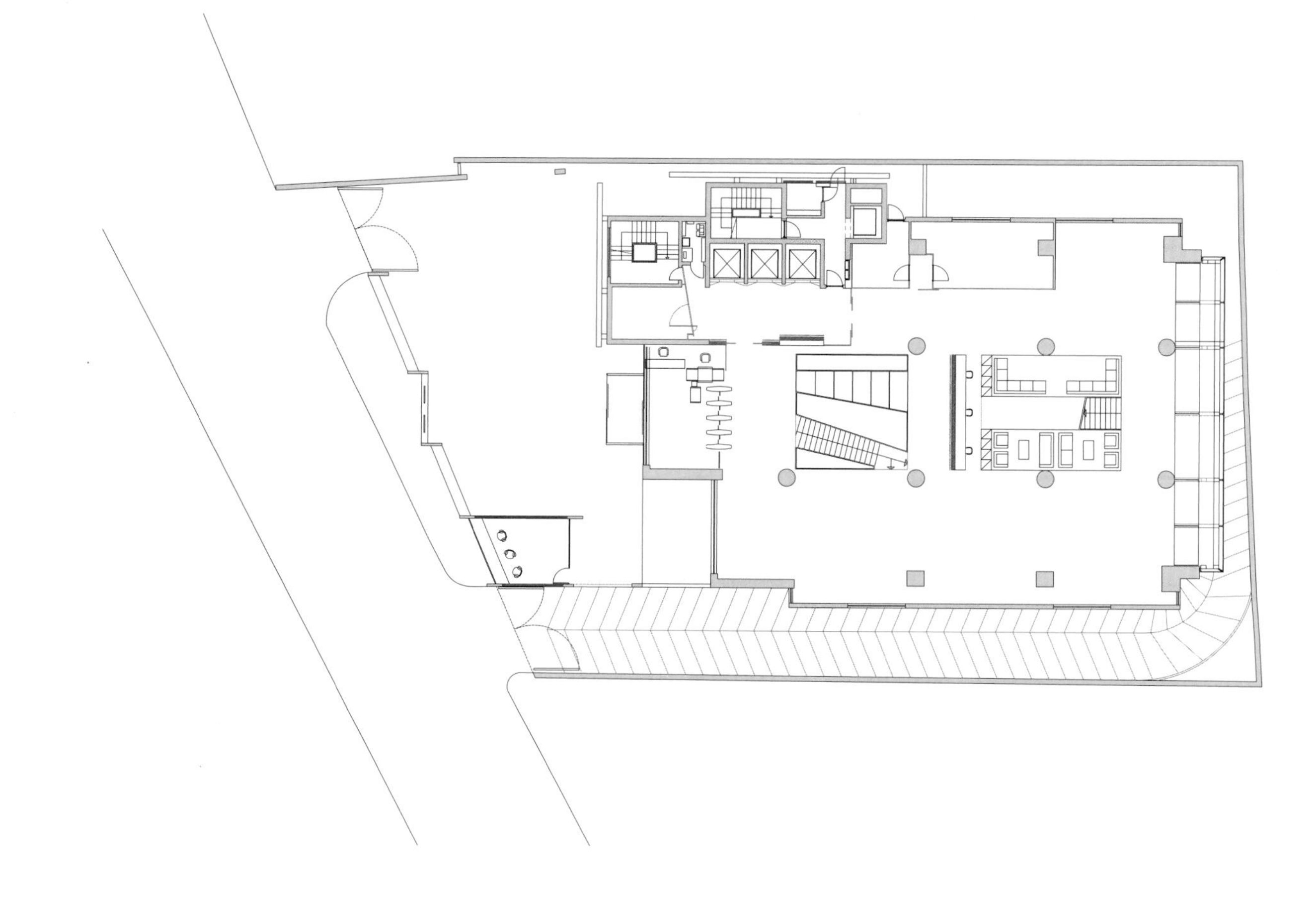

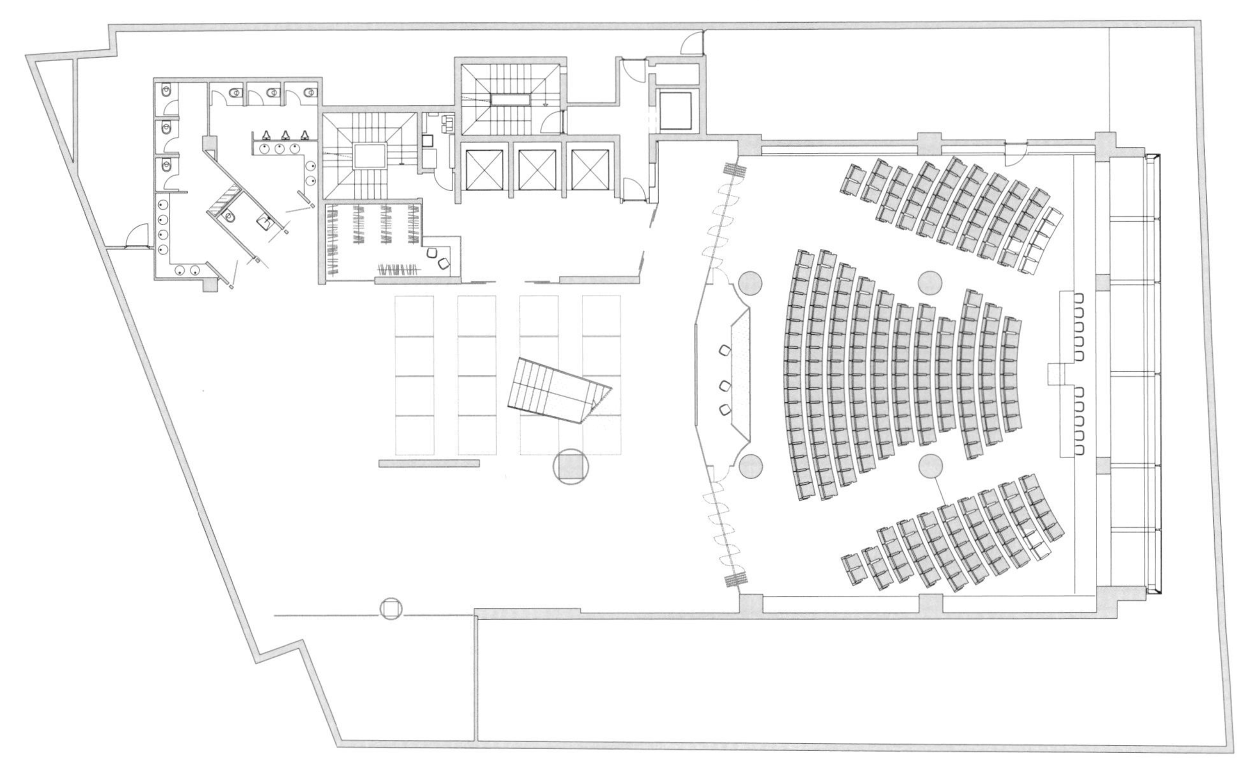

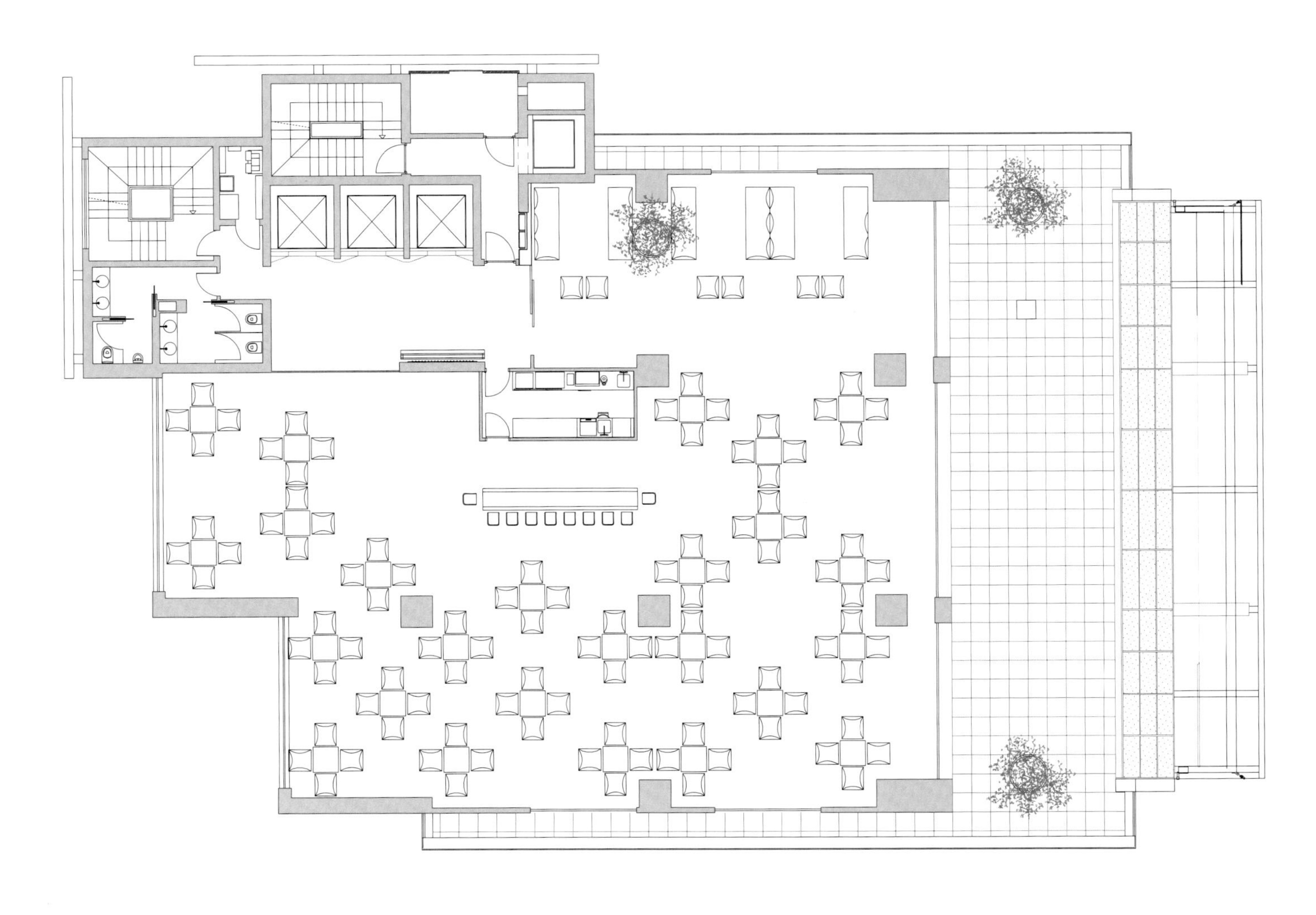

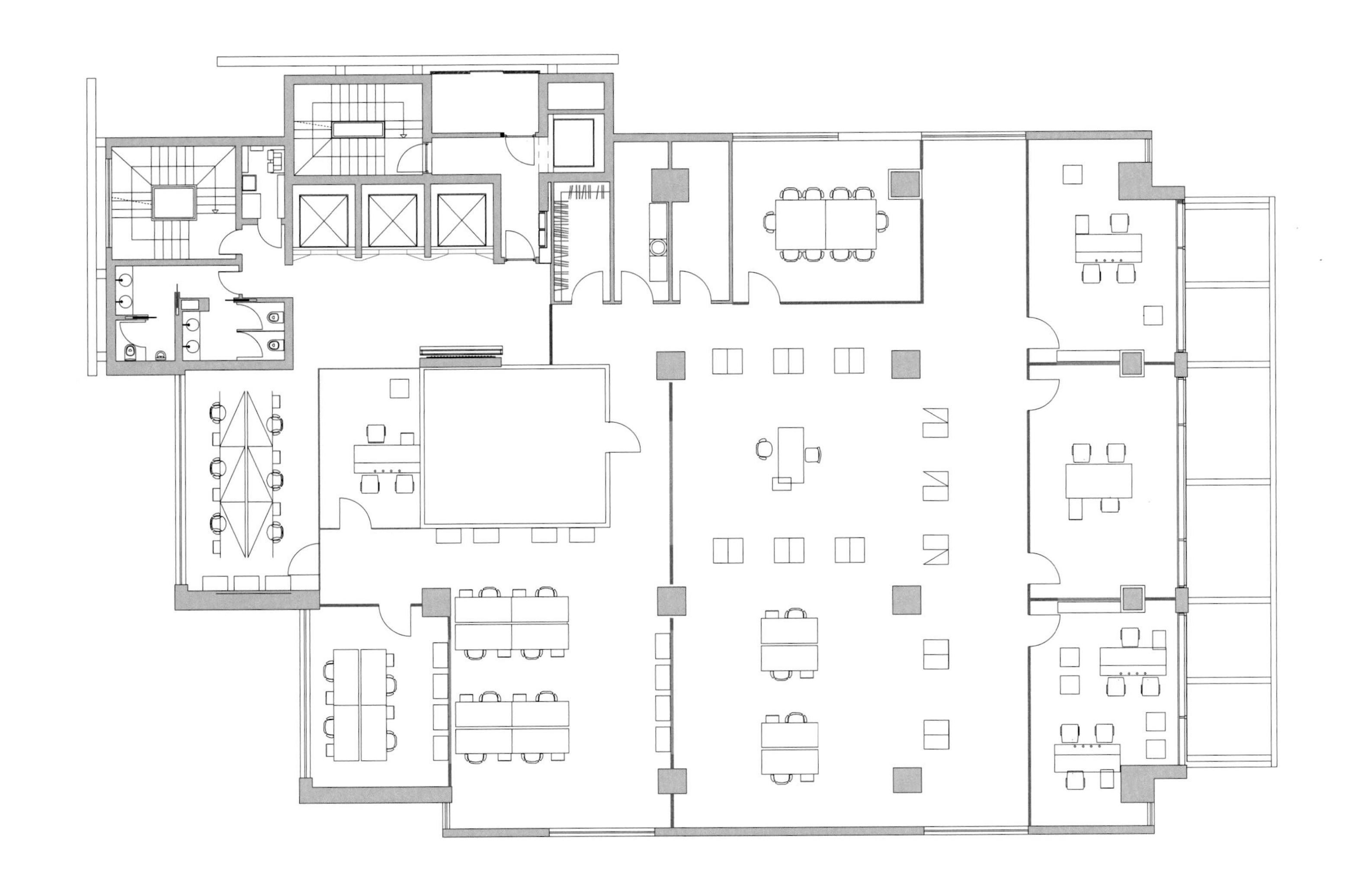

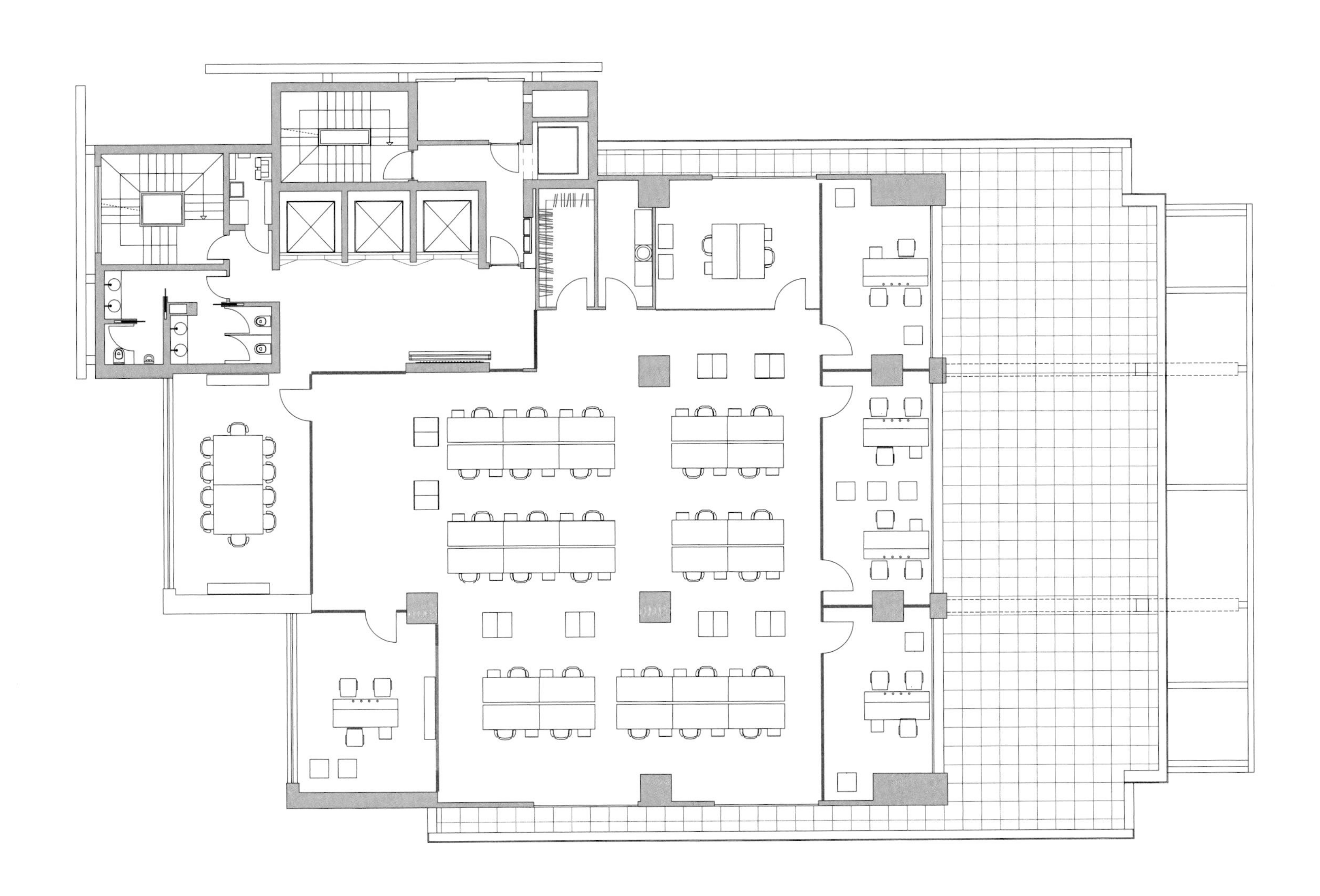

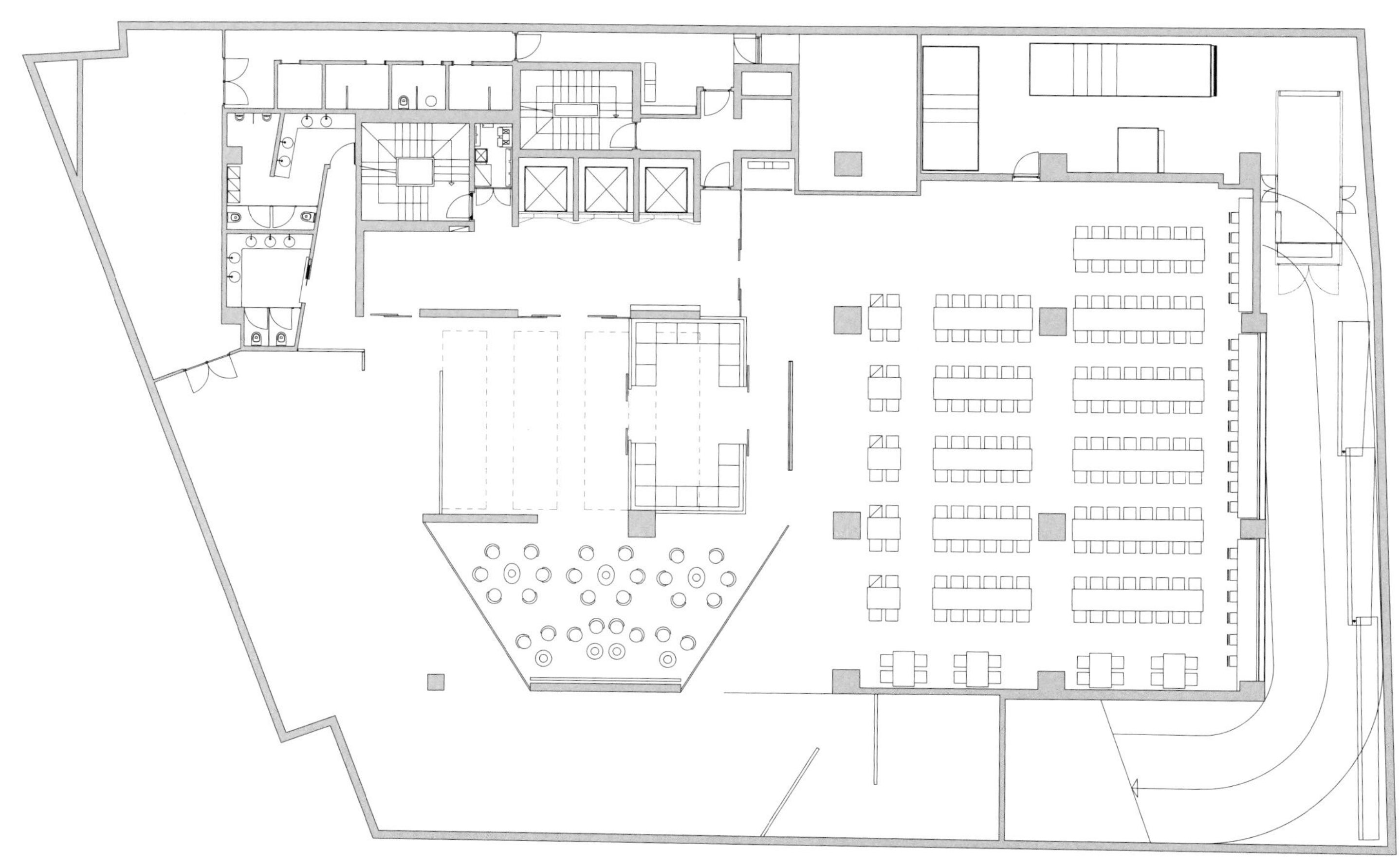

OFFICE TOWER AT 78 SHENTON WAY

ARCHITECT
Forum Architects

PROJECT DIRECTOR
Ho Sweet Woon

DESIGN DIRECTOR
Tan Kok Hiang

CLIENT
Shenton Singapore Holdings Pte Ltd

LOCATION
Singapore

PROJECT TEAM
Wong Chin Wah, Liew Chon Jack, Kai Wei Bin, Beverly Aquino, Florinio Pasco

PROJECT MANAGER
Davis Langdon & Seah Project Management Pte Ltd

AREA
45,928.93 m^2 (Gross Floor Area), 7,309.8 m^2 (Site Area)

PHOTOGRAPHER
Albert Lim

The project involves adding a new office tower extension to an existing high-rise office tower at 78 Shenton Way, being located on the edge of the Central Business District.

The existing building comprises a 34-storey tower sitting on a large 3-storey podium linked to a multi-storey carpark. The existing complex has site coverage of 65 %, leaving little space for the new tower without tearing down part of the existing building to make space for new extension.

To overcome the site constraints, the new tower comprising 8,000 square meters of gross floor area is built over and suspended above the existing multi-storey carpark deck. The new office floors are supported by five steel columns and one reinforced concrete core in strategic locations in order to allow the carpark below to remain in operation during construction.

The typical floor plate is designed with the service core on the west facade, opening up to a large unobstructed column free space with spans up to 21m and net areas of about 1,200 square meters per floor. The choice of non-centralized core allows a single large space shielded by the western sun with views out to the surrounding harbour beyond.

The overall facade expression of the new extension is based on an 1,100mm x 1,100mm square modular grid and the building is articulated as cubic volumes, which echoes subtly the existing solidly granite cladded tower. In deliberate contrast with the existing tower, the new extension is fully cladded in glass curtain wall, with the strong modular lines diffused by "dissolving" frit or perforation pattern on the glass or aluminium cladding.

To accentuate the height and the lightness, the top of the new tower is crowned with a "halo" roof, which floats above the tower.

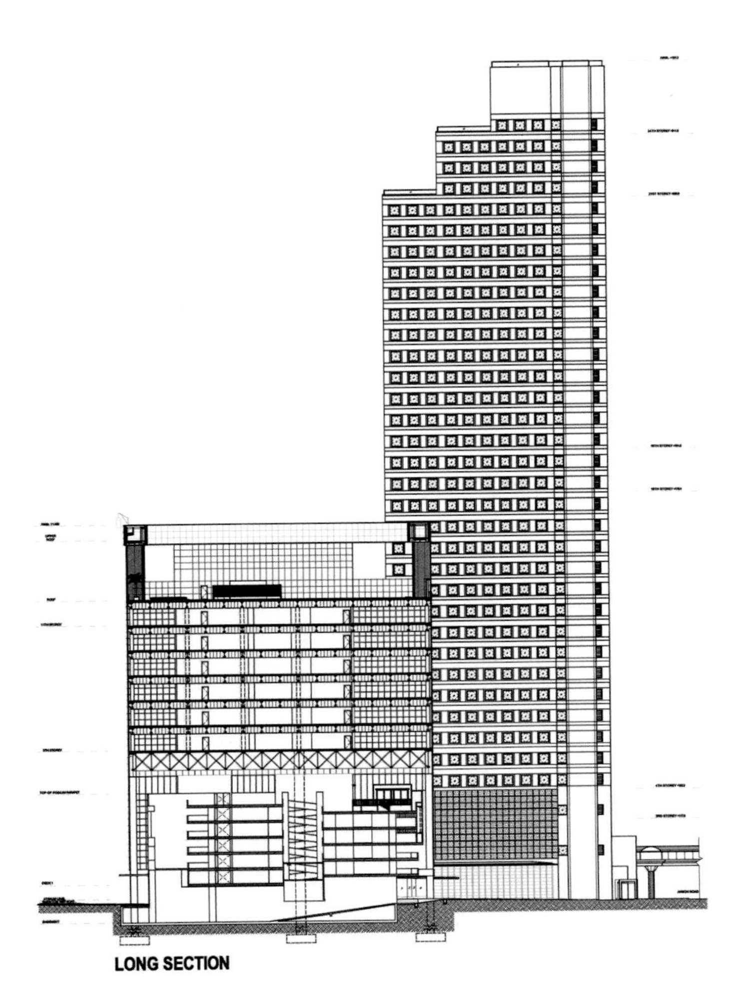

LONG SECTION

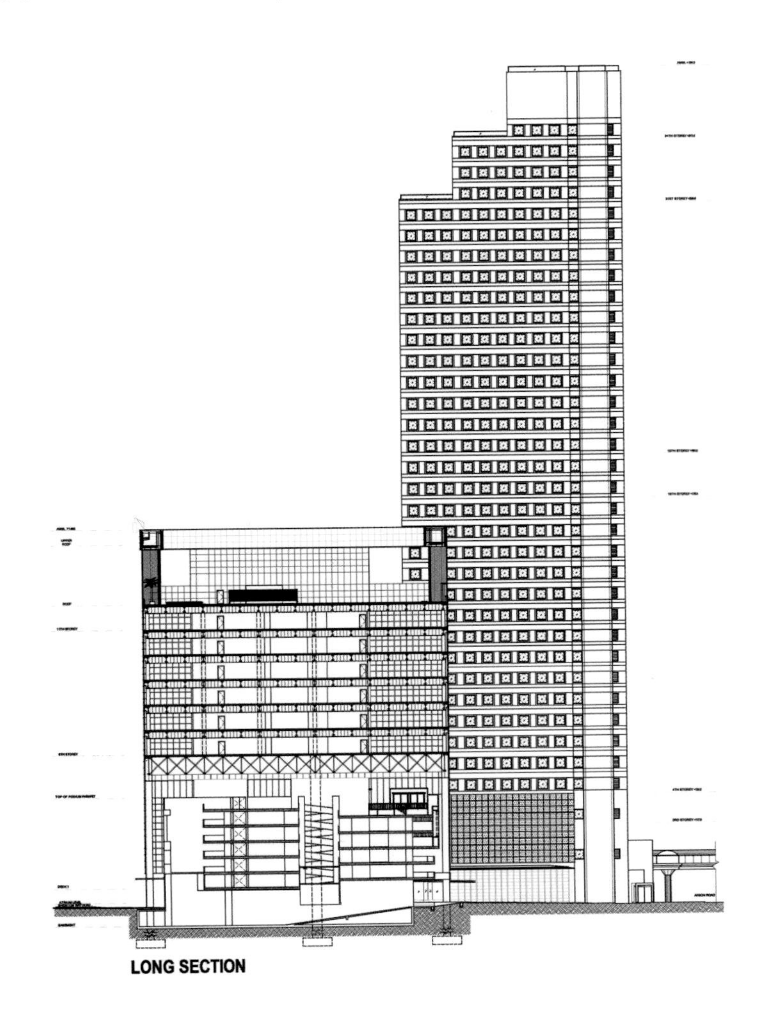

LONG SECTION

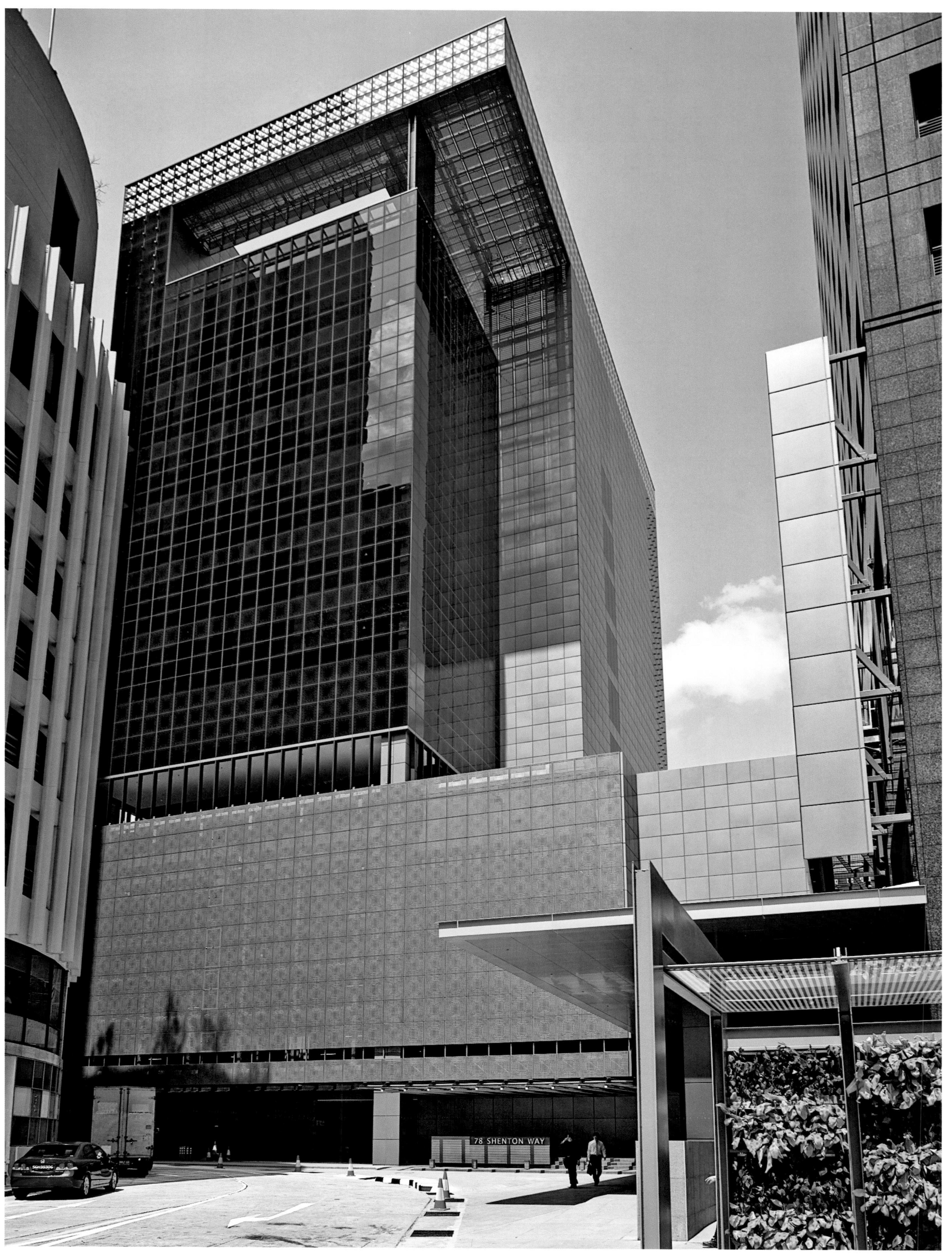
78 SHENTON WAY

78 SHENTON WAY

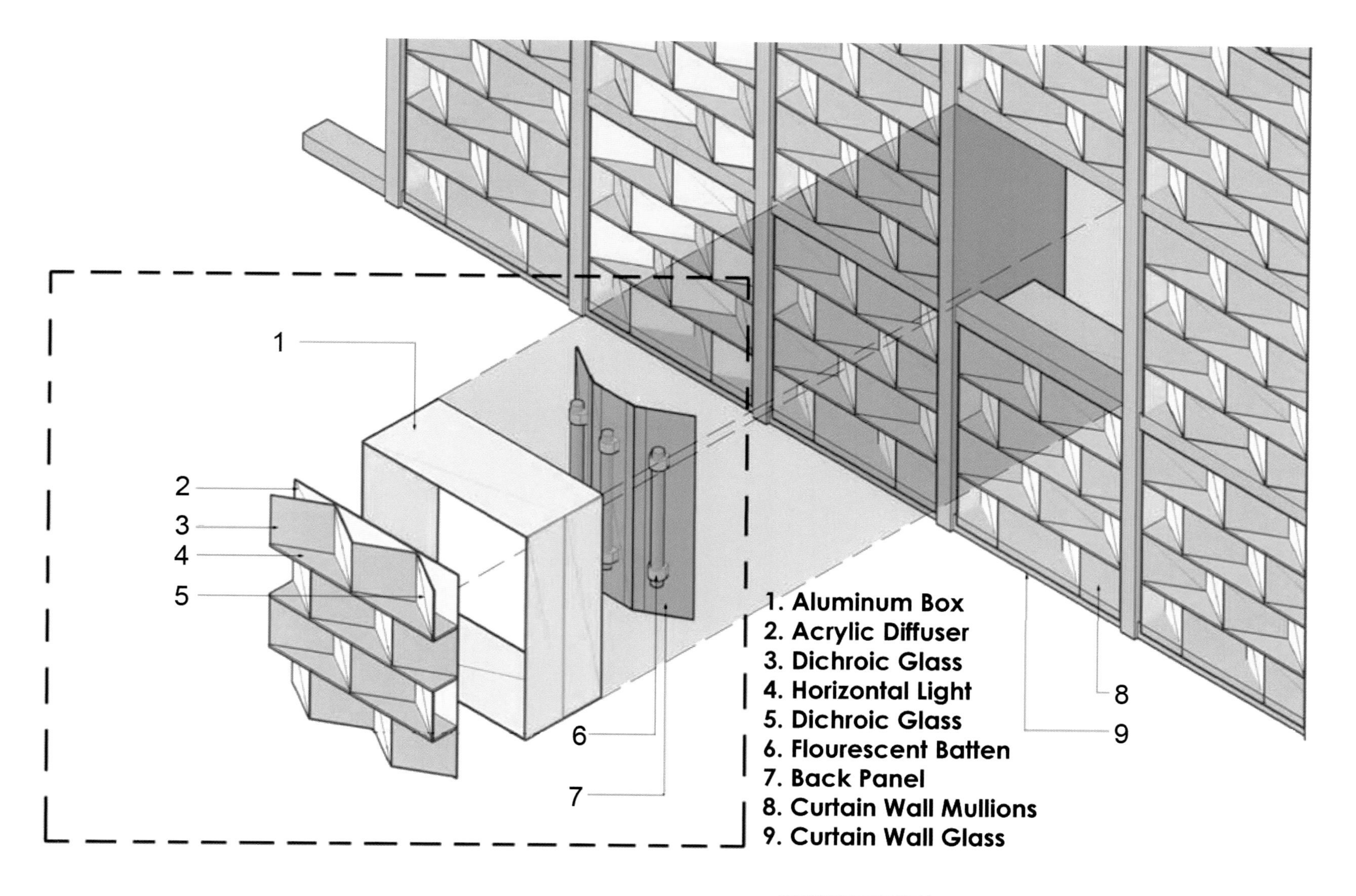

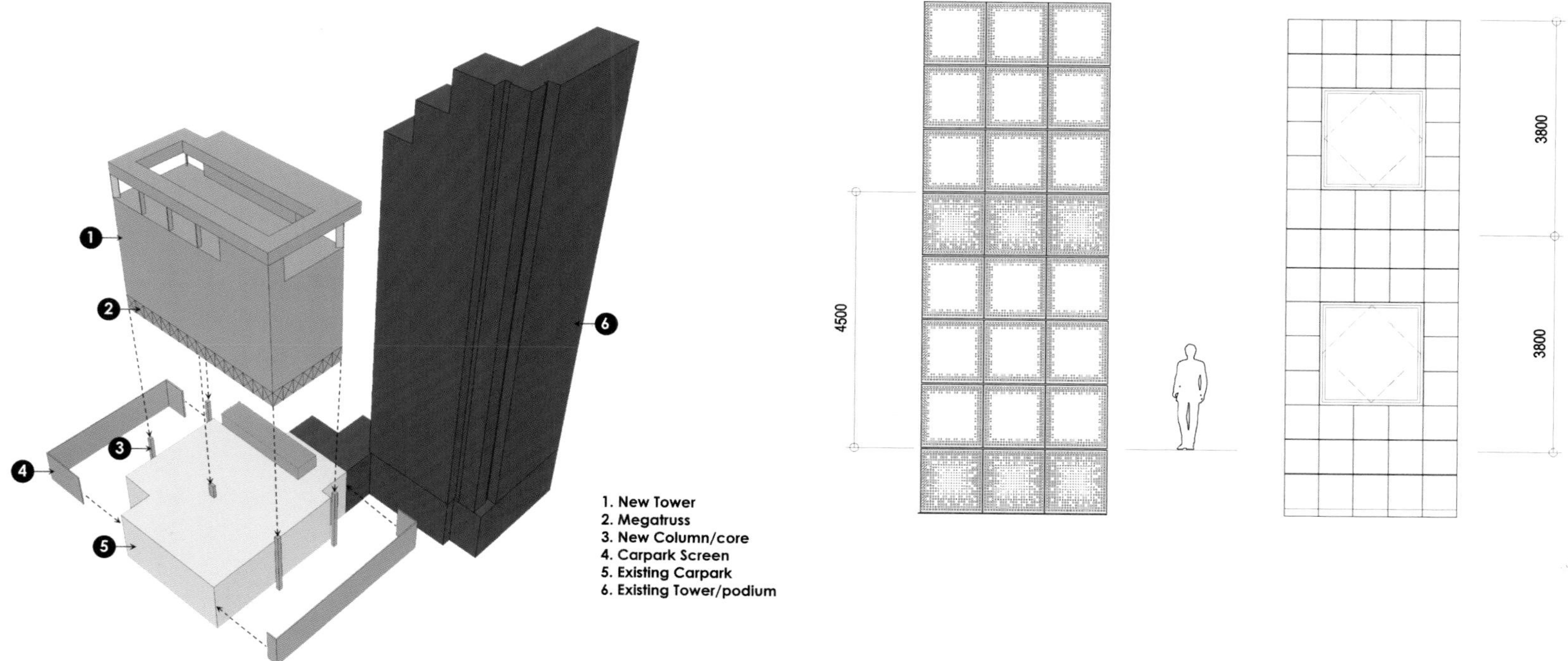

该项目涉及新建一座办公大厦，作为位于中央商务区边缘的珊顿大道 78 号高层办公大厦的扩建部分。

现有建筑包括一个 34 层的塔楼，坐落在一个 3 层高的底座上，与一个多层停车场毗连。现有建筑群的建筑密度已达 65%，由于没有拆除现有的一部分建筑来为新的扩建提供空间，所以仅留下一小部分空间给塔楼。

为了克服用地限制，建筑毛面积为 8000 平方米的新建塔楼将修建并悬浮在现有多层停车场露天平台上。为了让塔楼下面的停车场在施工期间正常运行，新修的办公楼层在大楼的核心位置由五根钢柱和一个加固钢筋混凝土芯支撑。

在西立面，典型的楼板被设计成服务中心，中心是一个跨度高达 21 米，净面积每层约 1200 平方米的巨大的通畅无阻的自由空间。非服务中心区是一个单一的大空间，这里可以观赏到落日的余晖及周围港口的美景。

扩建大楼的整个立面由若干个 1100 毫米见方的模块化网格构成。整个建筑呈立方体状，与现有的坚固的花岗岩包芯塔楼巧妙地融为一体。与现有塔楼形成鲜明对比的是，新扩建部分完全被玻璃幕墙覆盖，通过“溶解”玻璃或在玻璃或铝包层穿孔的方式在幕墙上形成醒目的模块线条。

为了突出高度和亮度，新建的塔楼顶端设计了一个带光环的屋顶，飘浮在塔楼之上，在阳光的照射下整座塔楼熠熠生辉。

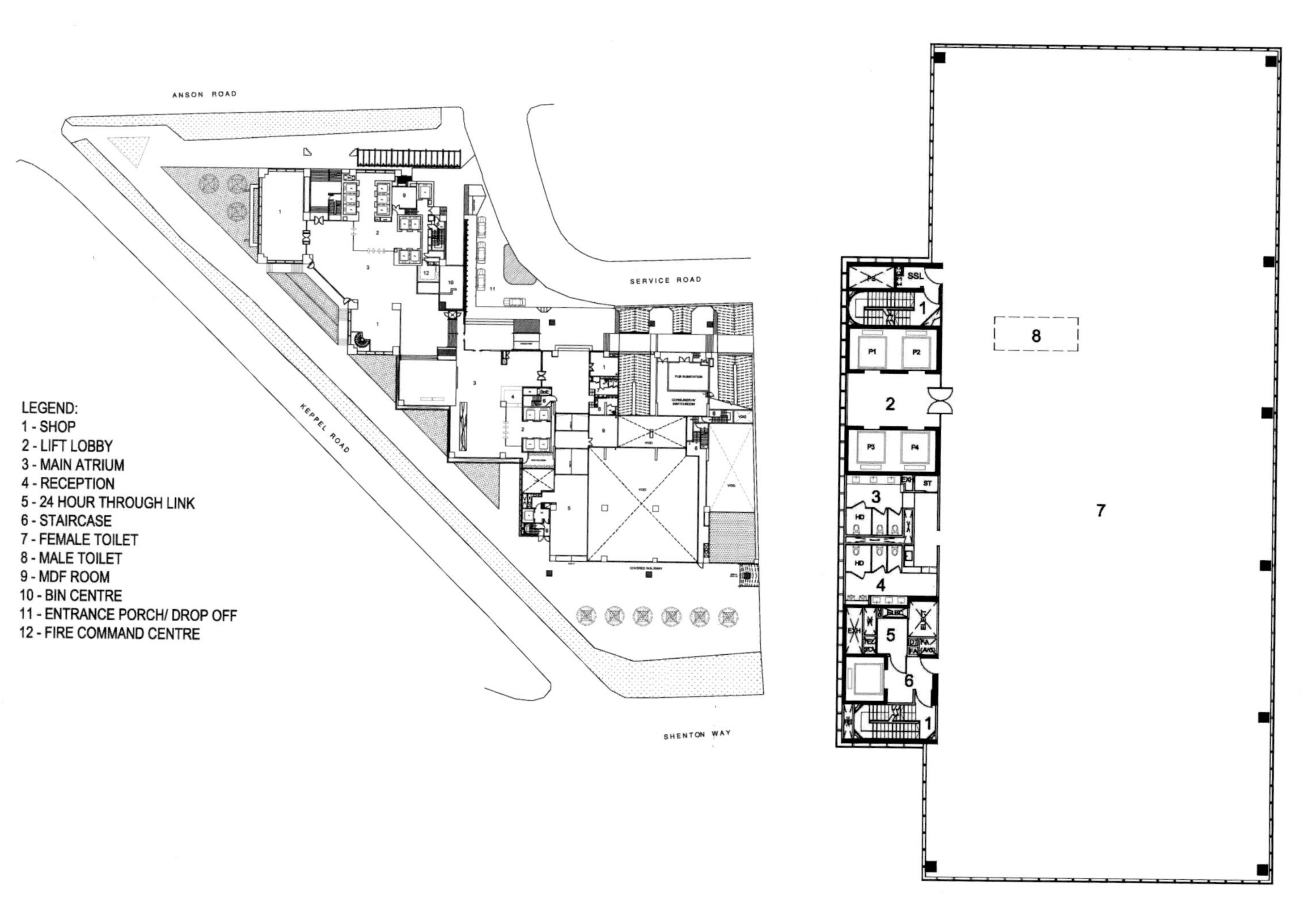
ANSON ROAD
SERVICE ROAD
KEPPEL ROAD
SHENTON WAY
LEGEND:
1 - SHOP
2 - LIFT LOBBY
3 - MAIN ATRIUM
4 - RECEPTION
5 - 24 HOUR THROUGH LINK
6 - STAIRCASE
7 - FEMALE TOILET
8 - MALE TOILET
9 - MDF ROOM
10 - BIN CENTRE
11 - ENTRANCE PORCH/ DROP OFF
12 - FIRE COMMAND CENTRE
LEGEND:
1 - STAIRS
2 - LIFT LOBBY
3 - FEMALE TOILET
4 - MALE TOILET
5 - M&E SERVICES
6 - FIRE FIGHTING LOBBY
7 - OFFICE UNIT
8 - KNOCK-OUT PANEL

ENTON WAY

78 SHENTON WAY

INDEX

Cox Rayner Architects

Cox Rayner Architects is the Queensland office of Cox Architecture, an Australian and international practice with projects in Singapore, China, India, Indonesia and the Middle East.
Cox Rayner is renowned as a practice which translates urban design into architecture which responds integrally to context, place and culture. The practice is especially distinguished in creating innovative solutions for tropical and subtropical climates, with environmental performance at the fore.
The practice is internationally recognized by winning two World Architecture Festival(Barcelona) Awards for the Helix Pedestrian Bridge in Singapore and the Kurilpa Pedestrian Bridge in Brisbane. Durability to create inventive ways to integrate structure and architecture has also results in a specialization in Exhibition and Convention Centers, major examples being Brisbane, Cairns North Queensland and Changi Singapore.

Paragon Architects

PARAGON ARCHITECTS

Paragon Architects is a dynamic and innovative architectural design business, based in Johannesburg / South Africa, and active since 1997. It is the originator of the Paragon Group of design businesses.
Their business has been built on their willingness to apply knowledge to almost any design task, and they have completed buildings and projects of many different types in almost all property industry sectors. Their current core competencies lie in the design of workplace environments, specifically head offices, along with retail, sports and leisure architecture. They have expanded their operational footprint beyond Southern Africa into Africa, and have specific current project interests in Angola, Ghana, Uganda and Rwanda. In 2009, they established a practice under the name of Paragon Arquitetura in Brazil.
Paragon Architects is known for being flexible and diverse in its approach to design. Their work is not style-driven, but lifestyle-driven. Elegant and efficient planning are at the core of their designs.

DaM

Architectural office DaM spol. s r.o. was founded in 1990 as a private Czech-Swiss architect´s office by two partner architects – Richard Doležal Dipl.Arch. ETH/SIA and Ing.Arch. Petr Malinský, these days with four more partners-architects: Petr Burian, Jiří Havrda, Jiří Hejda and Jan Holna.
The company office is in Prague. There are 30 regular collaborators/ employees (mainly architects and civil engineers) and also a range of external engineers-specialists of various professions. They have prepared and realized different kinds of important designs and constructions (including interior designs) both in Czech republic and Switzerland during 25-year existence of their Czech and Swiss office. The main types of their buildings are an office and housing construction and hotel buildings. Their newest buildings are a residential complex at Prague 5, and two new office blocks. At Prague 8-Karlin, Maint Point Karlin, and at Prague 4 BBC-Filadelfie.
They are able to provide complete service of architect, project and construction management (including all levels of technical supervision) and engineering together with their specialized collaborators.

WZMH Architects

WZMH ARCHITECTS

WZMH Architects is an award-winning architectural partnership established in 1961 in Toronto, Ontario, Canada. Since its inception, WZMH has built a reputation on innovative design, technical expertise and dynamic leadership.
The WZMH practice is comprised of eight principals, nine associate principals, eleven associates and a technical and administrative staff of approximately 100.
WZMH believes in a collaborative approach to all assignments, with all project participants working closely together in order to achieve a common goal: the delivery of the project to meet or exceed the expectations of the Client.
The firm's expertise encompasses all aspects of the architectural process from master planning, site evaluation and feasibility studies, through the design and construction documentation phases, to the final selection of finishes and fittings for building interiors.

Kurylowicz & Associates Architecture Studio

Kurylowicz & Associates Architecture Studio was founded in 1990 by Prof Stefan Kurylowicz. Today, it is the most renowned and titled architectural company in Poland. It employs 80 experienced architects in offices in Warsaw and Wroclaw.
The studio can boast many prestigious realisations in the whole country. It has experience in designing investments of every size and function. Throughout the time of its activity, the studio has gained the reputation of a reliable partner, able to present solutions of the highest quality within the fixed time and budget. Besides standard and complex architectural services, the company also offers counsel and support regarding advertisement and investment promotion. The quality of the solutions delivered by the company has been awarded many times both in Poland and abroad. But the biggest prize is the rich portfolio of satisfied customers. Many of which have been co-operating with them for many years.

Arditti+RDT/architects

Architect Mauricio Arditti graduated from the "Universidad Nacional Autonoma de Mexico" (UNAM) and his trajectory integrates over 50 years of professional experience. At the midpoint of his career, he received a second generation of Architects incorporating his two sons, Arturo (1984) and Jorge (1990), both "Anahuac University" graduates in México City. They all have complemented specialized studies abroad, in schools like "Harvard University" and the "Massachusetts Institute of Art", and have combined their knowledge and experience materialized in over 100 built projects, covering diverse types of buildings.
Academically, they've been professors abroad as part of the visiting faculty at "SCI-ARC" (Southern California Institute of Architecture) in Los Angeles, California and at the "Anahuac University" in Mexico City, where they are also members of the Advisory Board of its Architecture School.
Arditti+RDT/architects have been invited to lecture on their body of work and principles in Mexico and abroad, in events and institutions, They have been widely published in various national and international specialized magazines and books. Currently, the firm is working on a wide variety of projects that include corporate buildings, residential and tourist developments.

Erick van Egeraat

During his over 25 years of successful practice, Erick van Egeraat (Amsterdam, 1956) built a highly diverse portfolio. Containing ambitious and high-profile projects in the Netherlands, Europe and the Russian Federation. He has led the realisation of over 100 projects in more than 10 countries ranging from buildings for public and commercial use to luxury and social housing projects, projects for mixed use and master plans for cities and even entire regions. Each of these project represents his very personal and expressive vision on architecture and urban development. Both Erick van Egeraat and his work have been recipients of numerous international awards and citations. Erick van Egeraat graduated from Delft University of Technology, Department of Architecture, with honourable mention in 1984. In 1995 he established (EEA) Erick van Egeraat associated architects with offices in Rotterdam, Moscow, Budapest, London and Prague. In order to better meet the demands of a portfolio as diverse as he is, he successfully restructured his company early 2009 into what is now (designed by) Erick van Egeraat.

ingenhoven architects

Christoph Ingenhoven, Dipl.-Ing. Architect BDA, RIBA, AIA int.
In 1985, Christoph Ingenhoven founded the architectural studio of ingenhoven architects which has today become one of the world's leading architectural practices in sustainable design. Among his best known projects are the RWE headquarters in Essen, Germany – one of the world's first ecological high-rises; the Lufthansa Aviation Center at Frankfurt Airport; the European Investment Bank Building, Luxembourg, the headquarter of Daniel Swarovski Corporation Zurich and the high-rise tower 1 Bligh, the first building in Sydney rated 6 Star Green Star and 5 Star NABERS Energy and which has received several awards since its opening 2011: the CTBUH Best Tall Building Australasia and the International High-Rise Award 2012/13.
The Main Train Station in Stuttgart, Germany as well as the Office-and Apartment high rise tower at Marina One, Singapore, are both under construction.

Mario Cucinella Architects

Mario Cucinella Architects design architecture that, through research, the use of innovative technologies, and professional skills, embody an ideal of architectural quality integrating environmental sustainability, ethics and a positive social impact.
MCA, Mario Cucinella Architects is a company with a solid experience at the forefront of contemporary design and research. Sustainable building design and the rational use of energy is one of the central concerns in MCA's work and research.
Mario Cucinella's work has been internationally recognised and he was honoured with the MIPIM 2009 Green Building Award, the US Award 09 in the Architecture category, the MIPIM Architectural Review Future Projects Award, the "Energy Performance + Architecture Award" in 2005 (Paris), with the "Outstanding Architect" award by the World Renewable Energy Congress in 2004 (Denver) and in 1999 and he received the prestigious "Forderüngs Prize" for Architecture by the Akademie der Künste of Berlin.

Junglim Architecture Co.,Ltd

Two trees(木) have come together to form a lush forest(林).
JUNGLIM(正林) was founded in 1967 by brothers Jung-Chul Kim and Jung-Shik Kim.
The Founders' humanity, passion, and generosity formed the roots of JUNGLIM, whose two trees have grown into a firm of over 500 employees.
JUNGLIM's steadfast roots ensure that they exist as more than a successful corporate machine. They are a group of unique individuals who demonstrate a true love of architecture, creative problem-solving, innovative design methods, and social awareness. As JUNGLIM continues to expand in the future, they firmly believe that their roots will keep them grounded and committed to the philosophy that architecture is a cultural treasure as well as a humanitarian contribution.
JUNGLIM's social commitment reaches both within and outside the firm. Their mission is to build safe and beautiful spaces for the public while maintaining a productive and happy workplace for employees.

A-lab

A-lab is an Oslo-based, international architecture office that aims to create innovative and sustainable architecture. The office was founded in 2000 and now consists of 38 architects with diverse backgrounds and experience. With designers that have different skill sets they are able to operate in all phases of the design process. They also collaborate multidisciplinary with other companies.
They strive to set new standards in architecture and urbanism, in every project. Their project portfolio range from office buildings, residential projects and cultural centers and complex urban master plans, many of which are results of winning competitions. In 2012, a-lab has simultaneously led two of largest building projects in the country: DNB Barcode and Statoil Regional Head Office.
A-lab is also committed to the professional community and the current architectural discourse through honorary trusts and engagement in educational institutions.

Dam & Partners Architecten

Dam & Partners Architecten, founded in 1962 and led by Cees Dam and Diederik Dam, occupies itself with a wide range of projects differing in both nature and scale, from complex urban developments, offices and shopping centers, theaters and town halls, hotels, restaurants and leisure facilities, restoration of historical buildings, social housing and luxurious villas to interiors, furniture, carpets and glass. Dam & Partners Architecten has developed its own unique architectural philosophy and style that can be moulded to the concept and context of each commission. This results in a style that can undergo refreshing changes in appearance under the influence of the latest insights into sustainability, of new technologies, new opinions about design and the design process, and the intensive involvement of the firm and its employees in the arts and the cultural debate. Throughout this ongoing process, however, architecture in all its craftsmanship and with all its tradition and laws of dimensions, scale and rhythm remains the primary focus.

Broekbakema

Broekbakema connects people and society in relevant architecture with today's innovations. Established around 1910, Broekbakema has its roots in Rotterdam. Since its inception its buildings contribute to the wellbeing of people, to the creation of communities and to the development of their urban or rural environment. The firm designs houses, schools, museums, utility scale projects at all levels, interiors, urban development plans and transformations of recent monuments. A high-quality, accessible and more important, a relevant building is the result of co-creation with all stakeholders, something that is fundamental to Broekbakema. Broekbakema is a professional partner throughout the entire construction process, combining expertise and experience with innovation and sustainability. It was rewarded with the prestigious Dutch prize "Building of the Year 2011", awarded by the "BNA" Dutch Association of Architects.

HHD FUN

HHD FUN
HHD FUN
HHD FUN

HHD FUN is a design and research studio with interests in bringing knowledge from various fields outside of architecture and experimenting these means into the design of architecture. Parametric design and sustainability are their main research direction. The mathematics, geometric principles, algorithms, BIM, Artificial Intelligence and etc.are one portion of their approaches as the means in architecture generation. They collaborate with artist, fashion designer, mathematician, engineer, etc.and seeing these as opportunities of exploring new possibilities of design.

Van Aken Architecten (VAA)

Van Aken Architecten (VAA), founded in 1979, is located in Eindhoven and is active in the field of architecture, urbanism and interior architecture. Their working range includes high-rise buildings, care related buildings, offices, schools, town halls, residential buildings and architecture for the high-tech and leisure industries. Extensive transformations and renovations are also part of their expertise.
With an inventive and pragmatic approach, VAA looks further than just the design phase and delivers a total concept from design to delivery. The entire building process benefits from this.
With the knowledge and expertise of their 50 employees, VAA proves to be a reliable partner. In 2011 Van Aken Architecten proudly concluded a joint venture with St.Johnson International in Asia.

Boeri Studio

From 1999 to 2008, Boeri Studio has been a professional design agency specializing in architecture, urban design and city planning. During its last years of activity Boeri Studio has predominantly focused on the design of buildings and open spaces for European urban areas requiring regeneration or redevelopment. The sheer complexity of many of these situations has made strategic coordination among the various public institutions and countless market operators necessary from the outset of the development scenario. Boeri Studio undertook and carried through a series of projects regarding the transformation and re-use of several European waterfronts for the improvement of urban and tourist programs and developed similar projects within historical city centers. Since 2008 the professional activity of Boeri Studio has been divided in two different firms: Barreca&La Varra and Stefano Boeri Architetti.

Dark Arkitekter

DARK is one of Norway's largest planning, architecture and interior design practices. DARK is an Oslo-based studio consisting of seven companies working together to provide a broad range of services for their clients. Their work includes master planning, urban design, building design, landscape architecture and interior architecture, as well as furniture design, graphic design and visualization.
DARK was founded in 1988. Their highly qualified staff, currently 140 people, enables them to address all aspects of the building process, from Environmental Impact Assessments and Master Plans to completed buildings, from Project Management to detail design.
Their goal is to provide work of consistently high quality and standards, always answering to their clients' needs. These are basic components in their business strategy.

tvsdesign

tvsdesign

tvsdesign, a business corporation wholly owned by its employees, is an internationally recognized design firm that provides architecture, interior design and planning services. The firm has full service offices in Atlanta, Chicago, Dubai and Shanghai. They draw upon the knowledge and experience of many market segments including corporate/ commercial office, higher education, retail, public assembly, hospitality and sustainability. Founded in 1968, the firm has steadily expanded its practice and currently employs 125+ professionals.

Grupa 5 Architects

Grupa 5 Architects offer highly innovative and sustainable, architectural and engineering solutions suited to the individual needs of their clients. They believe that a successful cooperation with the client benefits both sides mutually and double fold. Their portfolio of completed projects built since 1998 includes a wide range of building types: residential, office, commercial, industrial, public, temples and buildings of culture. Grupa 5 has been part of the design team of such prestigious landmarks as the Metropolitan office building or the Royal Netherlands Embassy in Warsaw. Their award-winning built projects have set benchmarks of quality in Polish architecture. Grupa 5, managed by 6 partners currently employs around 30 architects in-house and cooperates on a permanent basis with around 100 consulting engineers and consultants. Their headquarters are located in the heart of green Mokotow, just outside the Warsaw CBD.

Forum Architects

Forum Architects is an architectural practice based in Singapore. It was founded by Tan Kok Hiang and Ho Sweet Woon in 1994. Kok Hiang and Sweet Woon both graduated from the National University of Singapore in 1987. Their early training under well known Singaporean architects, William Lim and Tay Kheng Soon instilled a deep concern about building in context, the environment and public engagement in architecture.
Forum has won more than 35 awards. More than 80 regional and international articles have been published on their works including three projects in the Phaidon Atlas of 21st Century Architecture.
Forum Architects are currently working on projects in Singapore, the Asia Pacific and the Middle East.

LVL Architettura

The architectural firm LVL Architettura, founded in 2006, is the association between:
Aurelio Galfetti (Switzerland, 1936) – graduated in Zurich – started in 1959 a long and successful architectural career making a lot of projects located in his Ticino region and all over Europe, that raised a great interest by critics. He founded with Mario Botta the Architecture Academy in Mendrisio.
Luciano Schiavon (Italy, 1962) – graduated in Padua – worked since 1989 on many projects and constructions concerning Architecture and Town Planning.
LVL Architettura designed Net Center in Padua, GHouse Tower in Jesolo, an urban square in San Donà, two urban developments in Padua and Treviso, a 130-meter high tower in Jesolo, hotel and dwellings in Venice and won a first prize in a competition for a resort on the Dolomites.

Herreros Arquitectos

Herreros Arquitectos is an architectural practice located in Madrid under the guidance of Juan Herreros, Chair Professor at the Madrid School of Architecture, and Full Professor at Columbia University.
Herreros Arquitectos has a solid international background with commissions in countries such as Korea, Panamá, France, etc. At present Herreros Arquitectos is developing the Munch Museum in Oslo, together with the Bjorvika residential area, and the Bogotá Conference Center both being results of in restricted international. Recently the office has completed the Bank of Panamá Tower in Panamá City where Herreros Arquitectos also is involved in the design and construction of a coastal park, and is also developing another big scale residential project in Casablanca Morocco.

3deluxe

The interdisciplinary design collective 3deluxe, consisting of about 30 individuals centred around Dieter Brell, Peter Seipp and Andreas and Stephan Lauhoff, has been creating groundbreaking impulses in the fields of architecture and interior design, graphic and media design.
Since the foundation of the practice in 1992 in Wiesbaden projects have been realised in various disciplines, which accounts for the broad spectrum of specialist knowledge of the highly qualified team. Meanwhile, 3deluxe maintains offices in Hamburg and Shanghai.
Established in 2005 as collaboration between Andreas and Stephan Lauhoff and communications designer Sascha Koeth, since 2008 located in Germany's media city Hamburg.

ARHI GRUP

Registered in 2004, ARHI GRUP is a company on the rise in the Romanian market. ARHI GRUP combines the talents and skills of its two managers and the other architects of the design staff, in a unique background of architecture, design, and knowledge of technical systems. ARHI GRUP benefits from trained and experienced personnel to offer personalized solutions for special design requirements.
Since 2004, ARHI GRUP has been committed to a number of medium and large projects, which challenged the company's architects and enriched their know-how and experience required for all major architectural programs, namely residential, office and retail segments.

Estudio Lamela

Estudio Lamela is one of the largest Spanish architectural firms, having an impressive past of over 55 years of extensive professional experience encompassing more than 1,600 projects. It offers design services for housing, offices, transportation facilities, sports complexes, retail and urban projects.
From offices in Madrid, Warsaw and México currently the company is staffed by a flexible and dynamic team of more than 70 professionals with different cultural backgrounds who are committed to the development of major projects and an ongoing search for innovative techniques, with the objective of achieving avant-garde results and high architectural quality. Additionally, the firm relies on the collaboration of appropriate specialists for each project and phase in the fields of structures, special and conventional services, environmental support, landscaping, health and safety, etc.

SAMOO Architects & Engineers

SAMOO Architects & Engineers is an architectural design firm headquartered in Seoul, Korea with diversified services including architectural design, urban planning, interior design, engineering, and construction management services. Since the firm's founding in 1976, SAMOO has developed into one of the world's largest architectural firms with a number of branch offices worldwide. Employing more than 1,000 professionals, SAMOO is involved in a diverse portfolio including office, civic, cultural, healthcare and biotechnology, residential, hospitality, academic, high-tech industrial, transportation, super-tall buildings and mixed-use projects. SAMOO was awarded the "red dot Design Award" in 2011 and ranked 9th worldwide by BD (UK) World Architecture TOP 100 and 2nd among architectural design firms of the Pacific Rim in 2012.

Dante O. Benini & Partners Architects

Student of Scarpa and Niemeyer, Dante O. Benini carried out building projects worldwide co-operating with Frank O. Ghery, Richard Meier, Arup and Daniel Libeskind too. He is registered at Rome Italian Architects Board (Ordine degli Architetti) and in the United Kingdom member of ARB (Architects Registration Board) and RIBA (Royal Institute of British Architects). In 1997 he founded "Dante O. Benini & Partners Architects" of which he is at present Leader Partner and Chairman together with Luca Gonzo Senior Partner and Managing Director. DOBP practice with Headquarters in Milan is present in London and Istanbul has specific departments dedicated to architectural project design, urban planning, interior design, yacht design. A staff of 60 people carries out, worldwide, complete projects of urban areas, major companies headquarters, industrial laboratories and shopping centers, exclusive clubs and private houses, yachts, as well as object design. Each project is designed on the basis of technical, economic and environmental sustainability, within a quality architectural context based on details care. DOBP practice has been awarded several architectural and design prizes and mentions (the latest has been the Chicago Athenaeum 2011 Good Design Award), won or was classified in international competitions and is invited to hold workshops and conferences both in Italy and abroad.

后记

本书的编写离不开各位设计师和摄影师的帮助，正是有了他们专业而负责的工作态度，才有了本书的顺利出版。参与本书的编写人员有：

Dominique Perrault, Cox Rayner Architects, Herreros Arquitectos, Mallol y Mallol, WZMH Architects, Cees Dam, Diederik Dam, Dam & Partners Architecten, ingenhoven architects, Erick van Egeraat, Jan van Iersel, Renze Evenhuis, Cees Schott AvB, SAMOO Architects & Engineers, Roman Dziedziejko, Mikolaj Kadlubowski, Michal Leszczynski, Krzysztof Mycielski, 3deluxe, Van Aken Architecten, Aurelio Galfetti, Luciano Schiavon, Carola Barchi, Paragon Architects, Forum Architects, Mario Cucinella, David Hirsch, Julissa Gutarra, Ho Sweet Woon, Stefan Kuryłowicz, Marcin Goncikowski, Tomasz Bardadin, Krzysztof Pydo, Katarzyna Pielaszkiewicz, Małgorzata Kowalczyk, Bogdan Stoica, George Mihalache, Barreca & La Varra, Estudio Lamela, Junglim Architecture, Arditti+RDT Arquitectos, A-lab, Dante O. Benini, Luca Gonzo, Hege Thorvaldsen, Vivian Steinsåker, Simona Ferrari, HHD FUN, Fernando Alda, Tom Arban, Mathieu van Ek, Giesecke, H.G.Esch, Keller, Maurizio Bianchi, Marco Bakker, Thea van den Heuvel, Luuk Kramer, Yum Seung Hoon, Marcin Czechowicz, Emanuel Raab, Sascha Jahnke, Enrico Cano, Sergio Cancellieri, Paolo Frizzarin, Albert Lim, Daniele Domenicali, Park Young Chea, Andrei Margulescu, Paolo Rosselli, Daniel Schäfer, Beppe Raso, Nils Petter Dale, Wei Gang, Wang Zhenfei

ACKNOWLEDGEMENTS

We would like to thank everyone involved in the production of this book, especially all the artists, designers, architects and photographers for their kind permission to publish their works. We are also very grateful to many other people whose names do not appear on the credits but who provided assistance and support. We highly appreciate the contribution of images, ideas, and concepts and thank them for allowing their creativity to be shared with readers around the world.